Methods in Cell Biology

VOLUME 45
Microbes as Tools for Cell Biology

Series Editors

Leslie Wilson

Department of Biological Sciences
University of California, Santa Barbara
Santa Barbara, California

Paul Matsudaira

Whitehead Institute for Biomedical Research and
Department of Biology
Massachusetts Institute of Technology
Cambridge, Massachusetts

Methods in Cell Biology

Prepared under the Auspices of the American Society for Cell Biology

VOLUME 45
Microbes as Tools for Cell Biology

Edited by
David G. Russell
Molecular Microbiology
Washington University School of Medicine
St. Louis, Missouri

ACADEMIC PRESS

San Diego   New York   Boston   London   Sydney   Tokyo   Toronto

Front cover photograph (paperback edition only): A platinum replica of the cytoplasmic face of a phagosome containing a single *Mycobacterium avium* bacillus. The phagosome was isolated and prepared as described in Sturgill-Koszycki, S., Schlesinger, P., Chakraborty, P., Haddix, P., Collins, H., Fok, A., Allen R., Gluck, S., Heuser, J. and Russell, D. G. (1994). Lack of acidification in *Mycobacterium* phagosomes produced by exclusion of the vesicular proton-ATPase. *Science* **263**, 678–681. The image was processed and colorized with Adobe Photoshop.

This book is printed on acid-free paper. ∞

Academic Press, Inc.
A Division of Harcourt Brace & Company
525 B Street, Suite 1900, San Diego, California 92101-4495

United Kingdom Edition published by
Academic Press Limited
24-28 Oval Road, London NW1 7DX

International Standard Serial Number: 0091-679X

International Standard Book Number: 0-12-604040-0 (comb)
International Standard Book Number: 0-12-564146-X (case)

PRINTED IN THE UNITED STATES OF AMERICA
94 95 96 97 98 99 EB 9 8 7 6 5 4 3 2 1

CONTENTS

Contributors   xi

Preface   xiii

Introduction   xv

1. Obtaining and Maintaining Microbial Pathogens

David G. Russell

    I. Introduction   1
   II. Biosafety Considerations   1
  III. Department of Health and Human Services Guidelines   2
  IV. Federal and Institutional Requirements   2
   V. Source of Pathogens   3
  VI. Culture and Maintenance of Pathogens   3
 VII. Conclusion   4

PART I   Manipulation of Pathogens

    2. Cultivation of Malaria Parasites

    William Trager

       I. Introduction   7
      II. Culture Medium   8
     III. Serum   8
     IV. Erythrocytes   10
      V. Culture Systems   10
     VI. Synchronization   16
    VII. Gametocyte Production in Culture   18
   VIII. Cloning Methods   20
    IX. Cryopreservation   20
     X. Serum Replacement   21
    XI. Axenic (Extracellular) Development of Erythrocytic Stages of
        P. falciparum   21
   XII. Cultivation of Erythrocytic Stages of Other Species of
        Malaria Parasites   21
  XIII. *In Vitro* Development of Preerythrocytic and Sporogonic Stages of
        the Life Cycle   22
        References   23

3. Molecular Tools for Genetic Dissection of the Protozoan Parasite
Toxoplasma gondii

*David S. Roos, Robert G. K. Donald, Naomi S. Morrissette,
and A. Lindsay C. Moulton*

I.	Introductory Overview	28
II.	*In Vitro* Culture of *T. gondii* Tachyzoites	32
III.	Molecular Transformation Systems for *Toxoplasma*	44
IV.	Summary and Outlook	59
	References	61

4. Transfection Experiments with *Leishmania*

Jonathan H. LeBowitz

I.	Introduction	65
II.	Stable and Transient Transfection	66
III.	Transfection	70
	References	76

5. Mutagenesis and Variant Selection in *Salmonella*

Renée Tsolis and Fred Heffron

I.	Introduction	79
II.	Mutagenesis	80
III.	Screening of Variants	92
IV.	Genetic Analysis	98
V.	Perspectives	101
	References	103

6. Mycobacterium: Isolation, Maintenance, Transformation, and
Mutant Selection

Nancy D. Connell

I.	Introduction	108
II.	Biosafety Considerations	108
III.	Culture Media and Conditions	109
IV.	Isolation of Mycobacteria	111
V.	Maintenance of Stocks	112
VI.	Genetic Techniques	112
VII.	Mutagenesis	118
VIII.	Strain Construction	120
IX.	Mutant Selection and Isolation	121
	References	123

PART II Microbial Adherence and Invasion Assays

7. Modulation of Murine Macrophage Behavior *in Vivo* and *in Vitro*

Gregory J. Bancroft, Helen L. Collins, Lynette B. Sigola,
and Caroline E. Cross

I.	Introduction	130
II.	Reagents and Solutions	131
III.	Animal Husbandry and Maintenance of Immunocompromised Mice	134
IV.	Eliciting and Harvesting Macrophages *in Vivo*	136
V.	Adherence of Macrophages to Solid Substrates	139
VI.	Assays for Macrophage Phagocytic Activity	140
VII.	Analysis of MHC Class II Antigen Expression	143
	References	145

8. *In Vitro* Assays of Phagocytic Function of Human Peripheral
Blood Leukocytes: Receptor Modulation and Signal Transduction

Eric J. Brown

I.	Introduction	147
II.	Phagocytosis Assays	148
III.	Phagocytic Receptors	156
IV.	Stimulation of Phagocytosis	159
V.	Phagosome Isolation	160
	References	162

9. Bacterial Adhesion and Colonization Assays

Per Falk, Thomas Borén, David Haslam, and Michael Caparon

I.	Introduction	165
II.	*In Situ* Screening of Host Receptor Distribution	166
III.	Bacterial Adherence to Cells in Culture	169
IV.	Biochemical Characterization of the Molecular Nature of Receptors *in Situ*	172
V.	Bacterial Inhibition Experiments *in Situ*	174
VI.	*In Vitro* Assays for Bacterial Adhesion	175
VII.	Probing Eukaryotic Cell Glycoconjugates with Purified Bacterial Adhesins	185
VIII.	Concluding Remarks	187
	References	188

10. Cytoadherence and the *Plasmodium falciparum*-Infected Erythrocyte

Ian Crandall and Irwin W. Sherman

I.	Introduction	193
II.	Ligands for Adherence	194
III.	*P. falciparum*-Infected Red Cell Adhesions	196

IV. Cytoadherence, an *in Vitro* Model of Sequestration: Practical
    Considerations of Cytoadherence Assays   198
    References   207

PART III   The Study of Intracellular Pathogenesis

11. Purification of *Plasmodium falciparum* Merozoites for Analysis of the
Processing of Merozoite Surface Protein-1

Michael J. Blackman

   I. Introduction   213
  II. Parasite Culture and Synchronization   214
 III. Merozoite Isolation   216
 IV. Assay for Secondary Processing of the Merozoite Surface
    Protein-1 (MSP-1)   217
    References   220

12. *In Vitro* Secretory Assays with Erythrocyte-Free Malaria Parasites

Kasturi Haldar, Heidi G. Elmendorf, Arpita Das, Wen Lu Li,
* David J. P. Ferguson, and Barry C. Elford*

   I. Introduction   222
  II. Release and Separation of Late Ring and Trophozoite Stage Parasites
    from the Erythrocyte Membrane (EM) and Tubovesicular
    Membrane (TVM) Network   227
 III. Synthesis and Secretion of Proteins by Intact,
    Ring/Trophozoite Parasites   230
 IV. Organization of Secretory Activities at Different Stages of the
    Asexual Life Cycle   234
  V. Release of Pigmented Trophozoits and Schizonts from
    Infected Erythrocytes by Osmotic Shock in Isoosmolar
    Dipeptide-Based Media   240
    References   245

13. Intracellular Survival by *Legionella*

Karen H. Berger and Ralph R. Isberg

    I. Introduction   247
   II. Laboratory Cultivation of *L. pneumophila*   248
  III. Tissue Culture of U937 Cell-Derived Macrophages   248
  IV. Intracellular Thymineless Death Enrichment   249
   V. Identification of Intracellular Growth Mutants from Enriched
    Bacterial Pools Using "Poke Plaque" Assays   256
  VI. Additional Remarks   257
    References   258

14. Isolation and Characterization of
Pathogen-Containing Phagosomes

Prasanta Chakraborty, Sheila Sturgill-Koszycki, and David G. Russell

    I. Introduction   261
    II. Choice of Pathogens and Particles   263
    III. Choice of Macrophage   264
    IV. Particle Adherence and Internalization Conditions   265
    V. Cell Lysis Conditions   266
    VI. Isolation of Phagosomes   269
    VII. Analysis of Phagosomal Constituents   271
    VIII. Storage and Handling of Two-Dimensional SDS–PAGE Data   273
    IX. Shortcomings   273
    References   275

15. Immunoelectron Microscopy of Endosomal Trafficking in
Macrophages Infected with Microbial Pathogens

David G. Russell

    I. Introduction   277
    II. The Host-Pathogen Interplay   278
    III. Intersection with the Endosomal Pathway   279
    IV. Processing of Infected Macrophages for
        Immunoelectron Microscopy   280
    V. Blocking Cryosections and Incubation with Primary Antiserum   281
    VI. Gold-Conjugated Second Antibodies.   282
    VII. Controls   283
    VIII. Final Preparation of the Grids   284
    IX. Routine Protocol for Analysis of Fluid-Phase Trafficking   286
    References   287

16. Measuring the pH of Pathogen-Containing Phagosomes

Paul H. Schlesinger

    I. Introduction   289
    II. Materials   299
    III. Procedures   302
    IV. When Things Are Not Perfect, or Even Very Close   307
    V. Conclusion   309
    References   309

17. Techniques for Studying Phagocytic Processing of Bacteria
for Class I or II MHC-Restricted Antigen Recognition by
T Lymphocytes

Clifford V. Harding

    I. Introduction   313
    II. Generating MHC-I- and MHC-II- Restricted T Cells to Detect
        Antigen Processing   315

    III. Antigen Presentation Assays   319
    IV. Observations and Implications   324
          References   324

Index   327
Volumes in Series   335

CONTRIBUTORS

Numbers in parentheses indicate the pages on which the authors' contributions begin.

Gregory J. Bancroft (129), Department of Clinical Sciences, London School of Hygiene and Tropical Medicine, London WC1E 7HT, England

Karen H. Berger (247), Department of Molecular Biology and Microbiology, Tufts University School of Medicine, Boston, Massachusetts 02111

Michael J. Blackman (213), Division of Parasitology, National Institute for Medical Research, London NW7 1AA, England

Thomas Borén (165), Department of Molecular Microbiology, Washington University School of Medicine, St. Louis, Missouri 63110

Eric J. Brown (147), Departments of Medicine, Molecular Microbiology, and Cell Biology and Physiology, Washington University School of Medicine, St. Louis, Missouri 63110

Michael Caparon (165), Department of Molecular Microbiology, Washington University School of Medicine, St. Louis, Missouri 63110

Prasanta Chakraborty (261), Department of Molecular Microbiology, Washington University School of Medicine, St. Louis, Missouri 63110

Helen L. Collins (129), Department of Clinical Sciences, London School of Hygiene and Tropical Medicine, London WC1E 7HT, England

Nancy D. Connell (107), Department of Microbiology and Molecular Genetics, University of Medicine and Dentistry of New Jersey, New Jersey Medical School, National Tuberculosis Center, Newark, New Jersey 07103

Ian Crandall (193), Department of Biology, University of California at Riverside, Riverside, California 92521

Caroline E. Cross (129), Department of Clinical Sciences, London School of Hygiene and Tropical Medicine, London WC1E 7HT, England

Arpita Das (221), Department of Microbiology and Immunology, Stanford University School of Medicine, Stanford, California 94305

Robert G. K. Donald (27), Department of Biology, University of Pennsylvania, Philadelphia, Pennsylvania 19104

Barry C. Elford (221), Department of Microbiology and Immunology, Stanford University School of Medicine, Stanford, California 94305

Heidi G. Elmendorf (221), Department of Microbiology and Immunology, Stanford University School of Medicine, Stanford, California 94305

Per Falk (165), Department of Molecular Biology and Pharmacology, Washington University School of Medicine, St. Louis, Missouri 63110

David J. P. Ferguson (221), Department of Microbiology and Immunology, Stanford University School of Medicine, Stanford, California 94305

Kasturi Haldar (221), Department of Microbiology and Immunology, Stanford University School of Medicine, Stanford, California 94305

Clifford V. Harding (313), Institute of Pathology, Case Western Reserve University, Cleveland, Ohio 44106

David Haslam (165), Department of Molecular Microbiology, Washington University School of Medicine, St. Louis, Missouri 63110

Fred Heffron (79), Department of Molecular Microbiology and Immunology, Oregon Health Sciences University, Portland, Oregon 97201

Ralph R. Isberg (247), Department of Molecular Biology and Microbiology, and Howard Hughes Medical Institute, Tufts University School of Medicine, Boston, Massachusetts 02111

Jonathan H. LeBowitz (65), Department of Biochemistry, Purdue University, West Lafayette, Indiana 47907

Wen lu Li (221), Department of Microbiology and Immunology, Stanford University School of Medicine, Stanford, California 94305

Naomi S. Morrissette (27), Department of Biology, University of Pennsylvania, Philadelphia, Pennsylvania 19104

A. Lindsay C. Moulton (27), Department of Biology, University of Pennsylvania, Philadelphia, Pennsylvania 19104

David S. Roos (27), Department of Biology, University of Pennsylvania, Philadelphia, Pennsylvania 19104

David G. Russell (1, 261, 277), Department of Molecular Microbiology, Washington University School of Medicine, St. Louis, Missouri 63110

Paul H. Schlesinger (289), Department of Cell Biology and Physiology, Washington University School of Medicine, St. Louis, Missouri 63110

Irwin W. Sherman (193), Department of Biology, University of California at Riverside, Riverside, California 92521

Lynette B. Sigola (129), Department of Clinical Sciences, London School of Hygiene and Tropical Medicine, London WC1E 7HT, England

Sheila Sturgill-Koszycki (261), Department of Molecular Microbiology, Washington University School of Medicine, St. Louis, Missouri 63110

William Trager (7), Rockefeller University, New York, New York 10021

Renée Tsolis (79), Department of Molecular Microbiology and Immunology, Oregon Health Sciences University, Portland, Oregon 97201

PREFACE

The field of microbial pathogenesis is currently enjoying a renaissance triggered by an increased interest shown by immunologists, cell biologists, and molecular biologists in the idiosyncrasies of these "primitive" organisms. This rapid expansion has deepened our appreciation of the complexities of the host–pathogen interface, and is revealing a myriad of strategies whereby microbes ensure the success of their infection. Studies initially intended to elucidate mechanisms of disease are also shedding light on basic cellular functions, and bacterial, protozoal, and fungal agents are becoming regarded, much as viruses were, as potential tools for dissection of cellular processes.

Unfortunately the majority of microbiological texts deal systematically with microbes and the diseases that they cause. This renders the field rather inaccessible to non-microbiologists, or even to microbiologists wishing to expand their interests. The editors of *Methods in Cell Biology* recognized that there was a need for a book to bridge the gap between microbiology and cell biology. This present volume represents a departure from tradition by substituting a thematic structure for the more usual phylogenetic organization. The volume describes a microbial toolkit for biologists who wish to use microbes as probes for basic cellular functions. I also hope that microbiologists will find the volume useful for its description of the cell biology of the host.

This volume provides basic information on culture and genetic manipulation of microbes (Chapters 1–6), then describes assays for analysis of the initial recognition stage between the pathogen and the host (Chapters 7–9), and culminates in a series of chapters describing analysis of that most intimate of relationships, intracellular parasitism (Chapters 10–17). Each chapter outlines practical procedures and describes the rationale behind their development. The volume should prove useful to anyone interested in the biology of infectious agents, or their exploitation as a new generation of cell biological reagents.

David G. Russell

INTRODUCTION

David Russell asked me to provide an introduction to *Microbes as Tools for Cell Biology*. From my standpoint, this volume just as easily could be called *Eukaryotic Cells as Tools for Microbiology*. By and large, the articles in this volume reflect the fact that pathogenic bacteria and protozoans are capable of invading eukaryotic cells, multiplying within them, and, in some cases, even persisting within them for long periods of time. Gaining access into the host cell has a number of advantages. In addition to avoiding the host immune system, intracellular localization places the pathogen in an environment potentially rich in nutrients and devoid of competing microorganisms. However, intracellular life is not free of difficulty. Bacteria or viruses internalized through the reorganization of the cytoskeleton find themselves within a membrane-bound vesicle, an acidic environment, which may be destined for fusion with potentially degradative lysosomes. Some viruses respond to the acidic environment by changing conformation, binding to the endosomal membrane, and releasing their nucleic acid into the cytoplasm. Bacteria such as *Shigella* (the cause of bacillary dysentery) and *Listeria monocytogenes* (a causative agent of meningitis and sepsis in the very young or very old) synthesize an enzyme that dissolves the surrounding membrane and permits them to replicate within the relative safety of the cytoplasm. These microbes become surrounded by a dense cloud of host cell actin filaments, which then rearrange to form a polarized "comet tail" that is associated with moving bacteria, which may drive through the plasma membrane to adjacent cells. Other organisms, such as the typhoid bacillus, the Legionnaire's disease agent, and the tubercle bacillus, parasitize professional phagocytes. These bacteria have evolved a specialized entry mechanism that bypasses the most fearsome of the macrophage's host defense machinery designed to kill infectious microbes. They apparently tolerate the initial endosome–lysosome fusion event. However, recent evidence suggests that these sophisticated, intracellular parasites directly modify this intracellular compartment into a privileged niche in which they can replicate optimally. Still other organisms, for example, the protozoan *Toxoplasma gondii,* inhibit the acidification of the endosomal vesicle and this, in turn, inhibits lysosomal fusion. The common theme is that certain pathogenic microorganisms have found a way to circumvent or to exploit normal host cell factors to suit their own purposes.

Clearly, the use of tissue culture cells and related methods has greatly facilitated the study of microbial invasion. By the same token, it is recognized that the study of microbial invasion facilitates the study of normal host cell function.

In the past, cell biologists exploited viruses and purified bacterial toxins to explore the biology of endocytosis and cellular trafficking. However, the use of living, intact pathogenic bacteria, protozoa, and fungi as probes to study the cell biology of a variety of cells is still a relatively new field. In my experience, cell biologists need to overcome two barriers in order to use microorganisms as experimental tools. The first is the fear of working with potentially infectious agents. Yet, the aseptic technique *does* work, and with a bit of reading, it will become quite apparent that one can usually employ avirulent mutant derivatives and still examine a cellular phenomenon of interest. In a parallel vein, it takes a brave cell biologist to overcome years of indoctrination and deliberately add microbes to sterile cell cultures. No doubt about it, living bacteria, pathogens or not, can wreak havoc on the homeostasis of a cell culture system. In addition to the toxins specifically created to affect eukaryotic function, microbial by-products of normal metabolism can be toxic to cultured cells. However, once these technical barriers are solved, the usual experimental tools of the cell biologist are readily applicable. Not unexpectedly, microscopy, in some form or another, was used in the initial studies of these host–parasite relationships. More recently, however, one sees intracellular trafficking studies with labeled components and ionic flux measurements—even the actin polymerization associated with *Listeria* has been reduced to a cell-free assay system. Vacuoles containing bacteria are now isolated from infected cells.

The awareness that microbial invasion and attachment systems offer enormous potential for examining essential host functions has forged collaborations between cell biologists and microbiologists. The union of these two previously isolated, independent scientific disciplines has yielded some lusty hybrid offspring. As is the case in other "mixed" marriages, there are new languages to be learned and new customs to be assimilated. Even at this early stage, the excitement is palpable. Cell biologists now appear at microbiology meetings and speak to overflowing audiences who wish to know more about their discipline. Similarly, cell biologists now organize symposia devoted to host–parasite interactions and they become enchanted with the sinister behavior of microbes that cause disease. They literally gasp when they see video images of *Listeria* motoring around a cell or the enormous cell ruffling that occurs when *Salmonella* enters an epithelial cell. *Microbes as Tools for Cell Biology* reflects this excitement and presents a variety of articles dealing with the biology of host–parasite relationships, and, one hopes, stimulates investigators to consider a new approach to understanding how cells work.

Stanley Falkow
Stanford University School of Medicine
Stanford, California

CHAPTER 1

Obtaining and Maintaining Microbial Pathogens

David G. Russell

Department of Molecular Microbiology
Washington University School of Medicine
St Louis, Missouri 63110

I. Introduction
II. Biosafety Considerations
III. Department of Health and Human Services Guidelines
IV. Federal and Institutional Requirements
V. Source of Pathogens
VI. Culture and Maintenance of Pathogens
VII. Conclusion

I. Introduction

This introductory chapter is intended to point the cell biologist in the right direction prior to embarking on any work involving pathogens. The chapter addresses two basic points: how does one obtain and culture the pathogen of interest, and how does one avoid becoming the culture vessel for the selected pathogen. These issues shall be discussed in reverse order.

II. Biosafety Considerations

Although there are certainly common precautions and practices applicable to the handling of all pathogenic organisms, each individual pathogen has its own specific hazards. For this reason, it is vital that a cell biologist wishing to work with pathogens contact, and preferably spend some time in, a laboratory that has an established expertise in handling that particular pathogen. Contact with clinical laboratories or infectious diseases laboratories may not be enough because these laboratories may be able to culture the pathogen for diagnostic purposes, but do not necessarily appreciate the precautions required for experimental manipulation of the pathogen.

Furthermore, for many microbial pathogens there are several avirulent or non-human pathogens that could be used as alternatives. Such possibilities should be explored prior to converting one's laboratory into a maximum security wing.

III. Department of Health and Human Services Guidelines

The Center for Disease Control and the National Institutes of Health jointly publish a booklet entitled *Biosafety in Microbiological and Biomedical Laboratories,* Stock No. 17-40-508-3, obtainable from the U.S. Government Printing Office, Washington, D. C. 20402 (tel. 202-275-3318). This booklet documents the various levels of biosafety containment required for each pathogen and outlines the requirements to satisfy each biosafety level (BL). The booklet contains numerous anomalies; for example, all parasites are categorized as BL2 pathogens irrespective of whether they cause transient or easily treated diseases, such as cutaneous leishmaniasis or amoebiasis, versus persistent diseases, with possibly fatal consequences, such as Chagas' disease (*Trypanosoma cruzi*). Despite these difficulties, which are the product of attempting to prescribe a universal solution to a myriad of pathogens, the booklet is an absolute requirement to outline the safety measures required and to inform one of one's legal responsibilities.

IV. Federal and Institutional Requirements

Although there is no direct stipulation under Federal Law that an institute must have a microbiological safety committee, the Occupational Safety and Health Administration regulations contain a general duty clause that requires all employers to provide a safe work environment. Many institutes have responded by forming their own microbiological safety committees that evaluate the relative hazard inherent to the use of a particular pathogen for a series of experiments. Some grant agencies require applications to be assessed by institutional microbiological safety committees. These committees should produce guidelines for the use of a pathogen and help in the design of adequate safety measures. Records must be kept regarding the education of laboratory staff to the potential dangers of the pathogen, and their understanding of the potential risks. A careful record of any accidental spillages and the decontamination procedures applied should also be kept.

For some pathogens, predominantly viral agents like vaccinia, it is necessary to vaccinate personnel prior to starting the research, whereas exposure to other infective agents is best evaluated by monitoring laboratory personnel for immune conversion (particularly toxoplasmosis and mycobacterial infections). Regardless, the investigator should establish the best mode of treatment for potential infections prior to importing the pathogen into the laboratory. It is vital to discuss the proposed studies with the institute's microbiological safety officer and committee prior to embarking on the experiments. Failure to conform to these regulations endangers one's staff and leaves one open to criminal prosecution.

V. Source of Pathogens

Once one has reorganized the laboratory to conform to the required safety standards, there a variety of possible sources of pathogens to be evaluated. For most cell biologists, I would recommend restricting oneself to a type-strain that has been extensively characterized in different laboratories. Obviously different isolates can exhibit substantial strain variation, therefore, wherever possible I would advise the use of "standardized" isolates. These may be obtained from a laboratory working actively on that particular pathogen; however, one should ask for the history of the culture in detail, and an assurance that the culture is exactly what it is reported to be.

There are several collections that may be used as potential sources of strains. These collections vary considerably in their level of regulation. The American Type Culture Collection (ATCC) (tel. 1-800-638-6597) is the repository for many bacterial, fungal, and protozoal strains; it is an invaluable resource. However, it is important to note that the strains are not being actively worked on at the ATCC, so some level of knowledge concerning the expected phenotype of the isolate requested is important to ensure that it is as described and has not been contaminated or attenuated with culture. The Centers for Disease Control, Atlanta (tel. 404-639-3883) retain many different collections of pathogens and are working with many of the isolates; however, these are not organized as a central resource and the heads of the various laboratories must be contacted individually to ascertain whether a particular strain or isolate is available.

There are many different laboratories in the United States and Europe that are supported by government or international agencies to type isolates and maintain an accessible collection of characterized strains. These sources are too diverse to list here, however, researchers in each field are usually aware of their existence and could point interested parties in the right direction.

VI. Culture and Maintenance of Pathogens

Many microbial pathogens will attenuate on constant culture and require passage through animals at relatively frequent intervals. To overcome this problem and to ensure that one retains a stock of uncontaminated, primary isolates it is vital to freeze down stabilates or aliquots of an early culture of your microbe. Freezing protocols vary with microbe and should be confirmed individually.

In this laboratory, I use two different books for the recipes for media to maintain microbes. For bacterial and fungal pathogens I use the *Handbook of Biological Media* by R. M. Atlas (L. C. Parks, Ed., 1993, CRC Press, Inc., International Standard Book No. 0-8493-2944-2). And for protozoal parasites I use *In vitro Methods for Parasite Cultivation* (A. E. R. Taylor and J. R. Baker, eds., 1987, Academic Press, International Standard Book No. 0-12-683855-0). In addition, for basic bacteriological techniques, the newly revised *Methods for General and Molecular Bacteriology* (P. Gerhardt, Ed., 1993, ASM Press, International Standard Book No. 1-55581-048-9) is very useful. I rarely have the need to look beyond these three comprehensive reference books.

VII. Conclusion

In this preliminary chapter I wish merely to direct any interested parties to the best source of information before embarking on any experiments using microbial pathogens. The risks involved vary tremendously with the different pathogens and must be assessed individually. For this reason there is no substitute for working in a laboratory that routinely handles the organism of interest. Laboratory infections do happen and have serious repercussions if recommended operating procedures have not been observed.

PART I

Manipulation of Pathogens

CHAPTER 2

Cultivation of Malaria Parasites

William Trager

Rockefeller University
New York, New York 10021

I. Introduction
II. Culture Medium
III. Serum
IV. Erythrocytes
V. Culture Systems
    A. Dish or Flask Cultures with Manual Change of Medium
    B. Semiautomated Methods
VI. Synchronization
VII. Gametocyte Production in Culture
VIII. Cloning Methods
IX. Cryopreservation
X. Serum Replacement
XI. Axenic (Extracellular) Development of Erythrocytic Stages of *P. falciparum*
XII. Cultivation of Erythrocytic Stages of Other Species of Malaria Parasites
XIII. *In Vitro* Development of Preerythrocytic and Sporogonic Stages of the Life Cycle
    References

I. Introduction

Continuous culture of any species of malarial parasite was first obtained in 1976 (Trager and Jensen, 1976) with *Plasmodium falciparum,* most important of the four species causing human malaria. The methods then reported have been widely used and have undergone relatively little modification (see also Trager and Jensen, 1980; Trager, 1987; Jensen, 1988). They depend on the maintenance of human erythrocytes under conditions that support intracellular development of the parasites. Accordingly, these cultures are not axenic (but see Section XI). Essential to this development, besides viable erythrocytes, are an appropriate tissue culture medium, which must be replaced at frequent intervals, and a gas phase with 3–5% CO_2 and 17% or less O_2. In this chapter, I will emphasize and give details of the techniques in use at present with which I have had direct experience. Reference will be made, however, to other relevant work, including cultivation of erythrocytic stages of species other than *P. falciparum,* and *in vitro* development of preerythrocytic and sporogonic stages of malaria parasites.

II. Culture Medium

The culture medium is RPMI-1640, originally developed for human white blood cells (Moore *et al.,* 1967) (Table I) supplemented with HEPES buffer and with hypoxanthine (Ifediba and Vanderberg, 1981; Divo *et al.,* 1985a). It is prepared as follows, for 1 liter of final medium.

Dissolve 50 mg hypoxanthine in 10 ml hot water and dilute to 900 ml. Glass redistilled water should be used throughout.

Add 10.4 g powdered RPMI-1640 (GIBCO), and 5.94 g *N*-12-hydroxy ethylpiperazine-*N'*-12-ethanesulfonic acid (HEPES). Dilute to a final volume of 960 ml and sterilize by filtration through a Millipore filter of 0.22-μm porosity. This solution can be stored up to one month at 4°C. We call this medium RP-C to signify RP without bicarbonate. To make RP (complete medium without serum), add 5% sodium bicarbonate solution (sterilized through a 0.22-μm Millipore filter) at the rate of 4 ml to 96 ml of RP-C. To complete this medium, add 10% human serum. We call the complete medium RPS.

III. Serum

This is an essential and important part of the medium (but see Section X for recent work on serum substitutes). Serum is best prepared from a unit of human blood of appropriate serotype collected without anticoagulant and allowed to clot at room temperature. It should then be stored at 4°C overnight (or up to 2 days but not longer) to permit shrinkage of the clot. The liquid portion is then transferred aseptically to centrifuge tubes and centrifuged at 500 × g for 10 min and the serum tubed or bottled. If stored at -20°C, it keeps indefinitely.

Fresh frozen plasma can be used in place of serum (Hui *et al.,* 1984), but this may give problems of partial clotting under some conditions.

The ABO type should of course be compatible with the cells being used. Type O cells can be used with any type serum, and AB serum with any type cells. AB serum is also useful if cultures are being initiated with infected *Aotus* monkey erythrocytes. We regularly use A+ cells with A+ serum whenever possible because this is what the New York Blood Center prefers to sell for research purposes.

Table I
Composition of RPMI-1640[a]

Component	Amount (mg/liter)
Inorganic salts	
Ca(NO$_3$)$_2$ · 4 H$_2$O	100.0
KCl	400.0
MgSO$_4$	48.84
NaCl	6000.0
NaHCO$_3$	2000.0[b]
Na$_2$HPO$_4$	800.0
Other	
Glucose	2000.0
Glutathione (reduced)	1.0
Phenol red	5.0
Amino acids	
L-Arginine (free base)	200.0
L-Asparagine	50.0
L-Aspartic acid	20.0
L-Cystine	65.0 (2 HCl)
L-Glutamic acid	20.0
L-Glutamine	300.0
Glycine	10.0
L-Histidine (free base)	15.0
L-Hydroxyproline	20.0
L-Isoleucine (allo-free)	50.0
L-Leucine (methionine-free)	50.0
L-Lysine-HCl	40.0
L-Methionine	15.0
L-Phenylalanine	15.0
L-Proline (hydroxy-L-proline-free)	20.0
L-Serine	30.0
L-Threonine (allo-free)	20.0
L-Tryptophan	5.0
L-Tyrosine	28.94 (Sodium salt)
L-Valine	20.0
Vitamins	
Biotin	0.20
D-Calcium pantothenate	0.25
Choline chloride	3.00
Folic acid	1.00
Isoinositol	35.00
Nicotinamide	1.00
P-Aminobenzoic acid	1.00
Pyridoxine-HCl	1.00
Riboflavin	0.20
Thiamine-HCl	1.00
Vitamin B$_{12}$	0.005

[a] From GIBCO catalog.
[b] Added before use.

Most commercially available human serum does not support good growth.

When cultivation is undertaken in malarious countries, it is necessary to test each serum in order to avoid using those with inhibitory antibodies or inhibitory levels of antimalarial drugs.

Animal sera have been used (Divo *et al.*, 1985b; Jensen, 1988) but all require a period of adaptation except for rabbit serum (Sax and Rieckmann, 1980).

IV. Erythrocytes

Human erythrocytes are the host cell of choice. In nature, *P. falciparum* is restricted to humans. Furthermore, human erythrocytes have a longer *in vivo* lifetime than those of other susceptible primates (as *Aotus trivirgatus*) and in most places are easier to obtain in quantity. We purchase a unit of blood collected in CPD (per liter 3.27 g citric acid, 26.3 g sodium citrate, 25.5 g glucose, 2.22 g sodium monobasic phosphate; used at 14 ml per 100 ml whole blood) or in CPD supplemented with adenine. This is thoroughly mixed and transferred aseptically to flasks in 50 to 100-ml amounts. Cells can be used for up to 5 weeks of storage at 4°C, and hence for about a week after expiry for transfusion purposes. The ABO type must be compatible with the serum being used.

Cells are prepared for culture in the following way. An appropriate volume (as 30–40 ml) is placed aseptically in a 50-ml conical graduated centrifuge tube (plastic with a screw top). It is centrifuged 10 min in the cold at 500–600 × *g*. The supernatant and buffy coat are removed and the cells are resuspended in approximately equal volume of RP (complete medium without serum) and recentrifuged. This wash procedure is repeated for a second wash. The twice washed cells are then resuspended in an equal volume of RPS (complete medium with serum). Such 50% erythrocyte suspensions can be used for culture after storage for up to about 4 days at 4°C.

V. Culture Systems

All systems are maintained aseptically, so that antibiotics are not normally used.

A. Dish or Flask Cultures with Manual Change of Medium

This is the simplest method; it requires little special equipment but it does require daily or nearly daily attention (Jensen and Trager, 1977). If 3.5-cm disposable Petri dishes are used, each dish receives 1.5 ml of erythrocyte suspension, typically with a 5% hematocrit and a 1% or less initial parasitemia (parasites per 100 erythrocytes). If dishes are being prepared from a previous culture, it is convenient to concentrate the cells of the culture by centrifugation and to resuspend them in an equal volume of fresh RPS. An appropriate volume of this suspension of known parasitemia can then be mixed with the necessary amount of 50% suspension of uninfected erythrocytes to give the desired initial parasitemia of 1% or less. A "0-time" smear is always prepared from this mixture and used to determine the actual starting parasitemia. For example, if a culture with a parasitemia of 8% is being subcultured, the cells would be diluted about 1:10 with uninfected cells. This infected 50% cell suspension would then be diluted with complete culture medium to give the desired volume of 5% suspension.

Alternatively, if, for example, different modifications of a medium are being tested, one would prepare a set of dishes each with 1.5 ml of medium and then add to each dish 150 μl of the 50% infected cell suspension.

Such dish cultures can be scaled up or down. To produce large numbers of parasites, one prepares a sufficient inoculum to start 50 or even 100 Petri dishes of 10-cm diameter, each with 10 ml of 5% infected cell suspension. When many permutations are to be tested in the same medium, as in drug tests, we use 24-well Linbro plates with 0.5 ml culture per well, or 96-well plates with still smaller volumes.

In any case the dishes or plates must be incubated at 37–38°C in an atmosphere with 3–5% CO_2 and 17% or less O_2. A "candle jar" provides typically 3% CO_2 and 15–17% O_2. This is a desiccator having a stopcock in the cover and a plain white candle inside. After the dishes have been set in, the candle is lit and the cover put on with the stopcock open. The moment the candle goes out (from accumulation of CO_2), the stopcock is closed. The jar is then placed in an incubator. The medium is changed daily. To do this, the dishes are removed carefully from the candle jar so as not to resuspend the cells, which are quite tightly settled. Old medium is then removed with a Pasteur pipet in such a way to avoid sucking up cells. About 1.2 ml can usually be removed from a 3.5-cm dish culture, and proportionally larger or smaller volume from larger dishes or from wells. Fresh RPS is then added, in appropriate volume, as 1.5 ml for a 3.5-cm dish or 10 ml for a 10-cm dish. The dishes are swirled gently to resuspend the cells and returned to the candle jar. The candle is burned as before and the jar returned to 37°C. Tissue culture flasks with the covers put on loosely can be used in place of dishes. They are handled in much the same way.

Although the atmosphere of 15–17% O_2 in a candle jar will support good growth of established culture lines, and supports initial development of some new isolates from patient blood, it is not ideal. Very early we had noted one strain that we were able to start in a flow vial culture, with an atmosphere of 5% O_2, but not under candle jar conditions (Jensen and Trager, 1978). Later work showed clearly that *P. falciparum* is microaerophilic (Scheibel et al., 1979). The optimal O_2 level is 3 to 5% with a CO_2 level of 3 to 5%. It is noteworthy that recent studies in South Africa and in New Guinea found that a large proportion of fresh isolates could be established in culture when a gas mixture with 5% CO_2, 5% O_2, 90% N_2 was used, but not when candle jars were used (Freese *et al.,* 1988; Southwell *et al.,* 1989).

Such gas mixtures can be provided in closed containers, as in desiccator jars equipped with an outlet as well as an inlet valve, or through the use of incubators in which the oxygen as well as the CO_2 level can be adjusted and automatically controlled. Flask cultures can also be individually gassed with a current of appropriate gas mixture passed through a filter to maintain sterility, and then tightly closed.

To monitor the growth, blood films are made at appropriate intervals, as every other day. This is best done, just after removal of the old medium, by taking up a droplet of the cells from the bottom of the dish. The extent of growth depends on the strain, the hematocrit, the starting parasitemia, the serum, and the cells. With a hematocrit of 5% and a starting parasitemia of 0.1%, there will generally be a 20- to 50- fold increase in 2 cycles (96 hr). The increase is usually greater in the first cycle than in the second. With established cultures and a good serum, a 10-fold increase at each cycle, as from 0.1 to 1% and from 1 to 10%, is not unusual. If one starts at 1%, however, the first cycle will indeed bring the parasitemia to 7–10% but a second cycle might not occur or at best might give only a small increase as to 12 or 15%. This is because the high lactic acid production by late stages of the parasite produces a low and unfavorable pH; the buffering system of the medium cannot handle it. High relative parasitemia can be obtained by using a low hematocrit. Thus with a 1% hematocrit, one can get parasitemias up to 25–30%, but this is the same absolute number of parasites as 5–6% parasitemia with a 5% hematocrit. With an initial low parasitemia and low hematocrit, the change of medium can be omitted on the first day.

A large-scale production system dependent on manual change of medium has been successfully applied (Palmer *et al.,* 1982).

It is generally convenient to maintain stock cultures on a small scale, since they can easily and quickly be scaled up to large volumes. The stock cultures have to be subcultured by dilution with fresh uninfected erythrocytes whenever the parasitemia reaches about 10–15%. Depending on the use to which the culture is being put, subculture is usually effected every 2 to 5 days.

A method using a 1% erythrocyte suspension in a medium with additional glucose in flasks kept on a shaker requires only to be subcultured every 3 days (Fairlamb *et al.,* 1985); daily change of medium is not needed.

B. Semiautomated Methods

Several methods have been described in which some mechanism replaces the human worker to effect a change of medium at desired intervals (for example, Jensen *et al.,* 1979; Ponnudurai *et al.,* 1983; Moloney *et al.,* 1990).

I have used a continuous slow flow of medium over a settled thin layer of erythrocytes. This was the method that gave the first successful cultures when corresponding material kept in suspension on a rocking device failed (Trager, 1976). The so-called flow vial originally used was soon replaced with a larger and more effective "flow vessel." This method (Trager, 1979) has been successfully used by technicians and research associates in my laboratory over many years with very little difficulty. Growth has generally been good and contamination rare, despite no use of antibiotics. This method will accordingly be described in some detail.

The glass flow vessel (Fig. 1) is made from rectangular 17 × 25-mm tubing with two vertical cylindrical necks 25 mm in diameter and a central side arm about 25 mm long and 11 mm in diameter attached at an angle. For sterilization this side arm is capped with aluminum foil and each vertical neck receives a gauze and cotton plug. The unit is wrapped in paper and dry-sterilized. To prepare it for use, the cotton plugs are replaced with silicone rubber stoppers, the one on the left bearing a medium delivery tube and a gas delivery tube with a cotton plug at its inlet, the one on the right provided with a medium outflow tube held in place by a packing of nonabsorbant cotton (see Fig. 1). This arrangement provides for exit of the flow of gas mixture and also enables a simple adjustment of the tip of the outflow tube to control the depth of the layer of medium. The outer end of both the medium delivery and the medium outflow tubes is capped with aluminum foil. These assemblies are wrapped in paper and sterilized by autoclaving, as is the vaccine cap, which is placed in the side arm.

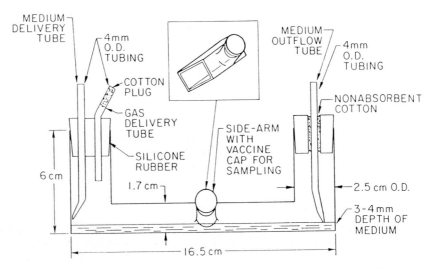

Fig. 1 Flow vessel designed to contain 12 ml of cell suspension. See text for full description. (From Trager, 1979, with permission).

Once the vessel is assembled, it is attached to a ring stand by a clamp on one neck and placed in an incubator that has access holes for the passage of tubing. Flexible Teflon tubing of $\frac{3}{16}$-inch inside diameter (I.D.) is used to connect the vessel to a reservoir flask of medium. The tubing has a short length of silicone rubber tubing at each end for ease of connection to the medium reservoir flask at one end and to the medium inlet tube on the left of the culture vessel. The tubing is also interrupted by a length of silicone rubber tubing of 4-mm I.D., which is inserted in a peristaltic pump. The lengths of tubing depend on the incubator used to hold the culture vessel and the refrigerator used to hold the reservoir flask of medium (see Fig. 2). The outflow tube at the right of the culture vessel is a length of silicone rubber tubing of $\frac{1}{8}$-inch I.D. interrupted at the part in the peristaltic pump with a stretch of silicone rubber tubing with internal diameter of $\frac{3}{16}$ inch (4.9 mm). By using slightly wider tubing for the outflow than for the inflow, both tubes can be in the same pump, yet the outflow will always be faster than the inflow, eliminating the possibility of the level of fluid in the culture vessel getting too high. The inflow and outflow tubings are

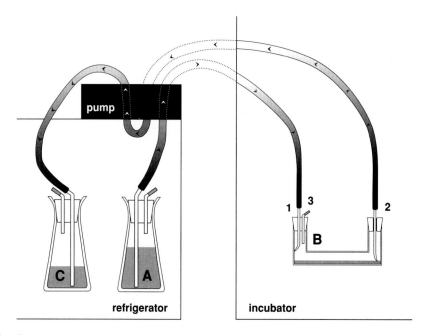

Fig. 2 Diagrammatic representation of the flow of medium in the continuous flow method. Medium in the reservoir flask (A), held in a refrigerator, passes through a peristaltic pump into an incubator where it enters the flow vessel (B) at tube 1. By the action of the same pump, medium is picked up at the other end of the flow vessel by tube 2 and delivered to a flask (C) for spent medium held in the refrigerator. A gas mixture enters the flow vessel at tube 3 and leaves through the cotton packing around tube 2.

capped at each end with aluminum foil and each is coiled, wrapped in a paper envelope, and sterilized by autoclaving.

For refrigeration of the reservoir flask and flask of spent medium, it is convenient to use a small Norcold refrigerator with four holes made in the top for passage of tubing, 2 in front and 2 in back. The peristaltic pump (Harvard Apparatus Model 1203) is set on top of the refrigerator, which is placed beside the incubator holding the flow vessel.

To set up a new culture system, a 500-ml flask holding 350 ml of sterile RP-C medium receives 14.7 ml sterile 5% sodium bicarbonate solution and 40 ml sterile serum. The flask is then provided with a silicone rubber stopper bearing a short cotton plugged air inlet tube and a delivery tube for medium that reaches to the bottom of the flask and has its outer end capped with aluminum foil. This has been separately sterilized by wrapping in paper and autoclaving. The flask is then set in the refrigerator. An empty sterile 500-ml flask equipped with a similar rubber stopper with air oulet and medium delivery tubes is also set in the refrigerator to serve for collection of spent medium. The inflow tubing is removed from its wrapper. One end is inserted into the incubator, the other end into the refrigerator, and the stretch of silicone tubing is placed in the pumping section of the peristaltic pump. By quickly removing the aluminum foil caps, the tubing is aseptically connected to the medium outlet tube of the reservoir flask in the refrigerator and to the inlet tube of the culture vessel in the incubator. The outflow tubing is similarly inserted in the pump and connected to the outflow tube of the culture vessel and the flask for spent medium (see Fig. 2).

A gas delivery tube of Tygon flexible tubing (I.D. $\frac{1}{8}$ inch, wall $\frac{1}{16}$ inch) is then attached to the gas inlet tube of the culture vessel. The gas comes from a tank of mixed gas having 3% CO_2, 2% O_2, and the balance N_2. The gas bubbles first through a large bottle half full of distilled water at a rate of about 1 bubble per second and then through a small flask of distilled water at about 2 bubbles per second from which it goes to a culture vessel. One tank can be used for two to four vessels, with the branching of the gas flow after passage through the large bottle. Appropriate clamps are used to control rate of flow of the gas.

With all the connections made, the peristaltic pump is set for a rapid rate of flow (as at "1" or "2") and turned on to fill the tubing and provide medium in the flow vessel to a depth of about 3 mm. The pump is turned off and the setting changed to "9." The inoculum of infected blood is then introduced aseptically through the vaccine cap on the side arm. Two milliliters of a 50% cell suspension with 1% parasitemia would give an initial hematocrit of 8% (1 ml cells in total volume of 12 ml). The vessel is gently rocked to distribute the blood evenly. The outlet tube of the culture vessel is adjusted so that its tip just touches the surface of the medium. After 1 hr the cells are fully settled. The pump is then turned on. The setting of "9" with the tubing size here used provides a delivery of about 50 ml medium per day, or the equivalent of four changes of the 12 ml in the culture vessel. The reservoir flask has to be changed after 6 days for a new set-up, but only once a week thereafter. The change is effected by mixing fresh complete medium in a 500-ml flask and, with the pump shut off, quickly transferring the rubber stopper with its delivery tube from the empty flask to the new one. The flask of spent medium is similarly exchanged for a new empty flask.

The frequency of subculture depends on the starting parasitemia. For a maximum production, subculture can be conveniently done on a thrice weekly schedule, as on Monday, Wednesday, and Friday. To subculture, the flow is shut off, the cells are resuspended, and a small sample is taken with syringe and needle through the vaccine cap and used for a stained smear from which the parasites are counted. If the parasitemia is, for example, 10%, one removes from the culture vessel (after thorough mixing) all but about 1 ml; hence a harvest of about 11 ml of infected blood. This is replaced with the corresponding volume of an 8% suspension of fresh uninfected erythrocytes in complete medium. After a thorough mixing, a small sample is withdrawn and used for a stained slide for a parasite count (which in this example would be a little below 1%). The cells are then allowed to settle for an hour before the pump is turned on.

If parasites are not needed so frequently, a very convenient schedule of subculture, which also encourages gametocyte formation, is one with intervals of 5, 5, and 4 days. For this the starting parasitemia after subculture should be 0.3 to 0.5%. With still lower parasitemia at subculture, as 0.1% or less, the cultures can be left unattended for 6–7 days. Some typical results with a gametocyte-forming clone are shown in Table II.

Such culture set-ups have been kept going continuously up to a year. Usually, however, it is better to replace it after 3 to 4 months, since a precipitate may form in the medium delivery tubing.

VI. Synchronization

The schizonic erythrocytic cycle of human malaria parasites requires either 48 hr, for *P. falciparum, P. vivax,* and *P. ovale,* or 72 hr, for *P. malariae.* In nature the cycle is typically synchronous but this is not true for most culture lines. Ordinary culture conditions evidently do not supply whatever it is that provides for synchrony and tunes the parasite cycle to the circadian rhythms of the host. As a result, all stages may be seen in a culture of *P. falciparum* at any one time, unlike the natural situation where rings are present in the peripheral blood on alternate days with often no parasites at all on the intervening days. This is because the late trophozoites and the schizonts are sequestered in the capillaries of deep organs, as brain or lungs, where they are attached to the endothelium by specialized structures on the erythrocyte surface called knobs. These knobs, as will be noted below, play a role in artificial methods for the synchronization of cultures.

Table II

A Series of Sequential Counts during a Period of 3 Months from a Stock Culture of Clone HB-3 Maintained in Two Flow Vessels (Exp. E-178)

Date and treatment[a]	Parasites[b] per 1000 erythrocytes									
	Vessel 1					Vessel 2				
	R	T	S	G	Total	R	T	S	G	Total
12/27/91	1	2	0	0	3	4	0	0	0	4
1/3/92 (Sub)	7	14	6	0	27 (6)	8	9	2	0	19 (3)
1/8 (Synch)	31	16	17	0	64 (12)	23	24	11	0	58 (11)
1/13 (Sub)	18	24	25	1	68 (4)	41	38	20	1	100 (4)
1/17 (Sub)	37	17	8	1	63 (6)	36	29	32	1	98 (9)
1/22 (Synch)	22	31	17	1	71 (9)	40	20	15	1	76 (9)
1/27 (Sub)	34	16	4	0	54 (4)	21	20	5	0	46 (4)
1/31 (Sub)	30	28	16	1	75 (7)	23	15	12	0	50 (4)
2/5 (Synch)	16	15	13	0	44 (10)	26	23	7	1	57 (8)
2/10 (Sub)	9	20	8	0	37 (4)	19	21	8	0	48 (4)
2/14 (Sub)	28	12	10	0	50 (7)	29	30	12	0	71 (7)
2/19 (Sub)	16	23	32	0	71 (5)	14	20	13	0	47 (4)
2/24 (Sub)	28	18	6	0	52 (3)	11	14	9	0	34 (3)
2/28 (Sub)	56	38	6	0	100 (3)	28	12	6	0	46 (6)
3/4 (Sub)	5	30	18	0	53 (13)	35	16	19	0	70 (10)
3/9 (Sub)	16	11	4	1	32 (3)	19	5	7	0	31 (2)
3/13 (Sub)	26	30	9	0	65 (10)	26	8	1	2	37 (5)
3/18 (Sub)	11	3	12	1	27 (3)	8	8	12	1	29 (4)
3/23 (Sub)	24	15	11	1	51 (2)	17	21	8	2	48 (9)
3/27 (Sub)	42	33	20	2	97 (6)	20	13	14	2	49 (5)
4/1	33	41	44	1	119	18	27	27	2	74

Note. The culture was initiated on 12/24/91 from a vial frozen 3/9/88.

[a] (Sub) indicates that the culture was subcultured by the addition of fresh erythrocytes. (Synch) indicates that the culture was subjected to synchronization by the procedure described in the text (Section VI).

[b] R, rings; T, trophozoites; S, schizonts; G, gametocytes. The number in parentheses under "Total" gives the total count immediately after subculture or synchronization. Accordingly, the extent of multiplication during each period can be obtained by dividing this figure into the total parasites at the next time of sampling. For example, Vessel 1 on 1/31 with a count of 75 had increased by almost 19 times from the count of 4 just after subculture on 1/27. During the next 5-day period, however, the count increased only 6-fold, from 7 to 44. For Vessel 1 the average increase per 4- to 5-day period during the 3 months was 16-fold; for Vessel 2 it was 11-fold. It will be noted that a low parasitemia just after subculture was usually followed by a larger-fold increase than a high parasitemia.

The first method for doing this resulted from an observation by Lambros and Vanderberg (1979) that infected cell suspensions exposed to a sorbitol solution would lose most of their late stages but not the rings. This led to an effective method of synchronization by a brief exposure to a 5% sorbitol (or mannitol) solution. Erythrocytes with trophozoites or late stages are permeable to sorbitol, which kills the parasites. Erythrocytes with rings (0 to about 16 hr old) are not permeable to these substances, just like uninfected erythrocytes. A single treatment with sorbitol does not give a very tight synchronization, but additional treatments given at appropriate times yield a much better result.

A second method depends on the separation of late stages by either sedimentation in Plasmagel (Pasvol *et al.*, 1978; Reese *et al.*, 1979) or centrifugation over 60% Percoll. In either case the synchrony is not very tight. It is much better to combine separation of late stages with a subsequent sorbitol treatment. We prefer Plasmagel separation for this purpose over Percoll because it also maintains, by phenotypic selection, the knob character (Magowan *et al.*, 1988). When normal erythrocytes are resuspended in Plasmagel or an appropriate gelatin solution (Jensen, 1978), they form rouleaux and settle quickly to the bottom of the tube (Reese *et al.*, 1979). This is also true of erythrocytes infected with rings of *P. falciparum* or with late stages of a line that does not form knobs. Such lines arise readily *in vitro* (Langreth *et al.*, 1979). However, erythrocytes with knobs on the surface do not enter into rouleaux and consequently remain suspended in the supernatant liquid. This provides a simple method for the separation of "knobby" late-stage parasites. Infected erythrocytes are made to a 25% suspension in RPS and mixed with an equal volume of Plasmagel, as for example, 4 ml and 4 ml. The mixture is placed in a test tube and left in a vertical position in a 37°C water bath for 45 min. The upper 5–6 ml is removed from the sedimented cells, centrifuged, and washed in RPS. This material has a 60–80% parasitemia of late trophozoites and schizonts of all stages. This is centrifuged and the brown pellet is mixed with an 8% washed erythrocyte suspension in complete medium at the rate of 0.05 ml pellet to 5-ml cell suspension. This is placed in a 50-mm Petri dish and incubated at 37°C for 3 hr in a candle jar. This material is then centrifuged, resuspended in 5% sorbitol for 15 min, including the 10 min for again centrifuging it, and washed and resuspended in complete medium. This suspension should contain only the rings 0 to 3 hr old that invaded during the 3-hr incubation, all late stages having been destroyed by the sorbitol. Depending on the composition of the starting material, one usually ends up with a parasitemia of 0.3–1%, all young rings. These are put back into culture, or into a flow vessel. They remain fairly synchronous through three subsequent cycles. This will vary somewhat with the isolate and with the particular culture conditions. Some isolates and clones when synchronized show a periodicity of less than 48 hr. For example, the time for a complete cycle for FCR-3 is about 44 hr.

VII. Gametocyte Production in Culture

This depends in the first place on the isolate. Under identical culture conditions, some produce more gametocytes than others, as first documented by Ponnudurai *et al.* (1982) and repeatedly confirmed by others (Bhasin and Trager, 1984; Graves *et al.*, 1984; Teklehaimanot *et al.*, 1987). This is presumably a genetic character. From an isolate (Honduras I CDC) that is itself a good gametocyte producer *in vitro,* we isolated three clones, of which two (HB-1 and HB-3) produced gametocytes as expected, but one (HB-2) has never under any conditions shown a gametocyte (Bhasin and Trager, 1984). Furthermore, not all gametocytes found in culture are equally infective to appropriate Anopheline mosquitoes (Ponnudurai *et al.,* 1982; Teklehaimanot *et al.* 1987). The clone HB-3 and clones (as 3D7) derived from NF54 of Ponnudurai *et al.* (an isolate obtained in The Netherlands from a traveler from Africa) have turned out to be particularly good for mosquito infection. These have been used for a most instructive genetic cross (Walliker *et al.,* 1987; Wellems *et al.,* 1987) and for infection of volunteers by sporozoite inoculation in vaccine trials (Davis *et al.,* 1992).

Extent of gametocyte production also varies with the culture conditions (see Alano and Carter, 1990, for review). If a culture is maintained at high level with frequent subculture for maximum production of asexual forms, gametocytes are rarely seen. They are, however, always present in small numbers and are sometimes noted in concentrates of schizonts prepared over Percoll. The usual practice to get larger numbers of gametocytes is to keep a culture for 1 to 3 weeks with change of medium but without provision of fresh erythrocytes (Ifediba and Vanderberg, 1981). *P. falciparum* gametocytes require about 10–14 days to reach full maturity.The continuous flow method described in Section V, B. lends itself very well to this purpose. For example, when three clones of the FCR-3 isolate from the Gambia (not a particularly good gametocyte producer) were tested by subculturing them to a parasitemia of 0.2% in a 6% erythrocyte suspension and keeping them in flow vessels without subculture, they showed on the 12th day total parasitemias of 6, 3, and 4%, with gametocytes constituting 11, 20, and 9%, respectively, of the parasites (Trager *et al.,* 1981). With the HB-3 clone, maintenance on a schedule of subculture at maximum intervals of 5, 5, and 4 days has shown a low but steady production of gameto-cytes over prolonged periods. For example, one such stock was kept for nearly a year, from August 10, 1986, to June 15, 1987, with the number of gametocytes ranging from 0 to 14% of the parasites and averaging 5–6% (Trager and Gill, 1989). There was no downward trend; the maximum of 14% occurred at 178 days and at the end of the period (309 days) 12% were gametocytes. It is important to note that in this series with subculture, that is, addition of a large proportion of fresh erythrocytes at 5, 5, and 4 days, the hematocrit remains approximately constant so that the relative parasitemia, expressed as parasites per 100 erythrocytes is a meaningful figure. This is not true with cultures kept over a week without addition of fresh erythrocytes. Despite daily provision of fresh medium, the erythrocytes in a dish culture show significant degeneration after the 6th day, and by 2 weeks the hematocrit will be much lower as a result both of this degeneration and of a continued cycling of asexual parasites. For example, we noted in such experiments a fall in the erythrocyte count from

500,000/μl at the start to 300,000/μl on the 11th day (Trager and Gill, 1989). Hence a count of 50 gametocytes per 10,000 erythrocytes on Day 11 is equivalent to about 30 on Day 5. After the 11th or 12th day, the number of erythrocytes drops off even more, a fact that many workers ignore.

Several specific substances have been observed to increase gametocyte formation *in vitro* (see Alano and Carter, 1990), but none sufficiently to be of much practical value. Similarly, the enhanced gametocyte formation in young erythrocytes (Trager and Gill, 1992), although of much theoretical interest, does not provide a practicable method for obtaining larger numbers of gametocytes.

VIII. Cloning Methods

The method of limiting dilution (Rosario, 1981) has been widely used. Since this depends on getting a single infected erythrocyte containing a single parasite, it is important to grow the parasites in such a way to minimize the number of multiply infected erythrocytes (ordinarily fairly high). When this method is used, recloning is probably advisable.

We have described a method permitting isolation and microscopic examination of a single infected cell with a single parasite (Trager *et al.*, 1981) without the use of a micromanipulator. Oduola *et al.* (1988) have obtained clones using a micromanipulator.

With any method the erythrocyte with a single parasite is inoculated into a small volume of 1% erythrocyte suspension. Fresh medium is provided daily after the first 48 hr and fresh erythrocytes at 5- to 6-day intervals. Clones can be expected to show a positive smear by 15–21 days after isolation.

IX. Cryopreservation

The method of Rowe *et al.* (1968) for erythrocytes has consistently given excellent results with cultures of *P. falciparum*. The cryoprotectant is a mixture of 70 ml glycerol with 180 ml 4.2% sorbitol in 0.9% NaCl solution, sterilized by filtration. To cryopreserve a culture, the infected cell suspension is centrifuged, the supernatant is discarded, and the cells are resuspended in an equal volume of the cryoprotectant. After 5 min for equilibration, the suspension is distributed in 0.5-ml amounts in appropriate small vials and frozen quickly by immersion in a dry ice–ethanol mixture. Frozen vials are stored in a liquid nitrogen refrigerator. To retrieve a culture, it is thawed in a 37°C water bath, transferred to a sterile centrifuge tube, and centrifuged 10 min at 500 × *g*. The supernatant is removed and replaced with an equal volume of sterile 3.5% NaCl solution. The suspension is again centrifuged and the supernatant removed and replaced with an equal volume of RPS, preferably with 15% rather than 10%

serum. This wash is repeated once. The cells are then mixed with a suspension of fresh uninfected erythrocytes in RPS and placed in culture.

X. Serum Replacement

Attempts to replace serum with defined substances have been unsuccessful. However, serum has been partially replaced by serum-derived materials. All require bovine albumin Fraction V at 5 g per liter. This was further supplemented with human high-density lipoprotein (HDL) at 1 mg/ml (Grellier *et al.*, 1991) to give growth equivalent to that with 10% human serum. Similar results have been reported using the bovine serum-derived lipids-cholesterol-rich preparation from Sigma at 10 ml per liter (Offula *et al.*,1993) or Nutridoma SP (a Boehringer Mannheim product from human blood) at 5% by volume (Lingnau, 1993). With this last supplement, good gametocyte formation was also reported.

XI. Axenic (Extracellular) Development of Erythrocytic Stages of *P. falciparum*

Under appropriate conditions 20–30% of merozoites will differentiate and develop into early rings extracellularly, but only 1 to 2% complete the schizogonic cycle through trophozoite and schizont back to merozoite (Trager and Williams, 1992; Trager *et al.*, 1992). Merozoites formed extracellularly were shown to be infective to erythrocytes. They were also able to initiate and complete a second extracellular cycle of development (W. Trager, J. Williams, and G. S. Gill, unpublished). The proportion undergoing full development, however, is too small to support continuous axenic culture.

XII. Cultivation of Erythrocytic Stages of Other Species of Malaria Parasites

Four species of malaria parasites of rhesus monkeys have been kept in culture by the same methods as for *P. falciparum* but using rhesus monkey erythrocytes in place of human erythrocytes. These are *P. knowlesi* with a 24 hr-cycle (Wickham *et al.*, 1980); *P. fragile*, a falciparum-like parasite (Chin *et al.*, 1979); *P. inui* with a 72-hr cycle (Nguyen-Dinh *et al.*,1980); and *P. cynomolgi*, a vivax-like parasite (Nguyen-Dinh *et al.*, 1981).

The rodent malaria *P. berghei* has been cultured with rat erythrocytes but requires reticulocytes (Janse *et al.*, 1989). Similarly, the requirement for reticulocytes is a limiting factor in the cultivation of *P. vivax* (Mons *et al.*, 1988), which has been maintained *in vitro* up to 3 weeks, but with decreasing numbers (Lanners, 1992).

XIII. *In Vitro* Development of Preerythrocytic and Sporogonic Stages of the Life Cycle

In nature, sporozoites of avian malaria parasites inoculated by the bite of an infected mosquito invade cells of the reticuloendothelical system to initiate a preerythrocytic cycle. For most species of avian parasites, merozoites formed in this cycle are able to invade either erythrocytes, initiating the erythrocytic cycle, or again reticuloendothelial cells to continue an exoerythrocytic cycle. Hence it is possible to maintain *in vitro* a continuous culture of exoerythrocytic schizogony and this has been done for *P. fallax* and *P. lophurae* using embryonic turkey brain tissue culture (see Huff, 1964, for a review). The parasites develop rapidly within the cells of the culture and often overgrow it. Infected cells from the fluid portion are used to seed new tissue cultures. This is an excellent method for the study of these forms; unfortunately, it has been little used in recent work.

No doubt one reason for its neglect is that it cannot be applied to the malaria parasites of humans and other mammals. All of these have only a single preerythrocytic cycle, which takes place within hepatic cells and produces merozoites capable only of infecting erythrocytes (Garnham, 1988). The single cycle of development has been reproduced *in vitro* by inoculating a tissue culture of hepatic cells with sporozoites removed from infected mosquitoes. The first success was with *P. berghei* in rat liver cells; infective merozoites were obtained in 44 hr (Foley *et al.*, 1978; Hollingdale *et al.*, 1981). Later, complete development of the preerythrocytic cycle of *P. vivax* (Mazier *et al.*, 1984a) and of *P. falciparum* (Mazier *et al.*, 1984b) was obtained from sporozoites inoculated to cultures of human hepatocytes. The preerythrocytic cycle of *P. malariae* has been cultured in chimpanzee hepatocytes (Millet *et al.*, 1988a) and those of *P. cynomolgi, P. knowlesi, P. coatneyi,* and *P. inui* on hepatocytes of *Macaca mulatta* (Millet *et al.*, 1988b). In all cases, only a small proportion of the inoculated sporozoites succeeded in developing into preerythrocytic schizonts.

The sporogonic cycle of development, which occurs in the mosquito, proved to be the most difficult to reproduce *in vitro*. Success was achieved only very recently, first with *P. gallinaceum* (Warburg and Miller, 1992) and soon after with *P. falciparum* (Warburg and Schneider, 1993).

Thus it is now possible to carry through *in vitro* the entire complex life cycle of *P. falciparum*. This provides a remarkable material for fundamental studies of cellular and molecular events associated with the differentiations that occur as the parasites switch from one developmental cycle to another.

References

Alano, P., and Carter, R. (1990). Sexual differentiation in malaria parasites. *Annu. Rev. Microbiol.* **44**, 429–449.

Bhasin, V. K., and Trager, W. (1984). Gametocyte-forming and non-gametocyte-forming clones of *Plasmodium falciparum. Am. J. Trop. Med. Hyg.* **33**, 534–537.

Chin, W., Moss, D., and Collins, W. E. (1979). The continuous cultivation of *Plasmodium fragile* by the method of Trager-Jensen. *Am. J. Trop. Med. Hyg.* **28**, 591–592.

Davis, J. R., Cortese, J. F., Herrington, D. A., Murphy, J. R., Clyde, D. F., Thomas, A. W., Baqar, S., Cochran, M. C., Thanassi, J., and Levine, M. M. (1992). *Plasmodium falciparum: In vitro* characterization and human infectivity of a cloned line. *Exp. Parasitol.* **74**, 159–168.

Divo, A. A., Geary, T. G., Davis, N. L., and Jensen, J. B. (1985a). Nutritional requirements of *Plasmodium falciparum* in culture. I. Exogenously supplied dialyzable components necessary for continuous growth. *J. Protozool.* **32**, 59–64.

Divo, A. A., Vande Waa, J. A., Campbell, J. R., and Jensen, J. B. (1985b). Isolation and cultivation of *Plasmodium falciparum* using adult bovine serum. *J. Parasitol.* **71**, 504–509.

Fairlamb, A. H., Warhurst, D. C., and Peters, W. (1985). An improved technique for the cultivation of *Plasmodium falciparum in vitro* without daily medium change. *Ann. Trop. Med. Parasitol.* **79**, 379–384.

Foley, D. A., Kennard, J., and Vanderberg, J. P. (1978). *Plasmodium berghei:* Infective exoerythrocytic schizonts in primary monolayer cultures of rat liver cells. *Exp. Parasitol.* **46**, 179–188.

Freese, J. A., Sharp, B. L., Ridl, F. C., and Markus, M. B. (1988). *In vitro* cultivation of southern African strains of *Plasmodium falciparum* and gametocytogenesis. *S. Afr. Med. J.* **73**, 720–722.

Garnham, P. C. C. (1988). Malaria parasites of man: Life-cycles and morphology (excluding ultrastructure). *In* "Malaria: Principles and Practice of Malariology" (W. H. Wernsdorfer and I. McGregor, eds.), Vol. 1, pp. 61–96. Churchill-Livingstone, Edinburgh.

Graves, P. M., Carter, R., Keystone, J. S., and Seeley, D. C., Jr. (1984). Drug sensitivity and isoenzyme type in cloned lines of *Plasmodium falciparum. Am. J.Trop. Med. Hyg.* **33**, 212–219.

Grellier, P., Rigomier, D., Clavey, V., Fruchart, J.-C., and Schrevel, J. (1991). Lipid traffic between high density lipoproteins and *Plasmodium falciparum*-infected red blood cells. *J. Cell Biol.* **112**, 267–277.

Hollingdale, M. R., Leef, J. L., McCullough, M., and Beaudouin, R. L. (1981). *In vitro* cultivation of the exoerythrocytic stage of *Plasmodium berghei* from sporozoites. *Science* **213**, 1021–1022.

Huff, C. G. (1964). Cultivation of the exoerythrocytic stages of malarial parasites. *Am. J. Trop. Med. Hyg.* **13**, 171–177.

Hui, G. S. N., Palmer, K. L., and Siddiqui, W. A. (1984). Use of human plasma for continuous *in vitro* cultivation of *Plasmodium falciparum. Trans. R. Soc. Trop. Med. Hyg.* **78**, 625–626.

Ifediba, T., and Vanderberg, J. P. (1981). Complete *in vitro* maturation of *Plasmodium falciparum* gametocytes. *Nature (London)* **294**, 364–366.

Janse, C. J., Boorsma, E. G., Ramesar, J., Grobbee, M. J., and Mons, B. (1989). Host cell specificity and schizogony of *Plasmodium berghei* under different *in vitro* conditions. *Int. J. Parasitol.* **19**, 509–514.

Jensen, J. B. (1978). Concentration from continuous culture of erythrocytes infected with trophozoites and schizonts of *Plasmodium falciparum. Am. J. Trop. Med. Hyg.* **27**, 1274–1276.

Jensen, J. B. (1988). *In vitro* cultivation of malaria parasites: erythrocytic stages. *In* "Malaria: Principles and Practice of Malariology" (W. H. Wernsdorfer and I. McGregor, eds.), Vol. 1, pp. 307–320. Churchill-Livingstone, Edinburgh.

Jensen, J. B., and Trager, W. (1977). *Plasmodium falciparum* in culture: Use of outdated erythrocytes and description of the candle jar method. *J. Parasitol.* **63**, 883–886.

Jensen, J. B., and Trager, W. (1978). *Plasmodium falciparum* in culture: Establishment of additional strains. *Am. J. Trop. Med. Hyg.* **27**, 743–746.

Jensen, J. B., Trager, W., and Doherty, J. (1979). *Plasmodium falciparum:* Continuous cultivation in a semiautomated apparatus. *Exp. Parasitol.* **48**, 36–41.

Lambros, C., and Vanderberg, J. P. (1979). Synchronization of *Plasmodium falciparum* erythrocytic stages in culture. *J. Parasitol.* **65,** 418–420.

Langreth, S. G., Reese, R. T., Motyl, M. R., and Trager, W. (1979). *Plasmodium falciparum:* Loss of knobs on the infected erythrocyte surface after long-term cultivation. *Exp. Parasitol.* **48,** 213–219.

Lanners, H. N. (1992). Prolonged *in vitro* cultivation of *Plasmodium vivax* using Trager's continuous flow method. *Parasitol. Res.* **78,** 699–701.

Lingnau, A. (1993). Serum-free cultivation of *Plasmodium falciparum* gametocytes *in vitro*. *Parasitol. Res.* **79,** 378–384.

Magowan, C., Wollish, W., Anderson, L., and Leech, J. (1988). Cytoadherence by *Plasmodium falciparum*-infected erythrocytes is correlated with the expression of a family of variable proteins on infected erythrocytes. *J. Exp. Med.* **168,** 1307–1320.

Mazier, D., Landau, I., Druilhe, P., Guguen-Guillouzo, C., Baccam, D., Baxter, J., Chigot, J.-P., and Gentilini, M. (1984a). Cultivation of the liver forms of *Plasmodium vivax* in human hepatocytes. *Nature (London)* **307,** 367–369.

Mazier, D., Beaudoin, R. L., Mellouk, S., Druilhe, P., Texier, B., Trosper, J., Miltgen, F., Landau, I., Paul, C., Brandicourt, O., Guguen-Guillouzo, C., and Langlois, P. (1984b). Complete development of hepatic stages of *Plasmodium falciparum in vitro*. *Science* **227,** 440–442.

Millet, P., Collins, W. E., Fisk, T. L., and Nguyen-Dinh, P. (1988a). *In vitro* cultivation of exoerythrocytic stages of the human malaria parasite *Plasmodium malariae*. *Am. J. Trop. Med. Hyg.* **38,** 470–473.

Millet, P., Fisk, T. L., Collins, W. E., Broderson, J. R., and Nguyen-Dinh, P. (1988b). Cultivation of exoerythrocytic stages of *Plasmodium cynomolgi, P. knowlesi, P. coatneyi,* and *P. inui* in *Macaca mulatta* hepatocytes. *Am. J. Trop. Med. Hyg.* **39,** 529–534.

Moloney, M. B., Pawluk, A.R., and Ackland, N. R. (1990). *Plasmodium falciparum* growth in deep culture. *Trans. R. Soc. Trop. Med. Hyg.* **84,** 516–518.

Mons, B., Collins, W. E., Skinner, J. C., Van der Star, W., Croon, J. J. A. B., Jr., and Van der Kaay, H. J. (1988). *Plasmodium vivax: In vitro* growth and reinvasion in red blood cells of *Aotus nancymai. Exp. Parasitol.* **66,** 183–188.

Moore, G. E., Gerner, R. E., and Franklin, H. A. (1967). Culture of normal human leukocytes. *JAMA, J. Am. Med. Assoc.* **199,** 519–524.

Nguyen-Dinh, P., Campbell, C. C., and Collins, W. E. (1980). Cultivation *in vitro* of the quartan malaria parasite *Plasmodium inui. Science* **209,** 1249–1251.

Nguyen-Dinh, P., Gardner, A. L., Campbell, C. C., Skinner, J. C., and Collins, W. E. (1981). Cultivation *in vitro* of the vivax-type malaria parasite *Plasmodium cynomolgi. Science* **212,** 1146–1148.

Oduola, A. M. J., Weatherly, N. F., Bowdre, J. H., and Desjardins, R. E. (1988). *Plasmodium falciparum:* Cloning by single-erythrocyte micromanipulation and heterogeneity *in vitro. Exp. Parasitol.* **66,** 86–95.

Ofulla, A. V. O., Okoye, V. C. N., Khan, B., Githure, J. I., Roberts, C. R., Johnson, A. J., and Martin, S. K. (1993). Cultivation of *Plasmodium falciparum* parasites in a serum-free medium. *Am. J. Trop. Med. Hyg.* **49,** 335–340.

Palmer, K. L., Hui, G. S. N., and Siddiqui, W. A. (1982). A large-scale *in vitro* production system for *Plasmodium falciparum. J. Parasitol.* **68,** 1180–1183.

Pasvol, G., Wilson, R. J. M., Smalley, M. E., and Brown, J. (1978). Separation of viable schizont-infected red cells of *Plasmodium falciparum* from human blood. *Ann. Trop. Med. Parasitol.* **72,** 87–88.

Ponnudurai, T., Meuwissen, J. H. E. T., Leeuwenberg, A. D. E. M., Verhave, J. P., and Lensen, A. H. W. (1982). The production of mature gametocytes of *Plasmodium falciparum* in continuous cultures of different isolates infective to mosquitoes. *Trans. R. Soc. Trop. Med. Hyg.* **76,** 242–250.

Ponnudurai, T., Lensen, A. H. W., and Meuwissen, J. H. E. T. (1983). An automated large-scale culture system of *Plasmodium falciparum* using tangential flow filtration for medium change. *Parasitology* **87**, 439–445.

Reese, R. T., Langreth, S. G., and Trager, W. (1979). Isolation of stages of the human parasite *Plasmodium falciparum* from culture and from animal blood. *Bull. W. H. O.* **57**, Suppl. 1, 53–61.

Rosario, V. (1981). Cloning of naturally occurring mixed infection of malaria parasites. *Science* **212**, 1037–1038.

Rowe, A. W., Eyster, E., and Kellner, A. (1968). Liquid nitrogen preservation of red blood cells for transfusion. *Cryobiology* **5**, 119–128.

Sax, L. J., and Rieckmann, K. H. (1980). Use of rabbit serum in the culture of *Plasmodium falciparum*. *J. Parasitol.* **66**, 621–624.

Scheibel, L. W., Ashton, S. H., and Trager, W. (1979). *Plasmodium falciparum:* Microaerophilic requirements in human red blood cells. *Exp. Parasitol.* **47**, 410–418.

Southwell, B. R., Brown, G. V., Forsyth, K. P., Smith, T., Philip, G., and Anders, R. (1989). Field applications of agglutination and cytoadherence assays with *Plasmodium falciparum* from Papua New Guinea. *Trans. R. Soc. Trop. Med. Hyg.* **83**, 464–469.

Teklehaimanot, A., Collins, W. E., Nguyen-Dinh, P., Campbell, C. C., and Bhasin, V. K. (1987). Characterization of *Plasmodium falciparum* cloned lines with respect to gametocyte production *in vitro,* infectivity to *Anopheles* mosquitoes, and transmission to *Aotus* monkeys. *Trans. R. Soc. Trop. Med. Hyg.* **81**, 885–887.

Trager, W. (1976). Prolonged cultivation of malaria parasites (*Plasmodium coatneyi* and *P. falciparum*). *In* "Biochemistry of Parasites and Host-Parasite Relationships" (H. Van den Bossche, ed.), pp. 427–434. North-Holland Publ., Amsterdam.

Trager, W. (1979). *Plasmodium falciparum* in culture: Improved continuous flow method. *J. Protozool.* **26**, 125–129.

Trager, W. (1987). The cultivation of *Plasmodium falciparum:* Applications in basic and applied research on malaria. *Ann. Trop. Med. Parasitol.* **81**, 511–529.

Trager, W., and Gill, G. S. (1989). *Plasmodium falciparum* gametocyte formation *in vitro:* Its stimulation by phorbol diesters and by 8-bromo cyclic adenosine monophosphate. *J. Protozool.* **36**, 451–454.

Trager, W., and Gill, G. S. (1992). Enhanced gametocyte formation in young erythrocytes by *Plasmodium falciparum in vitro. J. Protozool.* **39**, 429–432.

Trager, W., and Jensen, J. B. (1976). Human malaria parasites in continuous culture. *Science* **193**, 673–675.

Trager, W., and Jensen, J. B. (1980). Cultivation of erythrocytic and exoerythrocytic stages of plasmodia. *In* "Malaria" (J. P. Krier, ed.). Vol 2, pp. 271–319. Academic Press, New York and London.

Trager, W., and Williams, J. (1992). Extracellular (axenic) development *in vitro* of the erythrocytic cycle of *Plasmodium falciparum. Proc. Natl. Acad. Sci. U.S.A.* **89**, 5351–5355.

Trager, W., Tershakovec, M., Lyandvert, L., Stanley, H., Lanners, N., and Gubert, E. (1981). Clones of the malaria parasite *Plasmodium falciparum* obtained by microscopic selection: Their characterization with regard to knobs, chloroquine sensitivity, and formation of gametocytes. *Proc. Natl. Acad. Sci. U.S.A.* **78**, 6527–6530.

Trager, W., Williams, J., and Gill, G. S. (1992). Extracellular development *in vitro* of the erythrocytic cycle of *Plasmodium falciparum. Parasitol. Today* **8**, 384–387.

Walliker, D., Quakyi, I. A., Wellems, T. E., McCutchan, T. F., Szarfman, A., London, W. T., Corcoran, L. M., Burkot, T. R., and Carter, R. (1987). Genetic analysis of the human malaria parasite *Plasmodium falciparum. Science* **236**, 1661–1666.

Warburg, A., and Miller, L. H. (1992). Sporogonic development of a malaria parasite *in vitro. Science* **255**, 448–450.

Warburg, A., and Schneider, I. (1993). *In vitro* culture of the mosquito stages of *Plasmodium falciparum. Exp. Parasitol.* **76**, 121–126.

Wellems, T. E., Walliker, D., Smith, C. L., do Rosario, V. E., Maloy, W. L., Howard, R. J., Carter, R., and McCutchan, T. F. (1987). A histidine-rich protein gene marks a linkage group favored strongly in a genetic cross of *Plasmodium falciparum*. *Cell* (*Cambridge, Mass.*) **49,** 633–642.

Wickham, J. M., Dennis, E. D., and Mitchell, G. H. (1980). Long term cultivation of a simian malaria parasite (*Plasmodium knowlesi*) in a semi-automated apparatus. *Trans. R. Soc. Trop. Med. Hyg.* **74,** 789–792.

CHAPTER 3

Molecular Tools for Genetic Dissection of the Protozoan Parasite *Toxoplasma gondii*

David S. Roos, Robert G. K. Donald, Naomi S. Morrissette, and A. Lindsay C. Moulton

Department of Biology
University of Pennsylvania
Philadelphia, Pennsylvania 19104

I. Introductory Overview
    A. Genetics of *Toxoplasma gondii*
    B. The Parasite Life Cycle
II. *In Vitro* Culture of *T. gondii* Tachyzoites
    A. Growth of Parasites
    B. Replication Assays
    C. Safety Issues
III. Molecular Transformation Systems for *Toxoplasma*
    A. Vectors
    B. Transient Expression
    C. Stable Transgene Expression and Overexpression
    D. Nonhomologous Recombination, Insertional Mutagenesis, and Marker Rescue
    E. Homologous Recombination: Pseudo-diploids, Gene Knock-outs, and Perfect Gene Replacement
    F. Cloning by Complementation
    G. Special Considerations for Working with Pyrimethamine-Resistant Organisms
IV. Summary and Outlook
    A. Genetic Tools for the Future
    B. Outstanding Problems in the Cell Biology and Pathogenesis of *T. gondii*
References

═══════ I. Introductory Overview

A. Genetics of *Toxoplasma gondii*

Despite the importance of protozoan infections in human and veterinary disease worldwide, it has been difficult to unravel the mechanisms of pathogenesis for many of these organisms—particularly intracellular parasites. In part, this problem stems from the difficulty of culturing many such pathogens, and from the lack of suitable genetic systems. The Apicomplexan parasite *Toxoplasma gondii* provides a promising exception to this rule. *T. gondii* is convenient and safe to grow *in vitro* using standard cell culture techniques; the tachyzoite (merozoite) form is haploid and capable of indefinite asexual replication in culture. Beginning in the 1970s, pioneering work by Dr. E. R. Pfefferkorn and colleagues established the potential of *T. gondii* for genetic studies (Pfefferkorn, 1988). Parasites are readily mutagenized (Pfefferkorn and Pfefferkorn, 1979) and the clonal progeny of individual mutants can be plaque purified (Foley and Remington, 1969; Pfefferkorn and Pfefferkorn, 1976). In the wild, *T. gondii* is an unusually promiscuous parasite, infecting virtually any vertebrate tissue (Frenkel, 1973). *In vitro,* this wide host range facilitates experiments conceptually similar to somatic cell genetic studies, where existing mammalian cell mutants are used to determine what the parasite can do for itself and what it requires from its host (Pfefferkorn, 1981).

Within cells of the cat intestinal epithelium (but, alas, not yet in culture), *T. gondii* parasites are also capable of sexual differentiation and mating (Frenkel, 1973), permitting classical genetic crosses (Pfefferkorn and Pfefferkorn, 1980). The resultant progeny—8 haploid sporozoites within a single oocyst—are readily purified, and individual sporozoites can be isolated by micromanipulation (D.S. Roos and E. R. Pfefferkorn, unpublished). *Toxoplasma* exhibits normal Mendelian linkage patterns and chromosomal reassortment through the sexual cycle, with a level of reciprocal recombination suitable for genome mapping (Sibley *et al.,* 1992).

The molecular genetics of protozoan parasites have proved nearly as intractable as classical genetic approaches. Transformation of the Kinetoplastida (*Trypanosoma, Leishmania,* etc.) has only become possible within the past few years (Bellofatto and Cross, 1989; Laban *et al.,* 1990; Cruz and Beverley, 1990; Lee and van der Ploeg, 1990; ten Asbroek *et al.,* 1993), and there is still no practical system for transformation of such important pathogens as *Plasmodium* sp. (malaria) (Miles, 1988; Wellems, 1991). In contrast to many protists, the genetic structure of *Toxoplasma* is notable chiefly for being relatively conventional, i.e., similar to that of its mammalian host cells with respect to gene organization, codon usage, and nucleotide bias (Roos, 1993; Ellis *et al.,* 1993). These observations have led several investigators to examine the feasibility of molecular transformation in this parasite, and such studies have proved very successful (Soldati and Boothroyd, 1993; Donald and Roos, 1993; 1994; Kim *et al.,* 1993;

Sibley *et al.,* 1994). This review outlines the use of several of the molecular genetic tools that have recently been developed for the *T. gondii* system. Because the availability of these tools has resulted in a significant increase in the number of investigators interested in the basic biology of *Toxoplasma,* a brief introduction to parasite culture techniques is also provided.

B. The Parasite Life Cycle

The life cycle of *T. gondii* can be thought of as two independent cycles, intersecting in the form of the intracellular tachyzoite (Fig. 1). Asexual replication can occur in virtually any nucleated animal cell, and consists of the following steps (for reviews, see Joiner and Dubremetz, 1993; Sibley, 1993):

1. Attachment to the host cell. A variety of ligands, including laminins, may be involved.

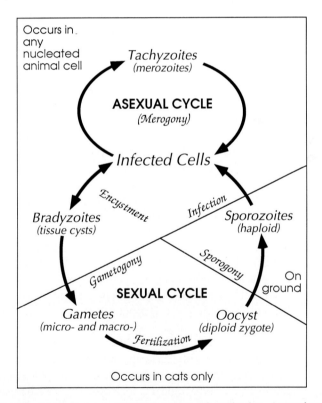

Fig. 1 Asexual and sexual cycles of *Toxoplasma gondii* replication. Asexual replication of *T. gondii* "tachyzoites" (merozoites) can be maintained indefinitely *in vitro,* in virtually any nucleated animal cell. Sexual differentiation occurs only within feline intestinal epithelium. See text for details on the genetic possibilities provided by this parasite.

2. Invasion into a specialized parasitophorous vacuole. The mechanism of invasion and precise origin of the vacuole is unknown, but this process is usually associated with clearing of one or more of the rhoptries—specialized organelles located at the apical end of the parasite. Secretion from dense granules is thought to play a role in establishing and/or maintaining vacuolar membrane function.

3. Multiple rounds of replication within the parasitophorous vacuole. *Toxoplasma* replicates by endodyogeny, producing two daughter parasites in each division. In contrast, many Apicomplexan parasites replicate multiple times before assembly of the mature daughter cells (merogony). Because all parasites within a given vacuole divide synchronously, the number of tachyzoites in any vacuole is invariably a power of two (Fig. 2).

4. Host cell lysis and parasite exit, facilitated by a twisting motility of the parasites. The motility of parasites during escape contrasts with their quiescence during replication, and is probably induced by the ionic changes that occur as the host cell membrane begins to give way under mechanical strain from the growing parasitophorous vacuole (Endo *et al.*, 1987; Endo and Yagita, 1990).

Within animal tissues, tachyzoites are also capable of differentiating into a more slowly replicating form—termed bradyzoites—which develop a periodic acid/Schiff-positive wall and may persist within the tissues for months or years, reemerging periodically to provide a natural boost to host immunity (McLeod and Remington, 1987). Given the high frequency of latent infection (10–90% in populations worldwide), bradyzoites represent a serious danger to immunocompromised individuals (Luft and Remington, 1992).

In contrast to the broad host range of asexual parasite forms, sexual differentiation is known to occur only in feline species (Frenkel, 1973). When bradyzoites infect intestinal epithelial cells (following ingestion of an infected mouse, for example), the parasites can differentiate into macro- or microgametes. The latter rupture out of infected cells and fuse with macrogametes to produce diploid oocysts, which develop a thick, impermeable wall and are shed with the feces. Upon exposure to oxygen and ambient temperature, the oocyst undergoes sporogony: two meiotic divisions and a single mitotic division produce eight haploid sporozoites, divided into two sporocysts within the oocyst wall. Sexual crosses can be carried out in cats under laboratory conditions (Pfefferkorn and Pfefferkorn, 1980), but although bradyzoite cysts have been observed in culture (Lindsay *et al.*, 1991; Bohne *et al.*, 1993; Soete *et al.*, 1993; and D. S. Roos, unpublished observations), it has not yet proved possible to reproduce the complete *Toxoplasma* sexual cycle *in vitro*. For further information on the biology of *T. gondii* infections see Frenkel (1973).

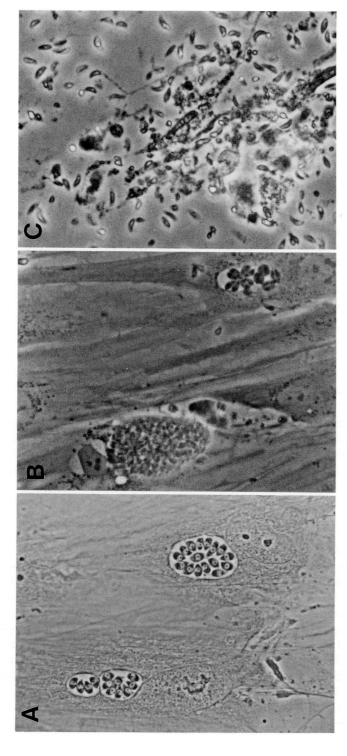

Fig. 2 *In vitro* culture of *T. gondii* in HFF cells. Tachyzoites replicate synchronously (by endodyogeny) within intracellular parasitophorous vacuoles. These vacuoles therefore always contain $2n$ parasites, where n indicates the number of divisions that have taken place since invasion. (A) Vacuoles at early stages of infection contain "rosettes" of 4, 8, or 16 parasites (fixed and stained with Giemsa). (B) Parasites continue to replicate within the parasitophorous vacuole, grossly distending the cell membrane. The large vacuole extends far out of the plane of focus in this phase-contrast micrograph. (C) The entire host cell monolayer is eventually destroyed, releasing highly refractile extracellular tachyzoites.

II. *In Vitro* Culture of *T. gondii* Tachyzoites

A. Growth of Parasites

1. Parasite Strains

Many strains of *T. gondii* have been studied in different labs, and although these differ in virulence and other respects, all strains are thought to constitute a single species (the only species within the genus *Toxoplasma*) (Levine, 1988). The virulent RH strain (Sabin, 1941) is commonly used because of its rapid replication rate, its high productivity, and its highly efficient lysis of host cells—facilitating the isolation of large numbers of tachyzoites relatively uncontaminated by host cell material. It should be noted that virulent strains identified as RH in several laboratories around the world appear to differ (Howe and Sibley, 1994), although it is not known whether these differences reflect heterogeneity within the original RH isolate, independent mutations over the past 50 years in culture, or laboratory contamination. To ensure strain homogeneity, it is advisable to work with strains of known provenance which have recently been clonally rederived (see below).

Toxoplasma RH forms few if any tissue cysts in mice—perhaps due to its virulence, or perhaps because of genetic alterations which prevent bradyzoite differentiation. RH strain *T. gondii* also fails to produce oocysts in cats. The ME49 strain has often been employed for studies on latent infection and sexual differentiation, and clonal isolates of this strain are available (Ware and Kasper, 1987).

Tachyzoites from virulent stains may be passaged either *in vitro* or by intraperitoneal infection of mice. Avirulent strains may be maintained in mice as latent tissue cysts, or in culture as tachyzoites. Only *in vitro* techniques are discussed below; for a discussion of *Toxoplasma* culture in animals, see Boothroyd *et al.*, (1994).

2. Host Cells

In clinical infections, *T. gondii* parasites are often found within macrophages and in the central nervous system (Frenkel, 1973), but these observations probably reflect the frequency with which macrophages encounter pathogenic organisms *in vivo,* and the immunologically privileged nature of the central nervous system, rather than any strong tissue tropism on the part of the parasite itself. Tachyzoites—the proliferative asexual form of *Toxoplasma*—may be maintained in virtually any mammalian cell type, including transformed cell lines (CHO, HeLa, LM, MDBK, Vero, 3T3, etc.). Parasites infect monolayer cultures more efficiently than suspension cultures, presumably because contact and invasion is easier to accomplish.

For routine cell culture, *T. gondii* tachyzoites are often grown in primary human foreskin fibroblasts (HFF cells), which offer several of advantages:

(i) The large, flat morphology of HFF cells provides an extensive plasma membrane, permitting multiple cycles of parasite replication within each cell before lysis. HFF cells therefore yield high parasite titers with minimal cell debris. (ii) HFF cells are strongly contact inhibited, permitting confluent monolayers to be prepared many weeks in advance and maintained until needed. (iii) Confluent HFF monolayers are highly resistant to many metabolic inhibitors (as a consequence of their strong contact inhibition), facilitating drug selection of parasites. (iv) As primary human cells, HFF cells provide a host comparable in certain respects to a clinical infection. Disadvantages of HFF cells include their limited replicative lifespan, slow growth rate (especially at high passage number), differences between HFF cultures available in different labs, and the possibility of contamination with latent human viruses.

HFF cells are maintained according to standard tissue culture techniques for monolayer cells and are released for passage using trypsin digestion. Care must be taken not to overtrypsinize HFF cells, as they are readily damaged. HFF cells are inoculated at moderate density (>2000 cells/cm^2) into tissue culture flasks (or plates, or wells) in modified Eagle's medium (MEM; ~0.3–0.4 ml/cm^2) containing 10% heat-inactivated newborn bovine serum (*HFF medium*), and incubated at 37°C in a humidified CO$_2$ incubator. Cells typically reach confluence (~3 × 10^4 cells/cm^2) in ~5–10 days, depending on passage number. Acid production is sufficiently low in HFF cells that flasks may be sealed after CO$_2$ levels have equilibrated (typically overnight), to minimize culture contamination. Confluent monolayers may be maintained for several weeks without replacing the original medium, but cultures differing in passage number, age, density, etc. may exhibit significant differences in their ability to support parasite growth. It is therefore advisable to use sister flasks of HFF cells set up in parallel whenever results from parallel parasite infections are to be compared.

3. Routine Parasite Culture

Because tachyzoite replication takes place intracellularly, serum growth factors are necessary primarily to satisfy the host monolayer rather than the *T. gondii* parasites themselves. (For labeling tachyzoite proteins with [^{35}S]-methionine, intracellular parasite replication appears to proceed normally for at least 12 hr in serum-free, methionine-free medium.) To minimize the possibility of exposing parasites to serum antibodies and complement, we typically replace the HFF medium with 1% heat-inactivated fetal bovine serum (FBS) in MEM (*Infection medium*) prior to inoculation with parasites. Certain transformed cell lines require higher serum content for maintenance in culture, but 1% FBS is sufficient for HFF cells once the cultures have reached confluence. For drug studies, dialyzed FBS may be employed.

T. gondii tachyzoites can be maintained indefinitely by serial passage, although continued *in vitro* culture may affect the ability of parasites to traverse the sexual cycle in cats or form tissue cysts in animals (as is apparently the case for the RH strain). Complete lysis of a 25-cm^2 T-flask (T25) containing confluent HFF cells typically produces $\sim 5 \times 10^7$ RH-strain tachyzoites. Because tachyzoites die with a half-life of ~ 10 hr unless they invade a new host, it is important that extracellular parasites be used shortly after lysis of the cell monolayer. Plaquing efficiencies for freshly "lysed-out" RH-strain parasites are typically ~ 20–40%, but other strains may exhibit lower viability, even if released from the host cells by syringe passage (see below).

Passage for routine parasite maintenance

1. Aspirate HFF medium from a T25 flask containing confluent HFF cells. [Transformed cell lines (CHO, J774, LM, Vero, etc.) may also be employed for parasite growth, but infection must be coordinated with cell growth so that parasite lysis peaks close to the time of cell confluence.]

2. Add 9 ml fresh Infection medium.

3. Infect with freshly lysed-out tachyzoites: a 1-ml inoculum of RH-strain parasites ($\sim 5 \times 10^6$ tachyzoites) will completely destroy the host cell monolayer in 2 days; a 0.1-ml inoculum will lyse out in 3–4 days.

4. Incubate at 37°C in a humidified CO_2 incubator. Once pH equilibrium has been attained (typically overnight, depending on the rate of gas exchange) flasks may be sealed to minimize the risk of contamination.

5. Monitor at least daily for parasite growth, using an inverted microscope equipped with phase-contrast optics. Intracellular parasites are visible as banana-shaped organisms within the relatively transparent, intracellular parasitophorous vacuole. As replicate replication proceeds, vacuoles containing 2, 4, 8, etc., parasites become apparent, often organized as "rosettes" with the parasites' apical ends directed outward (Fig. 2). As vesicles swell still further through continued parasite replication, cells become grossly distended and sausage-like. After cell lysis, emerging parasites often squirm vigorously. Extracellular parasites are markedly more refractile ("phase-bright") than intracellular forms; "phase-dull" extracellular parasites are lysed, and fail to exclude trypan blue and other vital stains.

6. When the host cell monolayer is destroyed, this entire procedure may be repeated from step 1, above. All waste material should be decontaminated with bleach, alcohol, or by autoclaving before disposal.

Once established within the intracellular parasitophorous vacuole, tachyzoite replication proceeds with a doubling time of ~ 7.5 hr (RH-strain). The time between infection and lysis of the host cell is variable, and probably depends chiefly on space constraints: vacuoles in smaller cells (or within narrow extensions of large cells) typically rupture after 2–4 parasite doublings (≤ 16 tachyzoites), whereas vacuoles within the body of larger cells (e.g., HFF) may contain as many as 256 (2^8) tachyzoites before lysing out of the host cell. No specific trigger for host cell lysis has been identified, and it is likely that lysis is simply a function of the cell's inability to contain an exponentially increasing parasite load. As noted above, emergence from the host cell is associated with a dramatic induction of parasite motility. Parasites may be at any stage in the replicative cycle when released from the host cell, and it is not uncommon to observe 'Siamese twins' afloat in the supernatant medium due to host cell lysis prior to the completion of endodyogeny (cytokinesis).

4. Purification of Tachyzoites

T. gondii tachyzoites may be purified by various means, although purification is not generally necessary for routine passage. The banana-shaped tachyzoites are ~6 μm in length, but only ~2 μm in diameter, and parasites may therefore be purified as a single cell suspension by filtration through 3-μm-pore-size polycarbonate filters (Nuclepore). Aggregated parasites and most host-cell debris remain behind. Filters and filter holders are available in a variety of sizes; a single 47-mm-diameter filter will generally suffice for filter purification of parasites from two 175-cm² T-flasks (T175s) of HFF cells. Small fragments of host cell debris remain in the supernatant following centrifugation for 20 min at 1500 × *g*. Tachyzoites may also be purified by filtration over glass fiber or cellulose columns (Bodmer *et al.*, 1972). Parasite density is determined using a standard hemocytometer.

Many parasite strains fail to emerge from the host cell monolayer as cleanly as the virulent RH laboratory strain. For such strains, tachyzoite yields may be improved by scraping the infected monolayer with a rubber policeman and forcing this material through a 27 gauge needle. Forcibly released parasites can be filtered as above.

5. Optimizing Production of Viable Parasites

Although tightly synchronous infection with *Toxoplasma* cannot be achieved at present (as noted above, the time between infection and cell lysis is variable, and host cells may lyse and release parasites at any stage during their replicative cycle), it is nevertheless possible to observe fluctuations in parasite titer in the supernatant medium, caused by alternating cycles of intracellular replication followed by parasite emergence. To maximize the production of viable parasites, it is necessary to consider the inoculating titer relative to total host cell number. If the penultimate cycle of infection results in lysing 50% of the available host cells, for example, the total yield of viable parasites will be suboptimal—residual extacellular tachyzoites from that round of infection will lose viability while parasites are replicating in the remaining host cells. For large-scale culture, it is therefore advisable to inoculate many flasks at high density, rather than inoculating at low density and waiting for the eventual destruction of the host cell monolayer. To favor the relatively synchronous emergence of parasites grown in multiple flasks, it is helpful to use host cell monolayers that are set up and maintained in parallel.

Large-scale culture protocol

1. Infect T25 flask as described above.
2. When parasites emerge, divide the entire 10-ml culture ($\sim$2–6 $\times$ 10^7 tachyzoites) between two T175s, each containing a total of 50 ml Infection medium.
3. When these parasites, in turn, 'lyse out' of the host cells ($\sim$2 days), the emerging parasites are used to infect 26 $\times$ T175s.
4. Follow cultures by inspection under an inverted microscope (at least twice daily), and when the host monolayer reaches >80% lysis filter supernatant medium through 3-μm polycarbonate membrane (see above) into 6 x 250-ml conical centrifuge bottles.
5. Vigorously knock flasks against palm of hand to dislodge residual parasites, rinse the residual monolayer with 5 ml phosphate-buffered saline (PBS), filter as above, and combine with the original filtrate. Repeat this procedure.
6. Remove a small aliquot from each centrifuge bottle for quantitation using a hemocytometer.
7. Pellet parasites for 20 min at 1500 $\times$ g in a refrigerated centrifuge maintained at 4°C. Yields are improved by slow deceleration.
8. Remove supernatant medium, resuspend pellets in a small volume of PBS, combine in a 15-ml conical centrifuge tube, and repellet as above. Yield for RH-strain: 5–10 $\times$ 10^9 tachyzoites ($\sim$0.1–0.2 ml packed cell volume); ME49 strain, 2–6 $\times$ 10^9. When working with highly concentrated parasites, it is particularly important to guard against possible infection through splashes to the eyes; see safety precautions discussed below.

It is also possible to scale-up parasite production using roller culture (Leriche and Dubremetz, 1991), although continuous agitation may inhibit effective invasion and therefore increase the percentage of dead extracellular tachyzoites harvested from the supernatant medium.

6. Cloning by Limiting Dilution or Growth under Agar

Both because extracellular tachyzoites may aggregate and because the residue of lysed host cells often contains many parasites enmeshed in cell debris, it is essential that tachyzoites be purified by filtration prior to cloning. Individual parasite clones may be isolated by limiting dilution in microtiter plates containing confluent host cells. Assuming 20–40% viability of the extracellular tachyzoites, inoculation of 96-well plates with 0.25 tachyzoites per well yields a predicted frequency of 5–9 wells containing a single parasite clone, and <0.5 wells containing more than one parasite. Wells containing single parasite plaques can be identified easily using an inverted microscope, and parasite clones re-moved to T-flasks for expansion. Depending on the rate of parasite growth it is generally possible to identify clones within 7–10 days of plating, and expand these to $>5 \times 10^7$ tachyzoites (i.e., the yield from one T25) in a total of 2–3 weeks.

As an alternative to cloning by limiting dilution, parasite clones may also be isolated by plaque purification. Serial dilutions of parasites are inoculated into confluent HFF monolayers grown in 6-well plates (or larger dishes), and incubated at 37°C for 8–16 hr to permit infection. The supernatant medium is then replaced with a 1:1 mixture of 2× Infection medium and 1.8% BactoAgar (agar melted at 56°C and cooled to <41°C before addition). After the agar has solidified at room temperature, infected dishes are incubated at 37°C until plaques are visible under the microscope—typically 7–10 days for RH-strain *T. gondii*. After marking the position of plaques under the microscope, plugs containing individual parasite plaques are removed using a sterile, plugged Pasteur pipette and inoculated directly into T25s containing confluent HFF cells.

7. Long-Term Storage

Toxoplasma tachyzoites may be frozen in liquid nitrogen using DMSO as a cryoprotectant. As noted above, parasites die rapidly in the extracellular environment, but intracellular tachyzoites fare very poorly during freezing. It is therefore preferable to prepare parasites for freezing by high-titer infection of HFF host cells, and to freeze large numbers of freshly lysed-out extracellular parasites (concentrated by centrifugation) in a small volume of cryoprotectant medium.

The protocol outlined below has been optimized with respect to time of parasite harvest (intracellular vs extracellular tachyzoites; parasites in the midst of host cell lysis vs completely lysed-out cultures), cryoprotectants (DMSO vs glycerol), cryoprotectant concentration, and serum concentration. Prolonged exposure to DMSO reduces viability, and it is therefore important that parasites be frozen as soon as possible after addition of the cryoprotectant. In contrast to others' experience, we have not found that more gradual addition of cryoprotectant improves parasite viability. In our hands, slow freezing protocols fail to improve upon simply placing of parasites in a precooled freezer box at −80°C, but direct freezing in either liquid nitrogen or a dry ice/ethanol bath is detrimental to parasite survival.

Freezing protocol

1. Infect a confluent T175 containing HFF cells with the entire contents of a freshly lysed-out T25 in a total of 50 ml Infection medium and incubate overnight at 37°C.

2. Observe the culture to ensure that virtually all cells are infected and follow until ~80% of the host cell monolayer has lysed.

3. Transfer the supernatant medium to a 50-ml conical centrifuge tube (filtration is not required), remove an aliquot for quantitation, and concentrate the parasites by centrifugation for 15 min at $1500 \times g$.

4. Aspirate the supernatant medium and resuspend parasites to a concentration of at least 10^8/ml in ice-cold Infection medium.

5. Add an equal volume of ice-cold Infection medium previously supplemented with 20% heat-activated FBS and 25% DMSO and mix by gently inverting (final concentrations: 10% FBS, 12.5% DMSO).

6. Pipette 0.5-ml aliquots (larger volumes do not thaw well) into 1.8-ml capacity round-bottom cryotubes and freeze at $-80°C$ in a precooled freezer box.

7. After checking viability by thawing one vial for each sample frozen, the remaining vials may be transferred to liquid nitrogen. (Viability at $-80°C$ is variable, probably because of temperature fluctuations introduced through use of most laboratory freezers; in liquid nitrogen, parasites remain viable for at least 15 years).

It is always advisable to confirm the viability of a frozen sample prior to discarding the parental parasite culture. As for mammalian cells, speed is essential in thawing parasite samples. The round-bottom cryotubes recommended above are preferable to tubes with a molded base because of their more rapid thawing rate; glass vials are still more effective vessels for parasite storage provided that an oxyacetylene torch is available for sealing.

Thawing protocol

1. Remove a frozen vial from liquid nitrogen (or $-80°C$ freezer) directly into a beaker of water at 37°C and agitate continuously until thawed (~40 sec).

2. Decontaminate the vial surface in 70% ethanol and wipe completely dry.

3. Transfer the thawed parasite sample directly to a 15-ml conical centrifuge tube containing 10 ml Infection medium and mix gently by inverting.

4. Pellet parasites for 15 min at $1500 \times g$.

5. Aspirate supernatant medium, gently resuspend parasite pellet in 10 ml fresh Infection medium, and add to a confluent monolayer of HFF cells in a T25 flask.

6. Observe cultures to verify that infection has been established and that no contamination is present.

Some investigators report improved parasite survival when the frozen aliquot is gradually diluted into Infection medium over the course of 15 min. Alternatively, thawed parasites may be directly inoculated into T25 flasks, and DMSO removed by replacing the medium after parasites have infected the monolayer. (It is not clear whether the presence of low concentrations of DMSO during the early stages of infection and replication is more detrimental to tachyzoites than centrifugation in the presence of the cryoprotectant.)

Using wild-type RH-strain tachyzoites it is commonly possible to rescue >50% of the parasite titer present before freezing (determined by plaque assay). Other strains and parasite mutants may freeze less successfully.

B. Replication Assays

Several techniques are available for following the survival of parasites under various conditions. As shown in Fig. 3A, all four of the assays described below yield comparable results in measuring the effects of pyrimethamine, an antifolate commonly used for inhibition of *T. gondii* both in culture and in the clinic.

1. Plaque Assays

Parasite viability may conveniently be assessed by plaque assay on HFF cell monolayers. Other cell types may also be used, but the slow replication rate of HFF cells (especially in the low serum Infection medium) permits plaque assays to be carried out even under conditions where parasite replication is greatly diminished. Although plaque assays may be performed in petri dishes or multiwell plates, plaque morphology is improved when the assays can be shielded from excess vibration (such as frequent opening and closing of incubator doors); T25 flasks offer the advantage of being easily sealed, after which they may be transferred to a standard (i.e., non-CO_2, nonhumidified) incubator or warm room for the duration of the assay.

Plaque assay protocol (adapted from Foley and Remington, 1969; Pfefferkorn and Pfefferkorn, 1976)

1. Inoculate T25 flasks, 35-mm dishes, or 6- or 24-well plates containing confluent HFF cells with serial 10-fold dilutions of freshly lysed-out tachyzoites in fresh Infection medium. For T25 flasks, up to ~200 plaques per flask can reliably be scored, so inocula of 200, 2000, 2×10^4, etc., are usually prepared (depending on the expected kill curves). For 35-mm dishes or 6-well plates, a maximum of ~50 plaques can be scored, vs ~15 per well in 24-well plates. Because of the narrower range of acceptable inocula using smaller plates or wells, statistical concerns mandate shallower dilutions (e.g., threefold) and inoculation of replicate wells.

2. In addition to the above, inoculate one or two control flasks/dishes/wells with a dilution of untreated parasites calculated to produce a reasonable number of plaques (e.g., 50 plaques per T25).

3. After pH equilibrium has been established but before any host cell lysis occurs (~8–24 hr), T25 flasks may be sealed and transferred to another incubator or warm room, if desired.

4. Incubate undisturbed for at least 6 days.

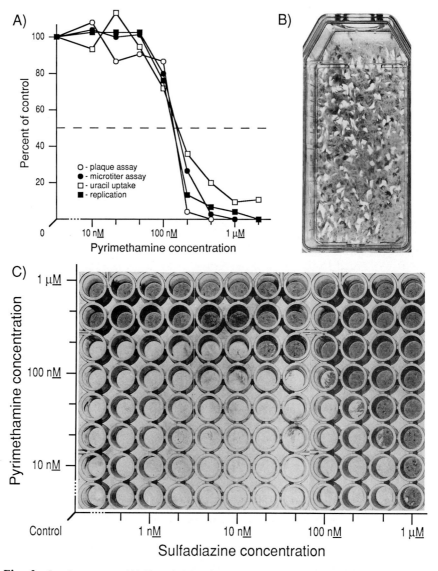

Fig. 3 *In vitro* assays. (A) Four independent assay techniques yield similar 50% inhibitory concentrations (IC_{50}) for pyrimethamine against RH-strain *T. gondii*. (○) Plaque assay (% control plaques); (●) microtiter assay (100—% maximum density); (□) uracil uptake assay (% control incorporation, 24–26 hr postdrug addition); (■) intracellular replication assay (% control doublings, 24–48 hr postdrug addition). See text for details on assay procedures. (B) Plaque assay. Eight days postinfection, the culture was fixed in methanol and stained with crystal violet to reveal clear, irregular parasite plaques against the mottled purple background typical of HFF cells. (C) Microtiter assay. HFF cells were infected with *T. gondii* tachyzoites in the presence of pyrimethamine and/ or sulfadiazine at the indicated concentrations. This highly virulent strain originally isolated from a patient with AIDS-toxoplasmosis is approximately twofold less sensitive to pyrimethamine than the RH-strain shown in (A) (parasites kindly provided by Drs. J. S. Remington and B. Danneman). As indicated in the text, such assays readily lend themselves to the analysis of multidrug interactions, such as the demonstration of approximately fivefold synergy between these two inhibitors of the folate pathway (note log–log scale).

5. Without disturbing experimental samples, remove one control flask or plate to assess the extent of plaque growth. Such controls can be examined for a maximum of 2–3 days after first being moved before satellite plaques begin to appear.

6. When controls indicate that plaques are of adequate size for visualization (usually ~9 days postinfection), experimental samples should be aspirated, rinsed in PBS, fixed in for 5 min in methanol, stained in crystal violet (5X stock: dissolve 25g crystal violet in 250 ml ethanol and add to 1000 ml 1% ammonium oxalate), and air-dried. Parasite plaques appear as irregular clear areas against the mottled violet background produced by confluent HFF cells (Fig. 3B).

To prevent formation of secondary plaques, especially if it is difficult to establish an undisturbed location for parasite incubation, medium may be aspirated 16–24 hr postinfection (i.e., after invasion has occurred, but before lysis of the host cells begins) and replaced with medium containing 1.8% BactoAgar, as described above. After plaques appear, the remaining host cells may be stained overnight by the addition of medium containing 0.01% neutral red dye on top of the agar. Parasites grown under agar typically take 1–2 days longer to form plaques than when grown in liquid culture.

2. Monolayer Disruption Assays

Parasite viability may also be determined by microtiter assays that assess the ability of parasites to lyse the host cell monolayer. These assays are particularly valuable for generating kill-curves following drug (or other) treatment. Although less quantitative than plaque assays, this approach is more rapid (results can often be obtained in 4–5 days), less dependent on precise estimation of inoculum titer, produces less plastic waste (and is commensurately less expensive), and is readily automated both for set-up and analysis. Microtiter assays also lend themselves to two-dimensional analysis (e.g., for drug interaction/synergy studies; Fig. 3C).

Microtiter assay protocol (one-dimensional assay)

1. 96-well microtiter plates containing HFF cells may be set up several weeks prior to use (typically inoculated with 10^3 cells per well in 250 μl HFF medium) and maintained until needed, but note that microtiter plates are unusually susceptible to desiccation—particularly at the edges and corners—and must be kept well humidified.

2. Aspirate medium from a 96-well microtiter plate(s) and replace with 200 μl Infection medium per well. (Aspirator manifolds and multiple-pipettors are helpful in working with microtiter plates.)

3. To the first well in each row, add 173 μl Infection medium containing 2.70X the top drug concentration desired.

4. Serially dilute 173 μl medium from one well to the next along the row. To provide an untreated control, do not add diluted drug to the last well (discard the 173-μl aliquot removed from penultimate well).

5. Dilute freshly lysed-out, filter-purified parasites to 2×10^4/ml in Infection medium and add 50 μl to each well of microtiter plate (final inoculum = 1000 parasites per well).

6. Incubate at 37°C in humidified environment containing 5% CO_2.

7. Observe daily and record cytopathic effect.

8. When the cell monolayer is destroyed in the control well (or later, to follow the progress in intermediate wells), aspirate and decontaminate medium, rinse in PBS, fix in methanol, and stain using crystal violet (see above). Quantitate monolayer destruction by densitometry, if desired.

As noted above, these assays readily lend themselves to two-dimensional analysis, producing a matrix of drug interactions (cf. Figure 3C). In setting up such assays, both drugs must be added prior to parasite addition. To ensure that drugs added in the second dimension do not affect the concentration of the first drug, increase the concentration of drug added to the first well in the first row *only* by 1.87-fold (beyond the 2.7-fold noted above; 5.03-fold above the desired final concentration). Addition of the second drug to this row (the first well in each column) will dilute the first drug to the proper final concentration.

3. Incorporation of [³H]-Uracil

Unlike its mammalian host cells, *T. gondii* tachyzoites are capable of taking up and utilizing uracil directly for pyrimidine salvage (Schwartzman and Pfefferkorn, 1981; Iltzsch, 1993). [³H]-Uracil therefore serves as a parasite-specific metabolic label for measuring viability (Pfefferkorn and Pfefferkorn, 1977). As a metabolic assay, uracil uptake does not require multiple cycles of parasite replication, and is therefore even more rapid than microtiter assays. Twenty-four-well plates provide a convenient level of labelling, but the protocol outlined below can readily be adapted for other configurations.

Uracil uptake assay protocol (adapted from Pfefferkorn and Guyre, 1984)

1. Inoculate confluent monolayers of HFF cells grown in 24-well plates with freshly harvested, filtered *T. gondii* parasites in 1 ml Infection medium and incubate under desired conditions.

2. Add 5 μCi [5,6-³H]-uracil (30–60 Ci/mmol) and incubate for 2 hr at 37°C. (Note: [³H]-uracil degrades within days in aqueous solution at 4°C and is stable for no more than a few weeks at −80°C. Aliquots may be stored for a period of many months in 70% ethanol at −80°C, however, and lyophilized and resuspended in culture medium immediately prior to use.)

3. Chill the tray(s) at −20°C for 2 min, add 1 ml ice-cold 0.6 N trichloroacetic acid (TCA) to existing medium, and incubate on ice for 1 hr to fix the monolayer.

4. Remove TCA solution to radioactive waste container, rinse, and immerse the plate(s) in running water bath overnight (or at least 4 hr).

5. Shake the plate(s) dry and add 0.5 ml 0.1 *N* NaOH per well. Incubate at 37°C for 1 hr to dissolve the TCA precipitate.

6. Count 0.25 ml from each well in scintillation fluid suitably acidified to neutralize NaOH (3.6 ml glacial acetic acid per gallon; use 3 ml scintillant per sample). Inoculation of plates with 10^4 RH-strain tachyzoites 24 hr prior to labeling typically results in incorporation of 10^4 cpm during the course of a 2-hr incubation (background from uninfected host cells is typically ~300 cpm).

Although uracil incorporated into extracellular parasites is precipitated as readily as for intracellular parasites, the TCA treatment does not necessarily fix extracellular parasites to the cell monolayer. If a substantial number of extracellular parasites are present, incorporation in the TCA supernatant may be collected by filtration through glass-fiber (Whatman GF/C) filters (Pfefferkorn and Pfefferkorn, 1977).

4. Direct Measurement of Doubling Time

As a more direct means of assessing the effect of various treatments, parasite replication rates may be determined by following the number of cell doublings as a function of time. Although somewhat laborious, these assays provide data at the level of individual parasites rather than a population average, without requiring the multiple cycles of infection necessary for plaque production.

Taking advantage of the fact that all parasites within a single vacuole replicate synchronously, it is possible to determine the number of mitotic divisions since infection as $\log_2$(parasite number). Although it is not currently possible to synchronize precisely parasite replication between different vacuoles (see above), a reasonable degree of synchronization can be established by inoculating cells in a small volume of medium and replacing the medium after ~4 hr—rinsing away parasites that failed to infect during the 4-hr window. The average number of cell divisions in the population (e.g., average of 100 randomly selected vacuoles) is scored at 4 to 8-hr intervals.

Alternatively, the replication rate of individual vacuoles may followed by video microscopy or scored by marking individual vacuoles and following the replication rate by visual inspection of the same exact vacuoles at 4-hr intervals.

C. Safety Issues

Thirty percent of the U. S. population is estimated to become infected with *T. gondii* in the course of their life (risk factors include gardening or farming, preference for rare meat, and cat ownership; Frenkel, 1973; McLeod and Remington, 1987). Infection in an otherwise heathly adult is normally asymptomatic (McLeod and Remington, 1987). Moreover, as opposed to the bradyzoite (tissue cyst) and oocyst forms, the tachyzoite parasite form discussed in this review is only infectious by direct introduction into the bloodstream or heavy inoculation of the mucous membranes (e.g., through an eye-splash). Nevertheless, *in vitro* culture probably selects for increased virulence, and regardless of the worker's antitoxoplasma serum titer, any needle stick or other potentially infectious exposure should be considered a presumptive case of toxoplasmosis and treated promptly with pyrimethamine–sulfadiazine accompanied by leukovorin rescue (Brooks *et al.*, 1987).

All work with live organisms should be carried out in disposable tissue culture ware using class-2 biosafety hoods. Eye protection should be worn when working outside of laminar flow hoods, particularly when using highly concentrated parasite cultures. The use of sharp implements (needles, Pasteur pipettes, etc.) should be limited, and flasks, pipettes, etc., that contact live parasites should be immediately decontaminated with alcohol, bleach, or acid, or by autoclaving or boiling. All laboratory personnel must be informed of the potential danger of *Toxoplasma* as a human pathogen and alerted to symptoms indicating possible parasite infection. Additional considerations apply to work with drug-resistant organisms (see below). Because of the special danger of toxoplasmosis for pregnant women and HIV-positive individuals, such individuals should not work with live parasites.

III. Molecular Transformation Systems for *Toxoplasma*

A. Vectors

A series of transformation vectors has been developed for *T. gondii*, based on regulatory and coding sequences derived from the parasite's fused dihydrofolate reductase–thymidylate synthase gene (DHFR-TS; Donald and Roos, 1993; 1994a). Figure 4 illustrates the basic structure of several constructs employed in the transformation schemes described below; DHFR-TS genomic and cDNA sequences are available from GenBank, with the Accession No. L08489. All plasmids are based on Bluescript pKS plasmid (Strategene), with polylinker cloning sites indicated. For transient transformation, a chloramphenicol acetyltransferase (CAT) reporter gene was engineered for expression under control of genomic DHFR-TS 5' and 3' noncoding sequences (p*dhfr*CAT*dhfr*). Stable transformation vectors employ these regulatory regions linked to DHFR-TS coding sequences, either as a cDNA-derived 'minigene' (pDHFR-TSc3) or in the native genomic configuration (pDHFR-TSg8). Selection is facilitated by the introduction of mutations in the DHFR domain predicted—by analogy with drug-resistant *Plasmodium falciparum* isolates (Hyde, 1990; Wellems, 1991)—to provide resistance to pyrimethamine (indicated by asterisks in Fig. 4). To test the expression of foreign genes linked to the DHFR-TS selectable marker, a CAT reporter construct under control of 5' and 3' noncoding sequences derived from the *T. gondii* major surface antigen P30 (SAG$^{1/2}$CAT

Soldati and Boothroyd, 1993) was introduced upstream of the DHFR-TS selectable marker (p*sag*CAT::DHFR-TSc3). Further details of vector construction have been reported elsewhere (Donald and Roos, 1993, 1994a).

B. Transient Expression

Flanking sequences from a variety of *T. gondii* genes have been successfully used to drive expression of CAT reporter plasmids introduced into parasites by electroporation (Soldati and Boothroyd, 1993; Donald and Roos, 1993). As shown in Fig. 5A, promoter sequences are essential for expression, whereas 3′ flanking sequences provide only modest enhancement of the signal (downstream sequences may be more important for other *Toxoplasma* promoters; Soldati and Boothroyd, 1993). Electroporation of HFF cells with *T. gondii* CAT constructs yields no detectable expression.

Basic electroporation protocol (Soldati and Boothroyd, 1993; Donald and Roos, 1993)

1. Pellet freshly harvested, filter-purified tachyzoites by centrifugation at $1500 \times g$ for 15 min and resuspend in filter-sterilized "intracellular" electroporation buffer (van den Hoff *et al.*, 1992) (*electroporation buffer:* 120 mM KCl, 0.15 mM CaCl$_2$, 10 mM K$_2$HPO$_4$/KH$_2$PO$_4$, pH 7.6, 25 mM HEPES, pH 7.6, 2 mM EDTA, 5 mM MgCl$_2$, 2 mM ATP, 5 mM glutathione). A sterile solution containing all salts and buffers may be prepared in advance and kept indefinitely at room temperature, to be supplemented with fresh ATP and glutathione (GSH) and resterilized by filtration through a 0.22-μm filter immediately prior to use.

2. Repellet parasites, resuspend in electroporation buffer to a concentration of 3.3×10^7/ml, and transfer 300 μl parasites (10^7) to a sterile 2-mm-gap electroporation cuvette.

3. Add 50 μg plasmid DNA resuspended in 100 μl electroporation buffer and mix gently by pipetting.

4. Electroporate parasites using a single 1.5-kEV pulse at a resistance setting of 24 Ω (pulse time ~0.25 msec; conditions optimized using a BTX model 600 Electro Cell Manipulator).

5. Leave cuvette undisturbed for 15 min at room temperature.

6. Inoculate parasites into a T25 flask containing a confluent monolayer of HFF cells in 10 ml infection medium and incubate at 37°C. If direct comparisons are desired, transfer only 0.3 ml (75%) of the electroporation mix, as it is difficult qantitatively to remove the entire sample from the cuvette.

Electroporation under these conditions diminishes parasite viability by ~25% (by plaque assay; but note that only 20–40% of extracellular parasites are viable at the time of harvest, as discussed above). CAT activity can be detected beginning ~8 hr posttransfection and continues to increase for at least 24 hr. (As

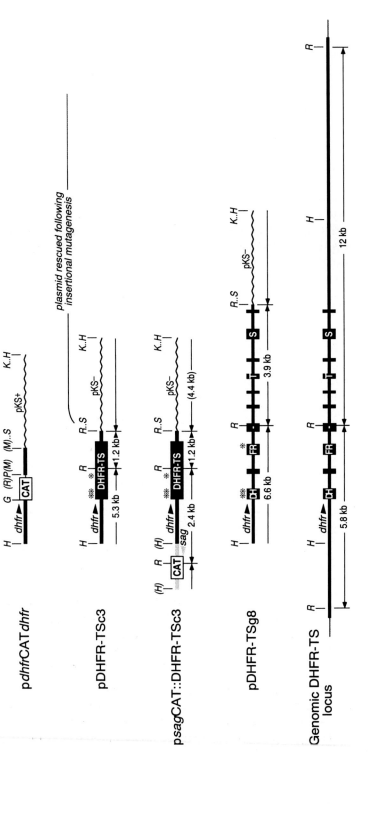

CAT is an unusually stable enzyme, these levels probably reflect accumulated product; the time of peak CAT transcription has not been determined.) Under these conditions, at least 50% of all viable parasites express CAT protein detectable by immunofluorescence (Fig. 5B).

Although the above conditions have been optimized for expression of p*dhfr*CAT*dhfr* in RH-strain tachyzoites, they are probably generally applicable, as acceptable electroporation parameters are rather broad: charging voltages from 0.25 to 1.5 kEV/mm and resistance settings from 0 to 64 Ω produce significant CAT activity. CAT activity can be readily detected using ≤ 1 μg plasmid and 10^6 parasites, and total expression increases as an approximately linear function of parasite number and plasmid concentration up to 10^7 parasites and 50 μg plasmid. Further increases (up to 5 x 10^7 parasites and 200 μg plasmid) enhance the CAT signal somewhat, but with decreased efficiency on a per parasite or per microgram basis. Linearized and intact circular plasmid DNA are equally effective in transient transformation assays.

Interestingly, of commonly used reporter genes, only CAT appears to function effectively in *Toxoplasma*. We have observed only trace expression from luciferase constructs and no expression whatsoever of β-glucuronidase. The possibility of transient expression is not limited to CAT, however—pyrimethamine-

---

Fig. 4 Transformation vectors for *T. gondii*. The DHFR-TS genomic map is shown at bottom; solid boxes indicate coding sequence—fragmented by 9 introns. The 5' end of mature mRNAs lies ~350 nt upstream of the translational start; polyadenylation can occur at >10 sites from ~250 to ~900 nt downstream of the stop codon. Genomic *Eco*RI sites (R) and *Hin*dIII sites (H) are indicated, along with the size of expected *Eco*RI restriction fragments.

All plasmid vectors are shown in linearized form p*dhfr*CAT*dhfr* contains 1.4 kb of 5' and 0.8 kb of 3' genomic sequence from the parasite's DHFR-TS gene fused to a bacterial CAT reporter cassette using synthetic *Bgl*II sites (G) introduced 7–12 nt upstream of the ATG codon in both CAT and the 5' DHFR-TS untranslated region. Wavy lines indicates pKS vector sequences; most of the additional restriction sites shown derive from the multiple cloning site: H, *Hin*dIII (ligated to genomic *Hin*dIII site); R, *Eco*RI; P, *Pst*I; M, *Sma*I (contains blunt-ligated DHFR-TS 3' end); S, *Sac*I; K, *Kpn*I. Restriction sites destroyed in the course of vector construction are indicated by parentheses. ".." denotes additional sequence derived from the pKS polylinker, which may be useful for vector manipulation. pDHFR-TSc3 is a "minigene" in which genomic 5' flanking sequences were fused to a full length cDNA clone. Restriction sites are abbreviated as above; internal *Eco*RI site corresponds to genomic site shown. The DHFR-TS 3' end terminates at an *Eco*RI site introduced during cDNA cloning. *Eco*RI restriction fragments expected for digestion of circular plasmid are indicated. Asterisks represent point mutations introduced to confer resistance to pyrimethamine (see text and Donald and Roos, 1993). Bracket shows plasmid sequence rescued along with flanking genomic DNA in insertional mutagenesis protocol (see text). p*sag*CAT::DHFR-TSc3 is identical to the above, except for the insertion of a CAT reporter gene in opposite orientation, under control of the SAG1 promoter (vector SAG ½ CAT kindly provided by Drs. D. Soldati and J. C. Boothroyd). The 4.4-kb *Eco*RI fragment does not hybridize with the DHFR-TS probe used in Fig. 6. Both the *dhfr* and the *sag* promoters are shown by arrows representing directionality. pDHFR-TSg8 is a genomic clone spanning the DHFR-TS gene from the endogenous *Hin*dIII site to a synthetic *Eco*RI site in the 3' flanking region.

(A)

pCAT

pCAT-*dhfr*

p*dhfr*-CAT

p*dhfr*-CAT-*dhfr*

No parasites

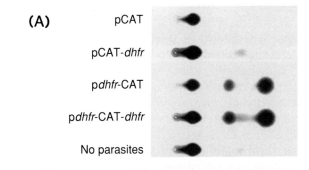

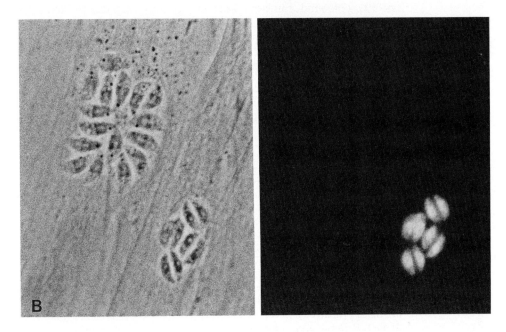

Fig. 5 Transient expression of recombinant CAT in electroporated *T. gondii*. (A) Extracellular parasites were transfected with a reporter cassette flanked by 5′ and/or 3′ sequences derived from the DHFR-TS gene, inoculated into HFF cells, and assayed for CAT activity 24 hr later. The 5′ DHFR-TS domain confers high levels of CAT expression; the 3′ domain is helpful but not essential. (B) 24 hr after transfection with p*dhfr*–CAT–*dhfr*, 55% of the parasitophorous vacuoles in infected HFF cells express CAT by immunofluorescence (right panel; compare with phase-contrast image at left). Each vacuole contains the progeny of a single parasite.

resistant DHFR-TS genes can also be detected in ~50% of viable electroporated parasites (Donald and Roos, 1993), and other groups have explored additional markers (unpublished).

Detection of CAT activity in transfected Toxoplasma tachyzoites

1. Harvest parasite-infected host cell monolayers ~24 hr postinfection using a rubber policeman, wash monolayer with PBS, and pellet combined mixture at 1500 × *g* for 15 min.

2. Resuspend in 0.5 ml ice-cold electroporation buffer (without ATP or GSH) supplemented with 15% glycerol and 10 m*M* β-mercaptoethanol and transfer to a microfuge tube. Samples may be stored frozen at −80°C at this point if desired.

3. On ice, add phenylmethylsulfonyl fluoride (PMSF) to 1 m*M* and sonicate 2 × 20 sec at ~55 W using a microtip (e.g., setting 4 on Branson Model 250 Sonifier). (PMSF may be prepared in advance as a 100 m*M* stock and stored in ethanol at −20°C. Thaw on ice immediately before use.)

4. Clarify by centrifugation for 10 min at 4°C in a microcentrifuge (at top speed) and transfer supernatant to fresh microfuge tube.

5. Pipette 50 μl of the sonic extract directly into the bottom of a microfuge tube and place the open tube(s) in the microfuge. The remaining sonic extract may be stored at −80°C for further use (e.g., for protein assay).

6. Carefully pipette a 5 μl drop of 40 m*M* acetyl coenzyme A and 3.5 μl [^{14}C]-chloramphenicol (30–60 Ci/mmol) on opposite walls of each tube. Spin briefly to initiate reactions synchronously.

7. Gently tap tubes to mix thoroughly and incubate for 30 min at 37°C.

8. Stop reactions by the addition of 300 μl ethyl acetate, vortex, and centrifuge 10 min in microfuge.

9. Remove the organic (upper) phase to fresh microfuge tube, lyophilize, resuspend in 10 μl ethyl acetate, and spot samples on glass thin-layer chromatography (TLC) plate (TLC plate should be dried for ~2 hr at >70°C before spotting).

10. Separate mono- and di-acetylated chloramphenicol from unacetylated substrate by ascending chromatography in an equilibrated chamber containing chloroform/methanol (95:5)

11. After plate has dried, cover with single layer of plastic wrap and expose to X-ray film or phosphor-imager cassette.

The relative strength of the various available *Toxoplasma* promoters remains unexplained. Surprisingly, the SAG1 promoter (from the P30 major surface antigen) appears to be the weakest in transient transformation assays, whereas promoters derived from DHFR-TS, ROP1 (a rhoptry protein), and TUB1 (β-tubulin) are approximately 5- to 10-fold more powerful. TUB1 is probably the strongest *Toxoplasma* promoter yet identified (Soldati and Boothroyd, 1993). In addition to the full-length 1.4-kb DHFR-TS promoter illustrated, a variety of comparably effective truncated promoters have also been engineered, including a fully active PCR cassette and *Nsi*I constructs spanning the ATG initiation codon.

C. Stable Transgene Expression and Overexpression

To develop a marker for selection of stable transformants, DHFR-TS mini-genes lacking introns were constructed containing 5' and 3' genomic flanking sequences fused to a cDNA-derived coding region (pDHFR-TSc3 in Fig. 4). Although DHFR-TS gene structure and expression appears to be unchanged in pyrimethamine-resistant mutants selected *in vitro,* several point mutations were introduced into the coding sequence by analogy with drug-resistant *P. falciparum.*

Mutation T83N (Thr^{83}→ Asn, formerly designated M3 in Donald and Roos, 1993; 1994) is analogous to Ser^{108}→ Asn in *Plasmodium,* which confers moderate pyrimethamine resistance.

F245S (Phe^{245}→ Ser, formerly M4) confers low-level pyrimethamine resistance to *P. falciparum* (Phe^{223}→ Ser) (known only from *in vitro* studies; Tanaka *et al.,* 1990).

S36R (Ser^{36}→ Arg, formerly M2; analogous to Cys^{59}→ Arg in *P. falciparum*) confers no phenotype in isolation, but dramatically enhances resistance when combined with Ser^{108}→ Asn in *Plasmodium* or either T83N or F245S in *Toxoplasma.*

For further discussion of these mutants, see Donald and Roos (1993); for reviews of DHFR-TS mutations in pyrimethamine-resistant malaria, see Hyde (1990) and Wellems (1991).

Upon transformation of pDHFR-TSc3/T83N or pDHFR-TSc3/F245S into *T. gondii* tachyzoites and selection in 1 μM pyrimethamine (approximately sixfold above the IC$_{50}$; see Fig. 2), transient replication above background levels was seen in ~50% of transfected parasites (Donald and Roos, 1993). Maintaining drug selection allowed the isolation of pyrimethamine-resistant parasites at a frequency of ~10^{-6}. These mutants remain drug-resistant even in the absence of continued selection. Hybridization with genomic DNA isolated from cloned, stably resistant mutants (Fig. 6) reveals the presence of multiple copies of transgenic DNA integrated into the parasite genome either as a tandem array (lane 2) or replicating as a stable extrachromosomal episome (lane 3). Selection in 600 nM pyrimethamine permitted isolation of drug resistant clones at somewhat higher frequency, and these clones harbor commensurately fewer DHFR-TS transgenes (lane 1).

DHFR-TS vectors containing double mutations T83N + S36R or F245S + S36R produced stable transformants at much higher frequency: ~5% of all viable transformants develop resistance to 1 μM pyrimethamine, even in the absence of drug selection (Donald and Roos, 1993; 1994). The majority

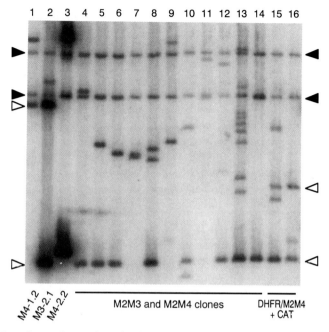

Fig. 6 Nonhomologous integration of stable DHFR-TS transgenes. DNA from drug-resistant parasite clones was digested with *Eco*RI and probed with DHFR-TS coding sequences. Each lane represents an independently cloned parasite line transfected with supercoiled circular plasmid pDHFR-TSc3 (see Fig. 4). Lanes 1–3 were isolated from clones transfected with single DHFR mutations M3 (T83N) or M4 (F245S); 4–14 were transfected with mutants M2M3 (T83N + S36R) or M2M4 (F245S + S36R). The endogenous 12- and 5.8-kb genomic bands are indicated by solid arrowheads, and were unaltered in any of these clones. Plasmid-derived bands are indicated by open arrowheads (5.3 and 1.2 kb for lanes 1–14). Most parasites transfected with the double mutants contain one plasmid band and one novel band, indicating that a single plasmid integrated by nonhomologous recombination. Resistance to 1 μM pyrimethamine is conferred by the M3 or M4 mutations only when present in high copy number (lanes 2 and 3). M4-1.2 was selected in 600 nM drug. Clones in lanes 15 and 16 were transfected with p*sag*CAT::DHFR-TSc3. These parasites stably express both pyrimethamine-resistant DHFR-TS and CAT. For further discussion see text. Reprinted from Donald and Roos (1993), with permission.

of these mutants contain single-copy insertions of the transfected plasmid, integrated into parasite DNA by nonhomologous recombination at sites dispersed throughout the genome (lanes 4–14).

Protocol for stable transformation (Donald and Roos, 1993)

1–5. Follow basic electroporation protocol, above.

6. Inoculate parasites into a T175 flask containing confluent HFF cells in 50 ml Infection medium *prepared with dialyzed serum* and supplemented with

1 μM pyrimethamine and incubate at 37°C. If multiple independent clones are desired, up to 10 T25 flasks may be inoculated. Pyrimethamine is stored at 4°C as a 10 mM stock in ethanol, and diluted in culture medium before use. As pyrimethamine is a "static" rather than "cidal" inhibitor, drug selection may be applied to parasites immediately after electroporation, or added up to 24 hr post-transfection.

7. When the host cell monolayer nears complete destruction, filter-purify extracellular tachyzoites and inoculate 96-well plates for cloning by limiting dilution as described above (maintaining drug selection, in medium prepared with dialyzed serum). The frequency of stable resistance among parasites that emerge from a first cycle of infection depends on the particular drug resistance allele employed and the precise conditions of selection. It is therefore advisable to inoculate microtiter plates at several dilutions. Alternatively, parasites may be passaged for another cycle in T-flasks prior to cloning.

Linear and circular DHFR-TS plasmids are equally effective at stable transformation. In most cases, integration of circular molecules occurs without significant loss or rearrangement of either host or plasmid genetic material—consistant with a model involving reciprocal crossing over at a single site (see below). Linear molecules also generally integrate at a single site without rearrangement, but are usually accompanied by the loss of a few (1–30) terminal nucleotides, and are sometimes ligated to short fragments derived from transfected plasmid DNA.

Selection of the particular pyrimethamine resistance allele to be used for transformation depends on the application at hand. Lanes 15 and 16 in Fig. 6 demonstrate that a foreign reporter gene (e.g., CAT) can be successfully integrated into the parasite genome in parallel with the DHFR-TS marker. These genes are stably maintained and expressed in the absence of specific selection. Insertion of the reporter gene upstream of DHFR-TS and in the opposite orientation (p*sag*CAT::DHFR-TSc3 in Fig.4) results in higher CAT expression than introduction in the same orientation as DHFR-TS. It is probable (but yet untested) that transfections using low-level pyrimethamine-resistance alleles coupled with a foreign gene would result in overexpression of both DHFR-TS and the gene of interest. By analogy with other systems (Kaufman, 1990), it is also likely that such mutants isolated (at higher frequency) in lower pyrimethamine concentrations could be "stepped-up" to higher drug levels with concomitant coamplification of the desired gene.

The surprisingly high frequency of stable integration using high-level pyrimethamine-resistance alleles permits many exciting genetic approaches to the study of *T. gondii*. It should be noted, however, that the traditional reliance on antifolates for clinical management of acute toxoplasmosis means that such mutants pose an unusual hazard (see discussion of special considerations for working with pyrimethamine-resistant organisms, below). Unfortunately, although other selectable markers are available for *Toxoplasma* (Kim *et al.*,

1993; Sibley *et al.*, 1994), the frequencies of resistance provided by these systems are ~100-fold lower than for the DHFR-TS double mutants, precluding some of the more promising applications described below.

D. Nonhomologous Recombination, Insertional Mutagenesis, and Marker Rescue

The high frequency of nonhomologous chromosomal insertion observed using cDNA-derived DHFR-TS minigenes suggested that it might be possible to knock out gene function efficiently by insertional mutagenesis, tagging the locus in the process. This approach has revolutionized the identification of important genes in other genetic systems (Berg and Spradling, 1991; Feldmann, 1991). Given that 5% of viable electroporated parasites stably integrate recombinant transgenes (Donald and Roos, 1993; 1994a), transfection of 10^7 parasites with a survival rate of 20% should yield ~10^5 independent random insertion events. Assuming completely random insertion (an unlikely prospect) this predicts an average of more than one hit in any 1-kb segment of the 8×10^7 base pair *Toxoplasma* genome.

The ability to identify insertional mutants of interest depends on a practical screen or selection strategy. Negative selectable markers such as uracil phosphoribosyltransferase (UPRT), hypoxanthine/guanine phosphoribosyltransferase (HGPRT), or adenosine kinase provide a suitable test (Donald and Roos, 1994b).

UPRT is a nonessential single-copy gene, whose loss confers resistance to fluorideoxyuridine (FUDR; Pfefferkorn, 1977). Eight transfections of RH-strain tachyzoites according to the procedure presented below yielded at least four independent FUDR-resistant clones. To ensure that mutants were independent, only one clone was isolated from each transfection, but the estimated frequency of >3×10^{-7} is within one order of magnitude of the theoretical value for truly random insertions. FUDR-resistant clones were unable to incorporate [^{3}H]-uracil and lack UPRT activity. Flanking genomic DNA was rescued along with the pKS vector by bacterial transformation and used to demonstrate that all of the FUDR-resistant mutants had integrated transfected plasmid into nearby (but distinct) sites within the parasite genome. Corresponding cDNA and genomic clones were isolated to provide the complete sequence of the *T. gondii* UPRT gene. To confirm the function of the putative UPRT locus and provide a mutant incapable of reversion to wild-type, homologous recombination technology (see below) was employed to replace the endogenous locus in wild-type parasites with a genomic clone from which essential coding sequences were deleted. Transfection of both FUDR-resistant insertional mutants and the UPRT "knock-out" clones with either reconstructed minigenes or the wild-type genomic locus restored UPRT activity.

Similar experiments have permitted identification of the *T. gondii* HXGPRT gene, based on selection for resistance to 6-thioxanthine. (Because the host cell enzyme exhibits no XPRT activity, thioxanthine is parasite specific; we are indebted to Dr. E. R. Pfefferkorn for suggesting this ingenious selection.) In addition to their value as a test for insertional mutagenesis, the UPRT and HXGPRT genes provide negative selectable markers for the development of positive–negative selection systems in *Toxoplasma* and candidates for alternative positive selection vectors (see below).

For reference in the insertional mutagenesis and marker rescue protocols outlined below, the linearized depiction of plasmid pDHFR-TSc3 shown in Fig. 4 can be considered to have integrated so as to disrupt the target gene of interest. The bracket indicates the rescued plasmid, including the bacterial vector and flanking genomic sequence from the target gene.

Protocol for insertional mutagenesis

1–6. Follow stable transformation protocol, above, using high-level pyrimethamine resistance vectors pDHFR-TSc3/T83N + S36R or pDHFR-TSc3/F245S + S36R *linearized* by restriction at the unique *Hind*III site. Although both linear and circular DNA produce insertional mutants, using linear molecules facilitates marker rescue and mapping of the resultant integrants (see below).

7. When the host cell monolayer nears complete destruction, filter-purify the pyrimethamine-resistant tachyzoites and apply desired secondary screen or selection (maintaining pyrimethamine selection as well). For identification of UPRT mutants, T175 flasks were inoculated with 10^7 tachyzoites in 5 μM FUDR. In some cases it may be desirable to force parasites out of the host cells by syringe-passage, to avoid selection against mutations that retard growth.

8. Isolate positive clones by plaque purification or limiting dilution and confirm phenotype.

Marker rescue protocol, for identification of the mutated genomic locus

1. Isolate parasite genomic DNA from putative insertional mutants.

2. Map the nature of transgene insertion by Southern analysis. In the case of a single insertion, restriction with polylinker enzymes that cut once in plasmid pDHFR-TSc3 will reveal a single-copy fragment extending into flanking genomic DNA (or two bands, depending on the location of the restriction site relative to the probe employed). Multiple independent insertions will produce multiple novel bands. Insertion may also occur as a tandem array, in which case multiple copies of the complete 6.5-kb vector will be observed, in addition to the single-copy flanking band(s). Head-to-head or tail-to-tail insertions have also been observed, producing more complicated (but predictable) restriction patterns.

As unique cutters that separate the selectable DHFR-TS marker from bacterial vector sequences, *Spe*I and *Xba*I are particularly useful for mapping. These enzymes may be used to map flanking restriction sites of suitable size for marker rescue. Additional enzymes that may be useful for marker rescue include *Eco*RI,

*Pst*I, *Xma*I, *Eag*I, *Sac*II, and *Sac*I. Note: restriction with enzymes expected to cleave close to the ends of the introduced plasmid (e.g., *Hind*III, *Cla*I) may yield confusing results due to deletion of these sites by intracellular exonucleases prior to integration, as noted above.

3. Restrict genomic DNA with an enzyme(s) predicted to generate fragments of suitable size for bacterial replication as a plasmid (up to ~20 kb, including the entire 3-kb pKS vector). Note: if mapping reveals multiple tandem insertions, double digestion may facilitate plasmid rescue from the tandem array. For example, *Spe*I/*Sph*I may be employed if the flanking *Spe*I site is of suitable size and is proximal to the flanking *Sph*I site; 4.6-kb fragments derived from the tandem array will have incompatible ends.

4. Ligate 500 ng restricted genomic DNA overnight in dilute solution (<1μg/ml), to promote intramolecular ligation.

5. Transfect competent *Escherichia coli* with 50 ng of the ligated genomic DNA, select on ampicillin plates, and isolate resistance plasmids. It is essential to use bacteria of high transformation competency for this procedure ("library"-competent or electrocompetent cells).

6. Confirm recovery of the expected size plasmid. Genomic DNA flanking the insertion point may be sequenced using standard pKS sequencing primers.

7. Probe Southern blots of DNA from wild-type parasites and mutant clones with flanking sequence fragment, to confirm that independent mutants all harbor insertions in the same region of genomic DNA.

8. Isolate cDNA clones and genomic clones corresponding to the tagged sequence. (Genomic and cDNA libraries for *T. gondii* are available from several sources, including the AIDS Research and Reference Reagent Program of the NIH.) Genomic DNA fragments or reconstructed minigenes may be introduced into mutant parasites according to the above transformation protocols, to confirm restoration of wild-type phenotype.

Because integration of electroporated plasmid DNA does not depend on pyrimethamine treatment, it should not be necessary to employ DHFR-TS vectors if direct selection for the desired mutants is sufficiently powerful—a decided advantage in terms of laboratory safety (see below). In other cases, pyrimethamine resistance is used to select for transgenic parasites, which can then be further screened by procedures appropriate to the gene of interest. As *Toxoplasma* tachyzoites are haploid, insertional mutagenesis of essential genes is lethal, but it should be possible to use the above procedure to tag and clone any nonessential gene for which a suitable selection or screen is available.

E. Homologous Recombination: Pseudo-diploids, Gene Knock-outs, and Perfect Gene Replacement

Although the cDNA-derived vectors discussed above appear to integrate into the *Toxoplasma* genome at random, vectors containing more extensive regions

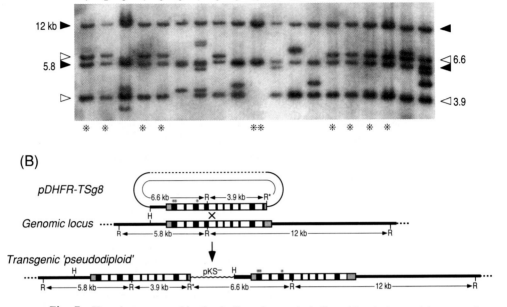

Fig. 7 Homologous recombination in *Toxoplasma*. As indicated by single asterisks, many drug resistant parasites transfected with alleles derived from the complete genomic DHFR-TS locus (plasmid pDHFR-TSg8 in Fig. 4) contain only endogenous and plasmid-derived bands of the expected sizes (solid and open arrows, respectively). As no free plasmid remains in these parasites, integration must have occurred by reciprocal crossing-over, as indicated in (B) and described in the text. Lane 10 (**) shows a parasite clone with no detectable plasmid sequences. Direct sequencing indicates that this clone is a perfect gene replacement, harboring the mutant DHFR-TS allele at its proper locus and no trace of the original wild-type allele. For further discussion, see Donald and Roos (1994) (reprinted with permission).

of continuous sequence homology (cf. pDHFR-TSg8 in Fig. 4) preferentially integrate by homologous recombination, at the same high frequency (Donald and Roos, 1994a). As shown in Fig. 7, when an 8-kb genomic clone spanning the DHFR-TS locus was transfected as a circular plasmid (according to the stable transformation protocol outlined above), approximately half of the drug-resistant parasites harbored transgenes integrated by homologous recombination. When a 16-kb genomic clone was employed, >80% homologous recombination was observed (Donald and Roos, 1994a).

Two forms of homologous recombination have been identified in *Toxoplasma*. Integration by reciprocal crossing-over at a single site produces a ''pseudodiploid'' duplication of the locus, as diagrammed in the bottom panel of Fig. 7. Integration by gene replacement results in loss of the wild-type allele by the mutant gene (cf. lane 10 in Fig. 7). Functional replacement could occur by several different mechanisms—double crossing-over, gene conversion, or recombination between the duplicated loci in a pseudodiploid (removing either the wild-type or the mutant gene copy, depending on the site of recombination). Unlike most eukaryotic systems (Orr-Weaver *et al.*, 1981; Kucherlapati and Smith, 1988; Cruz and Beverley, 1990; Hasty *et al.*, 1991; Finbarr-Tobin and Wirth, 1992) homologous recombination in *T. gondii* is far more common using circular plasmids than linearized DNA, suggesting that the "pseudodiploid intermediate" model is probably dominant (pseudodiploidy has never been observed in tachyzoites transfected with linear DNA). Depending on the nature of the mutant allele, gene replacement results in either allelic exchange or a functional gene knockout.

Protocol for pseudo-diploid formation or gene replacement

1–6. Follow stable transformation protocol, above, using plasmid DNA containing as large a segment of the targetted locus as feasible. See notes below for further considerations in plasmid design, the use of circular vs linear DNA, and selection schemes.

7. Isolate parasite clones and screen for presence of the mutant and wild-type alleles. In DNA hybridizations or PCR reactions, the pseudodiploid configuration results in the presence of both genomic and plasmid bands (and no novel bands); gene replacement yields only the genomic fragments (or fragments expected for the mutant allele, if different).

Linking a cDNA-derived pyrimethamine-resistant DHFR-TS gene to any target sequence of interest should permit pseudodiploid formation when *circular* plasmid DNA is transfected and parasites selected for pyrimethamine resistance. Pending the development of coupled positive–negative selection vectors such as those used in animal systems (Mortensen, 1993), production of gene knock-outs in *Toxoplasma* requires either selection for resistance encoded within the locus itself or extensive screening. Where negative selection is available, as for the loss of UPRT activity in FUDR-resistant mutants, nonfunctional genes can be transfected in either linear or circular form, and mutants screened for gene replacement. Deletion of a small internal fragment (for the production of irreversible knock-outs) does not appear to affect significantly the frequency of homologous vs nonhomologous recombination. Where no direct selection is available, gene knock-outs can be produced by inserting a selectable marker (e.g., CAT or a cDNA-derived mutant DHFR-TS allele) into the middle of the target gene), transfecting linear DNA to bias against pseudodiploid formation, and screening for loss of the targetted gene against the background of nonhomologous recombinants (Kim *et al.*, 1993). Alternatively, it may be possible to transfect circular plasmids containing both a positive selectable marker and the mutant gene of interest, and to screen pseudodiploids for subsequent loss of the wild-type gene once removed from selection. For this strategy it is essential to provide comparable lengths of genomic sequence on both sides of the introduced mutation, to maximize the frequency of pseudodiploid resolution into the mutant form.

F. Cloning by Complementation

The high frequency of stable transformation suggests that it should be feasible to clone *T. gondii* genes by complementation (as is now possible for *Leishmania;* Ryan *et al.,* 1993). To test this possibility, total genomic DNA was prepared from parasites harboring a drug-resistant DHFR-TS gene (lane 10 in Fig. 7), digested with various restriction enzymes, and electroporated into wild-type tachyzoites. Transfection of parasites with 50 μg of total DNA according to the procedures described above produced pyrimethamine-resistant plaques at a frequency of ~2.5 $\times$ 10^{-6}, regardless of the size of the genomic fragment spanning the DHFR-TS locus (9 to ~35 kb) or the position of the DHFR-TS gene within that fragment. Taking into account the ~10^4-fold difference in size between the 8 $\times$ 10^7 bp *T. gondii* genome and the plasmid vectors employed (6–19 $\times$ 10^3 bp), this frequency is surprisingly close to the 5% transformation rate measured for purified plasmid (Donald and Roos, 1993; 1994a). No plaques were observed when wild-type parasite DNA was used for transformation. These results demonstrate that molecular cloning by complementation is indeed feasible in *Toxoplasma*. Identification of unknown mutations would presumably require a *T. gondii* expression library in an appropriate bacterial shuttle vector. Alternatively, it may be possible to tag transfected genomic DNA by cotransfection with a bacterial plasmid (Kaufman, 1990).

G. Special Considerations for Working with Pyrimethamine–Resistant Organisms

Because classical therapy for acute toxoplasmosis involves treatment with antifolates, the use of pyrimethamine-resistant parasites poses a special hazard. Mice infected with transgenic RH-strain parasites harboring the high level pyrimethamine-resistance alleles T83N+S36R or F245S+S36R are completely refractory to pyrimethamine treatment (B. J. Luft, personal communication). Even very high inocula of these pyrimethamine-resistant parasites remain sensitive to azithromycin, clindamycin, atovaquone and sulfonamides, however. (Pyrimethamine-resistant parasites are actually hypersensitive to sulfonamides both *in vitro* and *in vivo*.) Any possible infectious contact with these mutants should be treated with clindamycin+sulfa or macrolide+sulfa.

As the tachyzoite form of *T. gondii* is infectious only by direct bloodstream introduction, release of drug resistant organisms into general circulation—even in the event of an accidental laboratory infection—is only possible if the subject is eaten by a cat. To preclude the production of infectious oocysts, pyrimethamine resistant parasites should not be passed through a feline host.

Development of new selectable markers that permit mutant selection at frequencies comparable to high-level pyrimethamine resistance DHFR-TS alleles via a folate-independent pathway is an important goal for the future.

IV. Summary and Outlook

A. Genetic Tools for the Future

Through the efforts of several laboratories, the development of molecular transformation techniques and strategies for *T. gondii* has proceeded with remarkable rapidity. In brief, recombinant molecules can be expressed either transiently or as stable transformants, as episomes or integrated into the parasite genome, and as single copy or multicopy transgenes. Stable integration can be produced by random nonhomologous recombination, single-site homologous recombination (producing a duplication at the transfected locus), or perfect gene replacement. Many of these outcomes can now be selected specifically, by the use of appropriate vectors and transformation conditions. The extraordinarily high frequencies of stable transformation observed permit cloning by complementation, insertional mutagenesis/marker rescue, gene knock-outs, and allelic replacement. In combination with available classical (Pfefferkorn, 1988; Pfefferkorn and Pfefferkorn, 1980) and "cell-genetic" (Pfefferkorn, 1981; 1988) possibilities and physical and genetic mapping strategies (Sibley *et al.,* 1992; Sibley and Boothoyd, 1992), these tools provide a powerful arsenal for investigations into the biology of intracellular parasitism.

Additional molecular tools would enhance genetic manipulation of *T. gondii* still further. The availability of negative selectable markers (e.g., UPRT, HGPRT) should readily permit development of positive–negative selection vectors for forced gene replacement (Mortensen, 1993). Episomal transformation vectors would be useful for the study of essential genes (and a variety of other applications) and for the production of bacterial shuttle vectors. The ability to establish pyrimethamine-resistant DHFR-TS transgenes in episomal form (cf. Fig. 6, lane 3) argues that such an approach is possible. As previously noted, cDNA libraries for transgenic expression in *T. gondii* would be useful for gene cloning by complementation.

Finally, the potential danger of pyrimethamine-resistant parasites in an accidental laboratory infection argues for use of alternative selectable markers. Three such markers have been developed to date, conferring resistance to chloramphenicol (Kim *et al.,* 1993), phleomycin (Perez *et al.,* 1989), or an ingenious complementation of tryptophan auxotrophy (Sibley *et al.,* 1994). Unfortunately, thus far none of these systems appears to provide transformation frequencies comparable to those observed using DHFR-based vectors, but the search for a suitable replacement remains a high priority. Identification of the *T. gondii* UPRT and, especially, HXGPRT genes (as noted above) provide possible positive—as well as negative—selection systems.

B. Outstanding Problems in the Cell Biology and Pathogenesis of *T. gondii*

Now that many of the necessary molecular genetic tools are available for studying *Toxoplasma,* the challenge for the future is to apply these techniques to the analysis of important biological problems. Among the many interesting questions of cell biological interest, we may wish to consider the following:

Lacking flagella or cilia, what structures and mechanisms provide motility to parasite tachyzoites (gliding motility during the "search" for host cells and twisting motility during cell invasion and exit)?

What structures and molecules are necessary for host cell invasion? What site(s) of attachment is involved?

How is the parasitophorous vacuole membrane established and maintained? What is the membrane composed of and what does it do?

How is the highly polarized parasite structure established and maintained? How are the components of these organelles targeted to the proper destinations?

What regulates the strictly synchronous division of parasites within the vacuole?

What is needed from the host cell; i.e., why is *Toxoplasma* an *obligate* intracellular parasite?

Are there specific signals for interconversion between bradyzoites and tachyzoites, and what regulatory and biochemical changes are involved?

Why does sexual differentiation occur only in feline intestinal epithelium, and what signals and controls are involved in this developmental pathway?

What essential features of the *T. gondii* are recognized by the immune system of the parasitized organism, and how might this information assist in the treatment of congenital infections and toxoplasmosis in AIDS?

Can novel metabolic pathways or structural elements be identified in *Toxoplasma* as potential targets for drug development, and how do such novelties broaden our appreciation of eukaryotic diversity?

What aspects of cell biology and biochemistry are conserved across the vast evolutionary gulf separating *T. gondii* from more familiar animal, plant, and fungal systems, and what do these similarities tell us about the essential elements of eukaryotic design?

Acknowledgments

The early studies of Dr. E. R. Pfefferkorn were instrumental in demonstrating the genetic possibilities afforded by *T. gondii,* and we thank the international community of toxoplasmologists for continuing to maintain an interactive and congenial research environment. Drs. J. C. Boothroyd, K. A. Joiner, K. Kim, B. J. Luft, E. R. Pfefferkorn, L. D. Sibley, D. Soldati, and L. G. Tilney and members of the Roos laboratory provided helpful discussion, access to manuscripts in press, and permission to cite unpublished research. Research described in this communication was supported by grants from the NIH. D.S.R. is a Presidential Young Investigator of the National Science Foundation (with support from the MacArthur Foundation), and a Burroughs Wellcome Fund New Investigator in Molecular Parasitology.

References

Bellofatto, V., and Cross, G. A. M. (1989). Expression of a bacterial gene in a trypanosomatid protozoan. *Science* **244,** 1167–1169.

Berg, C. A., and Spradling, A. C. (1991). Studies on the rate and site-specificity of P-element transposition. *Genetics* **127,** 515–524.

Bodmer, S. J., Voller, A., Pettitt, L. E., and Fleck, D. G. (1972). The purification of *Toxoplasma gondii* antigen from mouse peritoneal exudate. *Trans. R. Soc. Trop. Med. Hyg.* **666,** 530.

Bohne, W., Heeseman, J., and Gross, U. (1993). Induction of bradyzoite-specific *Toxoplasma gondii* antigens in gamma-interferon-treated mouse macrophages. *Infect. Immun.* **61,** 1141–1145.

Boothroyd, J. C., Kim, K., Pfefferkorn, E. R., Sibley, L. D., and Soldati, D. (1994). Forward and reverse genetics in the study of the obligate intracellular parasite *Toxoplasma gondii. Methods Mol. Genet.* **3,** in press.

Brooks, R. G., Remington, J. S., and Luft, B. J. (1987). Drugs used in the treatment of toxoplasmosis. *Antimicrob. Agents Annu.* **2,** 297–306.

Cruz, A., and Beverley, S. M. (1990). Gene replacement in parasitic protozoa. *Nature (London)* **348,** 171–173.

Donald, R. G. K., and Roos, D. S. (1993). Stable molecular transformation of *Toxoplasma gondii:* A selectable dihydrofolate reductase-thymidylate synthase marker based on drug resistance mutations in malaria. *Proc. Natl. Acad. Sci. U.S.A.* **90,** 11703–11707.

Donald, R. G. K., and Roos, D. S. (1994a). Homologous recombination and gene replacement at the dihydrofolate reductase-thymidylate synthase locus in *Toxoplasma gondii. Mol. Biochem. Parasitol.* **64,** 243–253.

Donald, R. G. K., and Roos, D. S. (1994b). In preparation.

Ellis, J., Griffin, H., Morrison, D., and Johnson, A. M. (1993). Analysis of dinucleotide frequency and codon usage in the phylum Apicomplexa. *Gene* **126,** 163–170.

Endo, T., and Yagita, K. (1990). Effect of extracellular ions on motility and cell entry in *Toxoplasma gondii. J. Protozool.* **37,** 133–138.

Endo, T., Tokuda, H., Yagita, K., and Koyama, T. (1987). Effect of extracellular potassium on acid release and motility initiation in *Toxoplasma gondii. J. Protozool.* **34,** 291–295.

Feldmann, K. A. (1991.) T-DNA insertion mutagenesis in *Arabadopsis:* Mutational spectrum. *Plant J.* **1,** 71–82.

Finbarr-Tobin, J., and Wirth, D. F. (1992). A sequence insertion targeting vector for *Leishmania enriettii. J. Biol. Chem.* **267,** 4752–4758.

Foley, V., and Remington, J. S. (1969). Plaquing of *Toxoplasma gondii* in secondary cultures of chick embryo fibroblasts. *J. Bacteriol.* **98,** 1–3.

Frenkel, J. K. (1973). Toxoplasmosis. A parasite life cycle, pathology, and immunology. *In* "The Coccidia (D. M. Hammond, ed.), pp. 343–410. University Park Press, Baltimore, MD.

Hasty, P., Revera-Pérez, J., Chang, C., and Bradley, A. (1991). Target frequency and integration pattern for insertion and replacement vectors in embryonic stem cells. *Mol. Cell. Biol.* **11,** 4509–4517.

Howe, D. K., and Sibley, L. D. (1994). *Toxoplasma gondii:* Analysis of different laboratory stocks of the RH strain reveals genetic heterogeneity. *Exp. Parasitol.* **78,** 242–245.

Hyde, J. E. (1990). The dihydrofolate reductase-thymidylate synthase gene in the drug resistance of malaria parasites. *Pharmacol. Ther.* **48,** 45–59.

Iltzsch, M. H. (1993). Pyrimidine salvage pathways in *Toxoplasma. J. Eukaryotic Microbiol.* **40,** 24–28.

Joiner, K. A., and Dubremetz, J. F. (1993). *Toxoplasma gondii:* A protozoan for the nineties. *Infect. Immun.* **61,** 1169–1172.

Kaufman, R. J. (1990). Selection and coamplification of heterologous genes in mammalian cells. *In* "Methods in Enzymology"(D. Goeddel *et al.,* eds.), Vol. **185,** 537–566. Academic Press, San Diego.

Kim, K., Soldati, D., and Boothroyd, J. C. (1993). Gene replacement in *Toxoplasma gondii* with chloramphenicol acetyl transferase as selectable marker. *Science* **262**, 911–914.

Kucherlapati, R., and Smith, G. R., eds. (1988). "Genetic Recombination." *Am. Soc. Microbiol.* Washington, DC.

Laban, A., Finbarr-Tobin, J., Curotto de Lafaille, M. A., and Wirth, D. F. (1990). Stable expression of the bacterial neo^r gene in *Leishmania enriettii*. *Nature (London)* **343**, 572–574.

Lee, M.G.-S., and van der Ploeg, L. H. T. (1990). Homologous recombination and stable transfection in the parasitic protozoan *Trypanosoma brucei*. *Science* **250**, 1583–1587.

Leriche, M. A., and Dubremetz, J. F. (1991). Characterization of the protein contents of rhoptries and dense granules of *Toxoplasma gondii* tachyzoites by subcellular fractionation and monoclonal antibodies. *Mol. Biochem. Parasitol.* **45**, 249–259.

Levine, N. D. (1988). Progress in taxonomy of Apicomplexan protozoa. *J. Protozool.* **35**, 518–520.

Lindsay, D. S., Dubey, J. P., Blagburn, B. L., and Toivio-Kinnucan, M. (1991). Examination of tissue cyst formation by *Toxoplasma gondii* in cell cultures using bradyzoites, tachyzoites, and sporozoites. *J. Parasitol.* **77**, 126–132.

Luft, B. J., and Remington, J. S. (1992). Toxoplasmic encephalitis in AIDS. *Clin. Infect. Dis.* **15**, 211–222.

McLeod, R., and Remington, J. S. (1987). Toxoplasmosis. *In* "Harrison's Principles of Internal Medicine" (E. Braunwald *et al.*, eds.), 11th ed. p. 791. McGraw-Hill, New York.

Miles, M. A. (1988). *Parasitol. Today* **4**, 28–29.

Mortensen, R. (1993). Overview of gene targeting by homologous recombination *Curr. Protocols Mol. Biol. Suppl.* **23**, 9.15.1–9.15.6.

Orr-Weaver, T. L., Szostak, J. W., and Rothstein, R. J. (1981). Yeast transformation: A model systems for the study of recombination. *Proc. Natl. Acad. Sci. U.S.A.* **78**, 6354–6358.

Perez, P., Tiraby, G., Kallerhoff, J., and Perret, J. (1989). Phleomycin resistance as a dominant selectable marker for plant cell transformation. *Plant Mol. Biol.* **13**, 365–373.

Pfefferkorn, E. R. (1977). *Toxoplasma gondii:* The enzymic defect of a mutant resistant to 5-fluorodeoxyuridine. *Exp. Parasitol.* **44**, 26–35.

Pfefferkorn, E. R. (1981). *Toxoplasma gondii* and the biochemistry of intracellular parasitism. *Trends Biochem. Sci.* **6**, 311–313.

Pfefferkorn, E. R. (1988). *Toxoplasma gondii* as viewed from a virological perspective. *MBL Lect. Biol.* **9**, 479–501.

Pfefferkorn, E. R., and Guyre, P. M. (1984). Inhibition of growth of *Toxoplasma gondii* in cultured fibroblasts by human recombinant gamma interferon. *Infect. Immun.* **44**, 211–216.

Pfefferkorn, E. R., and Pfefferkorn, L. C. (1976). *Toxoplasma gondii:* Isolation and preliminary characterization of temperature-sensitive mutants. *Exp. Parasitol.* **39**, 365–376.

Pfefferkorn, E. R., and Pfefferkorn, L. C. (1977). Specific labelling of intracellular *Toxoplasma gondii* with uracil. *J. Protozool.* **24**, 449–453.

Pfefferkorn, E. R., and Pfefferkorn, L. C. (1979). Quantitative studies on the mutagenesis of *Toxoplasma gondii*. *J. Parasitol.* **65**, 364–370.

Pfefferkorn, E. R., and Pfefferkorn, L. C. (1980). *Toxoplasma gondii:* Genetic recombination between drug resistant mutants. *Exp. Parasitol.* **50**, 305–316.

Roos, D. S. (1993). Primary structure of the fused dihydrofolate reductase/thymidylate synthase gene from *Toxoplasma gondii*. *J. Biol. Chem.* **268**, 6269–6280.

Ryan, K. A., Garraway, L. A., Descoteaux, A., Turco, S. J., and Beverley, S. M. (1993). Isolation of virulence genes directing surface glycosyl-phosphatidylinositol synthesis by functional complementation of *Leishmania*. *Proc. Natl. Acad. Sci. U.S.A.* **90**, 8609–8613.

Sabin, A. B. (1941). Toxoplasmic encephalitis in children. *JAMA, J. Am. Med. Assoc.* **116**, 801–807.

Schwartzman, J. D., and Pfefferkorn, E. R. (1981). Pyrimidine synthesis by intracellular *Toxoplasma gondii*. *J. Parasitol.* **67**, 150–158.

Sibley, L. D. (1993). Interactions between *Toxoplasma gondii* and its mammalian host cells. *Semin. Cell Biol.* **4,** 335–344.

Sibley, L. D., and Boothroyd, J. C. (1992). Construction of a molecular karyotype for *Toxoplasma gondii*. *Mol. Biochem. Parasitol.* **51,** 291–300.

Sibley, L. D., LeBlanc, A. J., Pfefferkorn, E. R., and Boothroyd, J. C. (1992). Generation of a restriction fragment length polymorphism linkage map for *Toxoplasma gondii*. *Genetics* **132,** 1003–1015.

Sibley, L. D., Messina, M., and Niesman, I. R. (1994). Stable DNA transformation in the obligate intracellular parasite *Toxoplasma gondii* by complementation of tryptophan auxotrophy. *Proc. Natl. Acad. Sci. USA* **91,** (in press).

Soete, M., Fortier, B., Camus, D., and Dubremetz, J. F. (1993). *Toxoplasma gondii:* Kinetics of bradyzoite-tachyzoite interconversion *in vitro*. *Exp. Parasitol.* **76,** 259–264.

Soldati, D., and Boothroyd, J. C. (1993). Transient transfection and expression in the obligate intracellular parasite *Toxoplasma gondii*. *Science* **260,** 349–352.

Tanaka, M., Gu, H.-M., Bzik, D. J., Li, W.-B., and Inselburg, J. (1990). Dihydrofolate reductase mutations and chromosomal changes associated with pyrimethamine resistance of *Plasmodium falciparum*. *Mol. Biochem. Parasitol.* **39,** 127–134.

ten Asbroek, A. L., Mol, C. A., Kieft, R., and Borst, P. (1993). Stable transformation of *Trypanosoma brucei*. *Mol. Biochem. Parasitol.* **59,** 133–142.

van den Hoff, M. J. B., Moorman, A. F. R., and Lamers, W. H. (1992). Electroporation in 'intracellular' buffer increases cell survival. *Nucleic Acids Res.* **20,** 2902.

Ware, P. L., and Kasper, L. H. (1987). Strain-specific antigens of *Toxoplasma gondii*. *Infect. Immunol.* **55,** 778–783.

Wellems, T. E. (1991). Molecular genetics of drug resistance in *Plasmodium falciparum* malaria. *Parasitol. Today* **7,** 110–116.

CHAPTER 4

Transfection Experiments with *Leishmania*

Jonathan H. LeBowitz

Department of Biochemistry
Purdue University
West Lafayette, Indiana 47907

I. Introduction
II. Stable and Transient Transfection
    A. Selectable Markers for Stable Transfection
    B. Vectors
    C. Transient Assays and Reporter Enzymes
III. Transfection
    A. Growth of Parasites
    B. Electroporation
    C. Selection of Stable Transformants
    D. Assay of Reporter Enzymes in Transient Transfectants
References

I. Introduction

Leishmania species, protozoan parasites of the family Trypanosomatidea, are medically important infectious agents that use a variety of novel processes such as RNA editing and *trans*-splicing of polycistronic precursor RNAs for gene expression. *Leishmania* cycle between an insect host and a mammalian host exhibiting distinct morphological and biochemical traits in the stage adapted to each host. In the mammalian host, *Leishmania* are intracellular parasites residing within the phagolysosome of macrophage. The complex nature of the biology of the infectious process, from the perspective both of host responses to parasite infection and of parasite responses to the host, presents a challenge to researchers that requires deployment of a large arsenal of techniques.

A number of powerful molecular genetic tools depend on the ability to introduce exogenous DNA into a host organism. These tools include stable expression of exogenous genes to generate cell lines with useful phenotypes, deletion of genes by homologous gene replacement to create null mutants, transient expression experiments to map *cis* elements governing expression of reporter genes, and screening of libraries by direct complementation of a scorable phenotype to isolate the relevant genes. With the advent of transfection methodologies for *Leishmania* and other trypanosomatids (Bellofatto and Cross, 1989; Laban and Wirth, 1989; Clayton *et al.*, 1990; Kapler *et al.*, 1990; Laban *et al.*, 1990; Rudenko *et al.*, 1990; Zomerdijk *et al.*, 1990; Bellofatto et al., 1991), these molecular genetic tools can be brought to bear on studies of *Leishmania* biology, sidestepping the inability to develop more traditional genetic methodologies.

This chapter describes the basic laboratory procedures required to introduce DNA by electroporation into *Leishmania*. Methods for generating stably transformed clonal parasite lines, as well as for transient transfection and analysis, will be presented. Attention will be paid to considerations of parasite culture as they pertain to transfection, DNA requirements, and other issues of specific applicability to *Leishmania* transfection. For generally applicable molecular biology protocols, the reader is referred to any of the excellent reference works available.

II. Stable and Transient Transfection

A. Selectable Markers for Stable Transfection

Transfection experiments can be divided into those in which the goal is to obtain parasite lines stably transformed with a particular piece of DNA, and those in which the goal is immediate analysis of the population of transfected cells for activities associated with the input DNA. The success of stable transfection experiments depends upon the availability of genes that confer a selectable phenotype to the host organism when the gene is expressed. Although this process can take weeks, the resultant stably transformed lines can be subjected to a variety of molecular, biochemical, and immunological analyses over a long period of time. Production of stable cell lines is essential for the analysis of phenotypes associated with expression of exogenous genes or for phenotypes associated with the disruption of endogenous genes. A number of selectable markers have been successfully used in *Leishmania* and other trypanosomatids (Table I). The availability of multiple selectable markers was essential for creation of null phenotypes at a given gene locus by sequential targeted gene replacement. With the exception of TK, each of the markers listed in Table I confers resistance to a drug or antibiotic that is toxic to wild-type parasites. *Leishmania* engineered to express the HSV-1 thymidine kinase gene die when grown in the presence of the nucleoside analog ganciclovir (LeBowitz *et*

Table I
Selectable Marker and Reporter Genes That Function in *Leishmania* and Other Trypanosomatids

Selectable marker gene	Selective drug	Reference
Neomycin phosphotransferase	G418	a
Hygromycin B phosphotransferase	Hygromycin B	b
Sh ble	Phleomycin	c
Puromycin phosphotransferase	Puromycin	d
N-Acetyl glucosamine-1-phosphate transferase	Tunicamycin	e
HSV thymidine kinase	Ganciclovir	f
Reporter enzyme	Detection method	Reference
β-Galactosidase	fluorescence	g
β-Glucuronidase	fluorescence	g
Chloramphenicol acetyltransferase	radiolabel	h
Luciferase	chemiluminescence	i

Note. a (Kapler *et al.*, 1990; Laban *et al.*, 1990; Lee and Van der Ploeg, 1990; Bellofatto *et al.*, 1991); b (Cruz *et al.*, 1991; Lee and Van der Ploeg, 1991); c (Freedman and Beverley, 1993; Jefferies *et al.*, 1993); d (Freedman and Beverley, 1993); e (Liu and Chang, 1992); f (LeBowitz *et al.*, 1992); g (LeBowitz *et al.*, 1990); h (Bellofatto and Cross, 1989; Laban and Wirth, 1989); i (Beverley and Clayton, 1993).

al.,1992). The ability to select against expression of the TK gene in *Leishmania* can be exploited in a variety of ways. For example one could select for mutations that inactivate *cis* elements used for expression of TK.

B. Vectors

A number of shuttle vectors that permit expression of exogenous genes in *Leishmania* have been described (Table II). Many of these vectors are functional in other trypanosomatid genera. For example, pX is capable of driving gene expression in *Crithidia fasciculata* and *Endotrypanum shaudinni* and pTEX functions in *Trypanosoma cruzi* (Coburn *et al.*,1991; Kelly *et al.*, 1992). When introduced into *Leishmania,* shuttle vectors are maintained as extrachromosomal multimers and therefore must provide all essential *cis* elements for transcription and RNA processing as well as any required *cis* elements for DNA replication. Production of functional mRNAs in *Leishmania* depends on processing of polycistronic precursor RNAs by *trans*-splicing of the miniexon to splice acceptor sites positioned upstream of all coding sequences (Borst, 1986; Agabian, 1990). Each splice acceptor site serves an additional function, also specifying the site of polyadenylation for the adjacent upstream gene (LeBowitz *et al.*, 1993). *Leishmania* expression vectors must provide segments of in-

Table II
Shuttle Vectors That Function in *Leishmania*

Vector	Locus[a]	Reference
pX, pX63NEO (HYG)	*L. major* DHFR-TS	b
pHM	*L. major* HMTX[r]	c
pTEX	*T. cruzi* gGAPDH	d
cLNEO (HYG)	Cosmid vectors	e
pALT-NEO	*L. enriettii* β-tubulin	f

Note. b (LeBowitz *et al.,* 1990, 1991); c (Freedman and Beverley, 1993); d (Kelly *et al.,* 1992); e (Ryan *et al.,* 1993); f (Laban *et al.,* 1990).

[a] Locus that supplies intergenic regions used in expression vector. DHFR-TS, dihydrofolate reductase-thymidylate synthetase; HMTX[r], H-region methotrexate resistance gene; gGAPDH, glycosomal glyceraldehyde phosphate dehydrogenase.

tergenic regions containing splice acceptor sites flanking the 5′ and 3′ ends of the selectable marker gene in order to ensure proper *trans*-splicing and polyadenylation. Also, evidence is accumulating that 5′ and 3′ untranslated regions (UTRs) contribute to expression of linked genes in Trypanosomatids (Pays *et al.,* 1990; Jefferies *et al.,* 1991; Aly *et al.,* 1993; Hug *et al.,* 1993; Hehl *et al.,* 1994). Surprisingly, neither transcriptional promoters nor replication origins have yet been identified in any of the current crop of *Leishmania* expression vectors, suggesting, perhaps, that these may be dispensable. This contrasts with the situation in *T. brucei* where DNA segments that stabilize episomal maintenance of vectors have been reported (Patnaik *et al.,* 1993) and where transcriptional promoter elements have been shown to be critical for high levels of expression of reporter genes (Clayton et al., 1990; Rudenko *et al.,* 1990; Zomerdijk *et al.,* 1990; Jefferies *et al.,* 1991).

A generalized scheme for construction of a *Leishmania* shuttle vector is shown in Fig. 1. The coding sequence for a *Leishmania* gene (gene 1) is excised from a segment of *Leishmania* genomic DNA and replaced with a selectable marker gene. The marker gene is flanked by intergenic region sequences that provide the splice site for miniexon addition at the 5′ end of the gene and the downstream splice site required for polyadenylation. Also, 5′ and 3′ untranslated regions from the replaced gene will be positioned appropriately relative to the marker gene in case they contribute to expression. The downstream splice site can then be utilized for expression of a gene of interest by placing a suitable polylinker site adjacent to the splice site in place of gene 2. It is also desirable to place an additional intergenic region downstream of this polylinker so that the gene of interest is flanked by, and will use, authentic *Leishmania* processing signals. Finally, bacterial plasmid sequences must be provided to

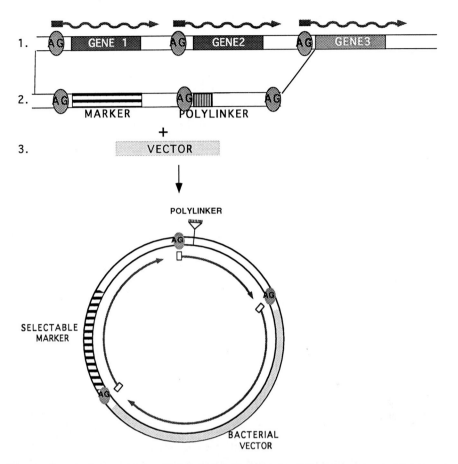

Fig. 1 Steps in the construction of a *Leishmania* shuttle vector: 1. Isolate a segment of *Leishmania* genomic DNA containing 2 genes and 3 intergenic regions making sure to retain splice acceptor site 3. 2. Replace gene 1 with coding sequence with a selectable marker coding sequence. Replace gene 2 coding sequence with a polylinker. 3. Add bacterial plasmid downstream of third intergenic region. mRNAs are represented by wavy lines in the genomic locus and by curved lines in the plasmid. Small rectangles and arrows bounding mRNAs denote miniexons and poly(A) tails, respectively. AG, location of splice acceptor sites.

allow propagation of the vector in *Escherichia coli*. It has become apparent that cryptic processing signals exist within the commonly used bacterial plasmids (Bellofatto *et al.,* 1991; Sherman *et al.,* 1991; Curotto de Lafaille *et al.,* 1992; LeBowitz *et al.,* 1993). When such bacterial sequences, appropriately oriented, form the 3′ flanking sequences for a gene, the resultant mRNA may utilize this fortuitous processing signal for 3′ polyadenylation.

C. Transient Assays and Reporter Enzymes

Transient transfection finds its most useful application in the analysis of potential regulatory sequences residing within a segment of DNA. A series of constructs, each containing defined deletions or mutations in the segment of DNA to be analyzed linked in *cis* to a reporter gene, can be assayed in a transient experiment. Mutations or deletions that correlate with loss of reporter enzyme activity indicate the location of essential *cis* elements. It is essential that each of the constructs be identical except for the particular mutations being assayed. The reporter enzymes listed in Table I, have been used successfully for transient transfections in trypanosomatids. We find a number of advantages in using β-galactosidase and β-glucuronidase as reporter enzymes. The assays are more sensitive than chloramphenicol acetyltransferase (CAT) assays, allowing quantitation of less than 0.1 pg of enzyme and do not require use of radioactivity. Although assay of β-galactosidase and β-glucuronidase with fluorescent substrates is less sensitive than luciferase assays, chemiluminescent substrates are commercially available that increase sensitivity beyond that of luciferase in the event that a luminometer is available. Both enzymes are stable in the moderate amounts of detergent required to lyse parasites and they can be assayed under identical conditions. These final features make it simple to control for transfection efficiency by cotransfecting a control plasmid containing one of the two reporters with a series of test constructs containing the other.

III. Transfection

A. Growth of Parasites

Leishmania transfection experiments can be divided into three phases: growth of promastigotes for electroporation, introduction of DNA into promastigotes by electroporation, and finally either plating transfectants on selective media for isolation of stable transformants or lysis of transfectants for assay of reporter enzyme activity in transient experiments. The electroporation phase and the plating or reporter enzyme assay phase are typically accomplished on consecutive days.

Promastigotes are maintained in culture at 26–27°C by serial passage of 10-ml cultures at dilutions of 1:100. In the M199 media that we use (Table III), doubling times for promastigote cultures range from about 12 to 20 hr and saturation densities range from 1 to 8×10^7 cell/ml depending on the particular species and isolate.

A typical transfection experiment will require the production of 100–500 ml of late long phase promastigotes depending on the number of DNA samples to be electroporated. Cultures can be grown either in roller bottles or in 225-cm^2 tissue culture flasks that accommodate up to 250-ml cultures. These flasks are laid almost flat in the incubator with necks propped up. Rolled-up paper towels are convenient for propping up the flasks.

Table III
M199 Media for Growth of *Leishmania* Promastigotes

Volume/500 ml	Reagent	Final concentration
318.5 ml	H_2O	
100 ml	5× M199	1×
25 ml	Heat-inactivated fetal calf serum	5%
25 ml	Bovine embryonic fluid	5%
5 ml	10 mM adenine in 50 mM HEPES, pH 7.5	100 μM
20 ml	1M HEPES, pH 7.5	40 mM
1 ml	0.25% hemin in 50% triethanolamine	0.0005%
0.5 ml	0.1% biotin in 95% ethanol	0.0002%
5 ml	5000 U/ml penicillin, 5 mg/ml streptomycin	50 U/ml; 0.05 mg/ml

Note. Complete M199 medium is conveniently prepared in 500-ml batches by combining the listed sterile solutions. Double distilled or deionized H_2O is aliquoted into 500-ml bottles and autoclaved. Indicated quantities of the sterile stock solutions are added to the sterile water. Bovine embryonic fluid, available from Sigma, is used as a partial replacement for fetal calf serum mainly as a cost-saving measure. If bovine embryonic fluid is unavailable, media should contain 10% fetal calf serum. Promastigotes of a variety of *Leishmania* species grow to equivalent or higher cell densities in the bovine embryonic fluid-containing media.

A good rule of thumb for estimating the required cell volume for a transfection experiment is to start at least 10 ml of cells for each electroporation point. Each sample to be electroporated requires 4×10^7 cells and we assume that the cells will be harvested at a density of about 5×10^6 cell/ml. We strongly recommend electroporating each sample in duplicate and performing relevant positive and negative controls. It is also a simple matter to time the growth of cells so that they reach late log phase at a convenient time. For example, if cells with doubling time of 12 hr are seeded at a density of 10^5 cells/ml, they will be ready for electroporation in just under 3 days. Of course, it is possible, by judicious choice of culture volume and initial cell density, to generate flexibility in the timing.

B. Electroporation

Introduction of DNA into *Leishmania* by electroporation proceeds in a similar fashion whether the desired goal is transient transfection analysis or production of stable transformants.

Sterility must be maintained for all of the steps outlined below. Each step should be performed in a laminar flow hood unless otherwise noted.

You will need the following: BioRad gene pulser and capacitance extender module, 0.2-cm electroporation cuvettes, M199 complete media, electroporation buffer (EPB): 21 mM HEPES, 137 mM NaCl, 5 mM KCl, 0.7 mM Na_2HPO_4, 6 mM glucose, pH 7.5. DNA purified on Quiagen columns or by cesium banding have been used successfully in transfection experiments in this laboratory. DNA purified by PEG precipitation has also been reported to give good results (Ryan *et al.*, 1993). We usually electroporate 20–50 μg of plasmid DNA for generation of stable transformants. When transient experiments are performed, 10 μg each of the test and control plasmid is sufficient. Homologous gene replacement experiments should be performed with 1–5 μg of the linearized targeting fragment, as higher levels are reported to yield transformants with undesired genotypes (Cruz and Beverley, 1990).

1. Prealiquot the DNAs to be transfected into sterile numbered eppendorf tubes using clean pipettors and sterile tips.

2. Put 10 ml of M199 media into 25-cm² tissue-culture flasks so that there is one flask for each transfection sample including controls.

3. Number sterile electroporation cuvettes (0.2-cm cuvettes) and place them on ice.

4. Count cells with hemocytometer or Coulter counter.

5. Transfer late log phase cells to sterile centrifuge bottles and collect the cells by centrifugation at $1000 \times g$ 5–10 min. Pour off the supernatant media.

6. Wash the harvested cells free of media by resuspending them in approximately $\frac{1}{10}$ vol of electroporation buffer.

7. Collect the washed cells by centrifugation and resuspend them in EPB at a cell density of 1×10^8/ml. Chill the cells on ice for 10 min.

8. We use a BioRad gene pulser for electroporation of *Leishmania*. The electroporation conditions used (500 μF, 2.25 V/cm) require use of a capacitance extender module with the gene pulsar unit. Place the gene pulsar and attached capacitance extender adjacent to the hood with the cuvette holder within the hood. Select the EXT. setting for capacitance on the gene pulsar unit and select the 500-μF setting on the capacitance extender module. Turn gene pulser unit on and select a voltage of 0.45 kV (2.25 kV/cm).

9. Mix 0.4 ml of the cell suspension (4×10^7 cells) with the DNA to be electroporated in the Eppendorf tube by pipetting up and down and transfer to a prechilled (4°C) electroporation cuvette.

10. Place the cuvette in the cuvette holder of the gene pulser and zap immediately by depressing the two buttons on the gene pulsar unit until it beeps. Afterward you should notice that the cells in the cuvette have a foamy head. Return the cuvette to ice.

11. Steps 9 and 10 are repeated until as many as 20 samples have been electroporated. By processing no more than 20 transfectants at a time, zapped cells will not sit on ice for an excessive amount of time. When more than 20 samples are being electroporated, we simply electroporate the first 20, transfer to media, then start on the next set of 20.

12. Using a sterile Pasteur pipette, transfer the samples to 10 ml of complete

M199 media prealiquoted into a 25-cm^2 flask. Use some of the M199 media to wash the cuvette. Incubate overnight at 26°C.

C. Selection of Stable Transformants

The ability to grow *Leishmania* promastigotes on agar plates provides experimenters with tremendous advantage in transfection experiments. Selection of parasites stably transformed with DNA containing a selectable marker is readily accomplished by growing the transfected parasites on agar plates containing the appropriate selective drug. In contrast to selection of transformants in solution, selection on plates permits isolation of clonal lines of transformed parasites and provides a ready means of quantitating the transformation efficiency in a particular experiment. The relative transformation efficiency of a particular DNA construct can provide clues to the presence of unforeseen effects on the expression of the selectable marker.

1. Preparation of Selective Plates

You will need the following: 2% noble agar, 2× complete M199 + biopterin, 100× selective drug.

Two percent noble agar is prepared by suspending the agar in water and autoclaving the suspension. The autoclaved agar can be used immediately or can be stored after it has hardened and used later by melting the agar in the microwave.

Prepare 500 ml of 2X complete M199 with biopterin by mixing the indicated quantities of the reagents and filter sterilizing. Store at 4°C (2 × M199 + biopterin: 133.5 ml H$_2$O, 200 ml 5× M199, 100 ml heat-inactivated fetal calf serum, 10 ml 10 m*M* adenine in 50 m*M* HEPES, pH 7.5, 40 ml 1 *M* HEPES, pH 7.5, 2 ml 0.25% hemin in 50% triethanolamine, 1 ml 0.1% biotin in 95% ethanol, 10 ml 5000 U/ml penicillin 5 mg/ml streptomycin solution, 2.5 ml 0.5 mg/ml biopterin).

Combine indicated stock solutions and filter sterilize.

To make plates:

1. Melt 2% noble agar.

2. Equilibrate agar and 2X M199 at 55°C. Each 100-cm^2 plate will require about 12.5 ml of 2% noble agar and about 12.5 ml of 2× M199. It is most convenient to transfer enough 2X M199 for all the plates to a sterile flask so that the selective drug can be added to it.

3. Mix equal volumes of 2X M199 supplemented with selective drug and 2% noble agar and dispense 25 ml into each plate. To avoid hardening of the agar, we usually mix enough agar and 2X M199 to prepare four plates and then repeat.

4. Let plates sit at room temperature overnight to dry.

5. Just prior to plating cells, equilibrate the plates with CO_2 by placing them in an incubator containing 5% CO_2 for about 2 hr. The plates should turn from a deep red to a red-orange color. At this point the plates should be wrapped in parafilm until cells are spread on them. If, however, the surface of the plates still contain condensation they need to be dried by opening them in the tissue-culture hood for 15–60 min.

Concentrations of the selective drug that are appropriate should be determined empirically for each *Leishmania* line. It is easiest to determine an effective drug concentration that inhibits growth of parasites in liquid culture, usually two to three times the EC_{50}. A good rule of thumb is to use a selective drug concentration on plates that is twofold higher than the drug concentration used in solution.

2. Plating Transfected Parasites

You will need the following. M199 plates with appropriate selective drug, no drug plates for estimating plating efficiency.

1. Transfer the transfected cells from the 25 cm^2 flasks into centrifuge tubes.

2. Centrifuge 5–10 min 1000 × g and decant media leaving 50–100 μl of media in the tube.

3. Resuspend cell pellet in the residual media.

4. Transfer cells to plate.

5. Flame a glass spreader that has been dipped into ethanol.

6. Spread the cells gently with the glass spreader.

7. Wrap plates in parafilm and place upright in 26°C incubator. Colonies will usually appear on drug-free plates in 1–2 weeks and on selective plates in 2–3 weeks. The number of colonies that will appear depends upon the transformation efficiency, which for *L. major* approaches 10^{-4} at saturating DNA concentrations (Kapler *et al.*,1990; Coburn *et al.*, 1991); the plating efficiency of the cells, which can vary greatly from line to line; and the DNA construct being used.

8. To expand colonies into liquid culture, pick well-isolated colonies with sterile toothpicks and transfer into 1-ml aliquots of M199 media containing selective drug aliquoted into 24-well microtiter dishes.

D. Assay of Reporter Enzymes in Transient Transfectants

In a transient transfection experiment, DNAs are electroporated into parasites on Day 1. The parasites are allowed to express the reporter gene overnight and are then lysed on Day 2 so that the reporter enzyme can be assayed. It should be emphasized that in transient experiments, population of cells, most of which have probably not taken up DNA, are being assayed. This protocol describes how to assay β-galactosidase (β-gal) and β-glucoronidase (gus) in parasites that have been electroporated with DNAs containing these reporter enzymes. As stated earlier, it is advantageous to cotransfect a control construct containing one of these reporters with test constructs containing the other reporter so that variations in transfection efficiency can be controlled for.

You will need the following. TPI solution (stock solutions are in parentheses): 50 mM Tris–HCl, pH 7.3, 150 μl/ml benzamidine (100 mg/ml in ethanol), 20μg/ml leupeptin (5 mg/ml), 200 μg/ml 1,10-phenanthroline (100 mg/ml in ethanol), 1 mM EDTA (0.5 M EDTA), 50μg/ml soybean trypsin inhibitor (10 mg/ml), 50 μg/ml BSA (10 mg/ml molecular biology grade), 1 mM PMSF (100 mg/ml in methanol).

Reaction mix: 23 mM Tris–HCl, pH 7.5, 125 mM NaCl, 2 mM MgCl$_2$, 12 mM β-mercaptoethanol. β-gal reaction mix additionally contains 0.1 mg/ml 4-methylumbelliferyl β-D-galactoside. Gus reaction mix additionally contains 0.34 mg/ml 4-methylumbelliferyl β-D-glucuronide. Substrates are added to reaction mixes from stock suspensions of 20 mg/ml 4-methylumbelliferyl β-D-galactoside or 72 mg/ml 4-methylumbelliferyl β-D-glucuronide in ethanol. To pipette substrate from these suspensions, make sure the substrate is well dispersed, then pipette the requisite amount using a pipette tip with the end cut off.

10% SDS; glycine carbonate solution: 133 mM glycine 83 mM Na$_2$CO$_3$, pH 10.7, filtered; Hanks' balanced salt solution (HBSS); Hoeffer TKO-100 minifluorimeter or other fluorimeter with filters or monochronometer allowing use of excitation wavelenghth 380 nM and emission at 460 nM.

1. Number 15-ml sterile centrifuge tubes and sterile Eppendorf tubes.

2. Prepare fresh Tris–protease inhibitor solution (TPI) minus PMSF. TPI is prepared from the indicated stock solutions all of which should be stored at −20°C with the exception of Tris and EDTA. Be sure to prepare enough extra for reagent blanks.

3. Transfer electroporated cells into sterile 15-ml centrifuge tubes and pellet the cells at 1000 × g, 10 min, room temperature.

4. Pour off media and resuspend the cell pellet in 1 ml of sterile HBSS.

5. Transfer the cell suspension to a sterile Eppendorf tube and pellet the cells by centrifugation for 1 min in a microfuge.

6. Add PMSF to freshly prepared TPI. Add 10% SDS to a final concentration of 0.1%. The concentration of SDS required to lyse parasites may vary with the species.

7. Resuspend the cells in 160 μl of the freshly prepared TPI/SDS solution.

8. Vortex vigorously to disrupt the cells and place on ice for 15 min.

9. Centrifuge 15 min in the microfuge to pellet the cell debris.

10. While the lysates are incubated on ice and centrifuged, it is convenient to prepare the reaction mix and transfer 320-μl aliquots to numbered assay tubes. Each transfectant should be assayed in parallel for β-galactosidase and β-glucuronidase. Each transfectant requires one tube with β-gal reaction mix

and one with gus reaction mix. Be sure to prepare enough extra for reagent blanks.

11. After the cell debris is pelleted, two 75-μl aliquots from each supernatant are transferred: one to a prealiquoted tube containing β-gal reaction mix and one to a tube containing prealiquoted gus reaction mix. This will start the reaction. When removing the supernatant be careful to avoid dislodging the pellet, which can be quite viscous. If the pellet does become dislodged it is often easier to remove it entirely before attempting to transfer the supernatant.

12. Incubate the reactions at 37°C. Reactions are linear for at least 20 hr, so choice of incubation time can be tailored to the convenience of the investigator. Obviously, longer incubation times produce greater signals so when assaying samples that are suspected of having low activity longer times are desirable. It can be convenient to allow the reaction to proceed overnight.

13. Reactions are terminated by the addition of 2 ml glycine carbonate solution.

14. Fluorescence is determined at 460 nM. The reaction product, 4-methylumbelliferone, fluoresces when it becomes ionized at alkaline pH. The excitation wavelength is 380 nM and fluorescence emission is maximal at 460 nM. This excitation and emission profile is matched by the characteristics of a fixed wavelength fluorimeter commercially available from Hoeffer Scientific, the TKO-100 minifluorimeter. Reagent blanks containing 75 μl TPI, 320 μl reaction mix,and 2 ml glycine carbonate are used to zero the instrument. Separate reagent blanks must be prepared for β-gal and gus.

15. Typically, we report results as the ratio of fluorescence units of the test reporter gene to fluorescence units of the control reporter gene. These ratios are then averaged for duplicate or triplicate sample points. This averaged ratio is a unitless number but is normalized in a way that controls for variability in transfection efficiency. In the event that one wishes to report quantitative data that describe the quantity of substrate hydrolyzed per unit time, the fluorescence units can be readily converted into moles of substrate hydrolyzed. This is done by generating a standard curve with known quantities of 4-methylumbelliferone (the reaction product that is commercially available) in glycine carbonate.

Acknowledgments

J.H.L. is supported by NSF Grant 9219767-MCB. This is paper 14152 of the Purdue University Agricultural Experminental Station. I thank Steve Beverley for allowing me to learn about transfection of *Leishmania* in his laboratory.

References

Agabian, N. (1990). Trans-splicing of nuclear pre-mRNAs. *Cell (Cambridge, Mass.)* **61**, 1157–1160.
Aly, R., Argaman, M., Pinelli, E., and Shapira, M. (1993). Intergenic sequences from the heat-shock protein 83-encoding gene cluster in *Leishmania mexicana amazonensis* promote and regulate reporter gene expression in transfected parasites. *Gene* **127**, 155–63.

Bellofatto, V., and Cross, G. A. M. (1989). Expression of a bacterial gene in a trypanosomatid protozoan. *Science* **244,** 1167–1169.

Bellofatto, V., Torres-Mūnoz, J. E., and Cross, G. A. M. (1991). Stable transformation of *Leptomonas seymouri* by circular extrachromosomal elements. *Proc. Natl. Acad. Sci. U.S.A.* **88,** 6711–6715.

Beverley, S. M., and Clayton, C. E. (1993). Transfection of *Leishmania* and *Trypanosoma brucei* by electroporation. *Methods Mol. Biol.* **21,** 333–348.

Borst, P. (1986). Discontinuous transcription and antigenic variation in trypanosomes. *Annu. Rev. Biochem.* **55,** 701–732.

Clayton, C. E., Fueri, J. P., Itzhaki, J. E., Bellofatto, V., Sherman, D. R., Wisdom, G. S., Vijayasarathy, S., and Mowatt, M. R. (1990). Transcription of the procyclic acidic repetitive protein genes of *Trypanosoma brucei*. *Mol. Cell. Biol.* **10,** 3036–3047.

Coburn, C. M., Otteman, K. M., McNeely, T., Turco, S. J., and Beverley, S. M. (1991). Stable DNA transfection of a wide range of trypanosomatids. *Mol. Biochem. Parasitol.* **46,** 169–179.

Cruz, A., and Beverley, S. M. (1990). Gene replacement in parasitic protozoa. *Nature (London)* **348,** 171–173.

Cruz, A., Coburn, C. M., and Beverley, S. M. (1991). Double targeted gene replacement for creating null mutants. *Proc. Natl. Acad. Sci. U.S.A.* **88,** 7170–7174.

Curotto de Lafaille, M. A., Laban, A., and Wirth, D. F. (1992). Gene expression in *Leishmania:* Analysis of essential 5′ DNA sequences. *Proc. Natl. Acad. Sci. U.S.A.* **89,** 2703–2707.

Freedman, D. J., and Beverley, S. M. (1993). Two more independent selectable markers for stable transfection of *Leishmania*. *Mol. Biochem. Parasitol.* **62,** 37–44.

Hehl, A., Vassella, E., Braun, R., and Roditi, I. (1994). A conserved stem-loop in the 3′ untranslated reion of procyclin mRNAs regulates expression in *Trypanosoma brucei*. *Proc. Natl. Acad. Sci. U.S.A.* **91,** 370–374.

Hug, M., Carruthers, V. B., Hartmann, C., Sherman, D. S., Cross, G. A. M., and Clayton, C. (1993). A possible role for the 3′-untranslated region in developmental regulation in *Trypanosoma brucei*. *Mol. Biochem. Parasitol.* **61,** 87–96.

Jefferies, D., Tebabi, P., and Pays, E. (1991). Transient activity assays of the *Trypanosoma brucei* variant surface glycoprotein gene promoter: Control of gene expression at the posttranscriptional level. *Mol. Cell. Biol.* **11,** 338–343.

Jefferies, D., Tebabi, P., Le Ray, D., and Pays, E. (1993). The ble resistance gene as a new selectable marker for *Trypanosoma brucei:* Fly transmission of stable procyclic transformants to produce antibiotic resistant bloodstream forms. *Nucleic Acids Res.* **21,** 191–195.

Kapler, G. M., Coburn, C. M., and Beverley, S. M. (1990). Stable transfection of the human parasite *Leishmania* delineates a 30 kb region sufficient for extrachromosomal replication and expression. *Mol. Cell. Biol.* **10,** 1084–1094.

Kelly, J. M., Ward, H. M., Miles, M. A., and Kendall, G. (1992). A shuttle vector which facilitates the expression of transfected gene in *Trypanosoma cruzi* and Leishmania. *Nucleic Acids Res.* **20,** 3963–3969.

Laban, A., and Wirth, D. F. (1989). Transfection of *Leishmania enreittii* and expression of chloramphenicol acetyl transferase gene. *Proc. Natl. Acad. Sci. U.S.A.* **86,** 9119–9123.

Laban, A., Tobin, J. F., de Lafille, M. A. C., and Wirth, D. F. (1990). Stable expression of the bacterial neo^r gene in *Leishmania enriettii*. *Nature (London)* **343,** 572–574.

LeBowitz, J. H., Coburn, C. M., McMahon-Pratt, D., and Beverley, S. M. (1990). Development of a stable *Leishmania* expression vector and application to the study of parasite surface antigen genes. *Proc. Natl. Acad. Sci. U.S.A.* **87,** 9736–9740.

LeBowitz, J. H., Coburn, C. M., and Beverley, S. M. (1991). Stimultaneous transient expression assays of the trypanosomatid parasite *Leishmania* using beta-galactosidase and beta-glucuronidase as reporter enzymes. *Gene* **103,** 119–123.

LeBowitz, J. H., Cruz, A., and Beverley, S. M. (1992). Thymidine kinase as a negative selectable marker in *Leishmania major*. *Mol. Biochem. Parasitol.* **51,** 321–325.

LeBowitz, J. H., Smith, H. Q., Rusche, L., and Beverley, S. M. (1993). Coupling of poly (A) site selection and trans-splicing in *Leishmania*. *Genes Dev.* **7,** 996–1007.

Lee, M. G.-S., and Van der Ploeg, L. H. T. (1990). Homologous recombination and stable transfection in the parasitic protozoan *Trypanosoma brucei*. *Science* **250**, 1583–1587.

Lee, M. G.-S., and Van der Ploeg, L. H. T. (1991). The hygromycin B resistance encoding gene as q selectable marker for stable transformation of *Trypanosoma brucei*. *Gene* **105**, 255–257.

Liu, X., and Chang, K. P. (1992). The 63-kilobase circular amplicon of tunicamycin-resistant *Leishmania amazonensis* contains a functional N-acetylglucosamine-1-phosphate transferase gene that can be used as a dominant selectable marker in transfection. *Mol. Cell. Biol.* **12**, 4112–4122.

Patnaik, P. K., Kulkarni, S. K., and Cross, G. A. (1993). Autonomously replicating single-copy episomes in *Trypanosoma brucei* show unusual stability. *EMBO J.* **12**, 2529–2538.

Pays, E., Coquelet, H., Tebabi, P., Pays, A., Jefferies, D., Steinert, M., Koenig, E., Williams, R. O., and Roditi, I. (1990). *Trypanosoma brucei:* Constitutive activity of the VSG and procyclin gene promoters. *EMBO J.* **9**, 3145–3151.

Rudenko, G., LeBlancq, S., Smith, J., Lee, M. G. S., Rattray, A., and Van der Ploeg, L. H. T. (1990). Procyclic acidic repetitive protein (PARP) genes located in an unusually small alpha-amanitin resistant transcription unit. *Mol. Cell. Biol.* **10**, 3492–3504.

Ryan, K. A., Dasgupta, S., and Beverley, S. M. (1993). Shuttle cosmid vectors for the trypanosomatid parasite *Leishmania*. *Gene* **131**, 145–150.

Sherman, D. R., Janz, L., Hug, M., and Clayton, C. (1991). Anatomy of the parp gene promotor of *Trypanosoma brucei*. *EMBO J.* **10**, 3379–3386.

Zomerdijk, J. C. B. M., Oullette, M., ten Asbroek, A. L. M. A., Kieft, R., Bommer, A. M. M., Clayton, C. E., and Borst, P. (1990). Active and inactive versions of a promoter for a varient surface glycoprotein gene expression site in *Trypanosoma brucei*. *EMBO J.* **9**, 2791–2801.

CHAPTER 5

Mutagenesis and Variant Selection in *Salmonella*

Renée Tsolis and Fred Heffron

Department of Molecular Microbiology and Immunology
Oregon Health Sciences University
Portland, Oregon 97201

I. Introduction
II. Mutagenesis
    A. Spontaneous Mutants
    B. Chemical Mutagenesis
    C. Deletion or Replacement of Large DNA Fragments
    D. Transposon Mutagenesis
III. Screening of Variants
    A. General Considerations
    B. *In Vitro* Models of Infection
    C. Parameters Affecting Tissue Culture Assays
    D. Assay Protocols
IV. Genetic Analysis
    A. Cloning and Sequencing DNA Flanking the Transposon
    B. Mapping Strategies
V. Perspectives
    References

I. Introduction

In this chapter we will describe some frequently used methods for mutagenesis of the *Salmonella* genome, as well as potential methods for screening a bank of mutants and genetic analysis of interesting variants. The section on mutagenesis focuses on frequently used systems for transposon mutagenesis, while the section on variant selection describes *in vitro* cell culture assays designed to enrich a pool of mutants for those with defects in genes related to pathogenesis. In the following section on genetic analysis, some of the tools available for the characterization of interesting mutants will be discussed. Although the review will focus on methods developed in *S. typhimurium,* many of these are also applicable to other *Salmonella* serovars.

II. Mutagenesis

A. Spontaneous Mutants

With some pathogens, avirulent variants arise spontaneously after prolonged passage in the laboratory. Although these mutations are now regarded merely a nuisance to researchers, the impact that this had on pathogenesis research should not be underestimated. In 1921, a spontaneous variant of *Mycobacterium bovis* was obtained after repeated passages in culture—this mutant turned out to be the basis for one of the first bacterial vaccine strains Bacille Calmette-Guérine (BCG). Vaccines for a variety of other pathogens are also based on spontaneous mutants.

B. Chemical Mutagenesis

For some pathogens it was found that spontaneously arising mutants always acquired the same defect. Other pathogens, including *Salmonella,* are not as easily attenuated simply by passage *in vitro.* To faciliate variant selection, chemical mutagens were therefore used to create a wider range of avirulent microbial variants. An advantage of chemical mutagenesis is that a wide range of mutations can be created, including single nucleotide changes, nonsense mutations, and frameshift mutations. Many of these mutations will exert specific effects on proteins, such as a single amino acid change, and will not affect the expression of downstream genes in an operon. Such mutants are, however, difficult to analyze since the nature of the mutation is often not clear. Thus, they may require extensive genetic analysis in order to identify the mutation. In addition, the occurrence of multiple mutations is likely after chemical mutagenesis, which further complicates identification of the genes responsible for avirulence. For example, the first *S. typhi* vaccine strain, Ty21a, created by chemical mutagenesis was thought to be attenuated because of an lipopolysaccharide (LPS) defect, due to a *galE* mutation. However, it was later shown that the strain contains several other mutations, all of which have not been defined (Silva *et al.,* 1987). Furthermore, a defined *galE* mutant constructed by site-directed mutagenesis of *S. typhi* was found to be fully virulent (Hone *et al.,* 1988).

C. Deletion or Replacement of Large DNA Fragments

Variants that have alterations in a defined region of the genome have been created in two independent ways: plasmid curing and construction of hybrids between different *Salmonella* strains. Most Salmonellae contain large plasmids of ~90 kbp, which have been designated virulence plasmids and are considered part of their genome. This plasmid, however, is missing in *S. typhi*. Curing of the plasmid causes a reduction in virulence of most *Salmonella* serovars. This large "deletion" has been studied extensively, and it appears that the main virulence determinants of the plasmid are localized in the 8-kb *vir* region. Complementation of plasmidless derivatives of *S. dublin* and *S. typhimurium* with this region has been shown to restore nearly full virulence to these serovars (Gulig *et al.*, 1992; Krause *et al.*, 1991). Methods for curing Salmonellae of large virulence plasmids have been developed by Stojiljkovic *et al.* (1991; Table I) and Tinge and Curtiss (1990).

A more random approach for selection of variants was the construction of hybrids between *S. typhimurium* SR11, which is fully virulent in the mouse, and the less virulent strain LT2 by Hfr matings (Benjamin *et al.*, 1986). Some hybrids had a degree of virulence intermediate to LT2 and SR11. Due to the defined nature of the genetic exchange, it was possible to determine the endpoints of the exchanged fragment. Although the large size of the exchanged region made the analysis of these hybrids labor-intensive, this type of approach will be of particular value in the future. Another approach that could be useful in generating hybrids between *Salmonella* strains utilizes *mutLS* strains of *S. typhi*, which are defective in mismatch repair (Zahrt *et al.*, 1994). This defect overcomes the recombination barrier between *S. typhi* and foreign but related DNA and allows introduction of markers from *S. typhimurium* into *S. typhi*. These approaches of generating hybrids will allow us to address questions such as that of which genes are responsible for the host specificity of the different *Salmonella* serovars.

D. Transposon Mutagenesis

1. Properties of Transposons

a. General Considerations

The breakthrough in creating large numbers of defined mutants that are easy to analyze came with the introduction of transposon mutagenesis. These mobile genetic elements insert more or less randomly in the genome, thus disrupting the function of the gene into which they have inserted. Transposons used for mutagenesis generally contain selectable markers. This important feature allows transfer of the mutation into a clean background, ensuring that an observed phenotype can be attributed to a single mutation. At some point after isolating transposon mutants, each single mutation should be retransduced into a clean wild type background and its phenotype retested. This ensures that any phenotypes observed are not artifacts of the mutagenesis procedure.

In addition, presence of the transposon provides several possibilities of cloning and mapping the affected gene, which greatly facilitates further analysis of mutants. One disadvantage of this mutagenesis method, however, is that mutants are generally polar, thus raising the possibility that not the transposon's target gene but one of the affected downstream genes may be responsible for the observed phenotype.

In choosing a transposon for mutagenesis, it is advantageous to consider subsequent genetic manipulations beforehand—some transposons include features such as unique restriction sites for cloning flanking DNA or restriction mapping, which may prove useful for later analysis of mutants. A wide assortment of transposons is available for use in enteric bacteria. Some of these, such as Tn3 and γδ, are useful for mutagenesis of cloned genes, but cannot be used for mutagenesis of the *Salmonella* genome because the *Salmonella* chromosome is immune to insertion of these transposons. We will therefore restrict our discussion to a few transposons used most frequently in studying *Salmonella* pathogenesis.

b. Features of Transposon-Induced Mutations

Insertional inactivations are the oldest use of transposons. Mutants contain an insertion in the genome that can be selected using markers on the transposon. These insertions generally lead to loss of function of the target gene but may also affect expression of downstream genes in an operon.

Fusion mutants have the properties of insertion mutants with the additional feature that they contain a marker gene whose expression can be studied in the context of the insertion site. Generally these markers are not expressed in the transposon's donor molecule, so that expression of the marker gene provides an additional possibility for screening for transposition events. Screening for the marker gene will, however, yield only a fraction of all possible transposon insertions, as not every transposition will result in expression of the marker gene. Transposons are available for construction of three different types of fusions: transcriptional, translational, and promoter fusions. Transcriptional fusions result when a transposon inserts such that the marker gene (for example, *lacZ*) is transcribed from a promoter close to the insertion site. In order to be expressed, the transposon must be inserted in the proper orientation to the chromosomal promoter. These transcriptional, or "operon," fusions can be used to study the regulation of the gene that has been inactivated by the transposon insertion. Uses of operon and gene fusions have been reviewed by Slauch and Silhavy (1991).

Translational fusions lead to synthesis of a hybrid protein. These result when a transposon carrying a marker gene inserts in the correct orientation and reading frame so that the marker gene is translated in frame with the amino terminus of the target gene (e.g., *lac*Z, *pho*A, or *kan*). The resulting fusion proteins may have the function of the marker gene, or, if the essential domains of the target gene are included, the fusion protein may be bifunctional. Since these fusions are subject to the additional constraint of protein stability, it is not surprising that only a small fraction of all transposon insertions yield a functional fusion protein. However, the hybrid protein may yield a great deal of information about the target gene. For example *pho*A fusions indicate that the target protein is either secreted, contains periplasmic domains, or is localized to the bacterial envelope.

Promoter fusions place the target gene under the control of a promoter (e.g., P_{lac}) carried on the transposon. This approach is analogous to cloning in an expression vector in that regulated expression of the target gene is possible, with the difference that the target gene is present in only one copy in the genome. Transposon-induced promoter fusions have been used to acheive conditional expression of virulence genes in *Bordetella* (Cookson *et al.*, 1990).

c. Stability of the Insertion

The wild-type transposons originally used for mutagenesis are often unstable. Since they encode a functional transposase, these transposons may undergo secondary transposition events or catalyze deletions or inversions of adjacent sequences (particularly after storage), confounding subsequent genetic analysis. The difficulties encountered in the analysis of mutants containing transposon-induced genetic rearrangements can be avoided by using minitransposons. These transposon derivatives are not only smaller, as their name implies, they also lack a functional transposase, which is provided in cis or in trans. Loss of the transposase subsequent to insertion of the transposon into the genome yields a stable transposon mutation that cannot undergo secondary re-arrangements. For this reason, minitransposons should be used whenever possible. In the names of transposons, defects (for example, in transposition) are indicated by adding the letter *d* (Tn*10d* = mini-Tn*10;* Mu*d* = mini-Mu).

d. Specificity of Insertion

Although no transposon inserts completely at random, the degree of specifity of most transposon insertions is sufficiently low to isolate insertions in almost any gene of interest. For addressing statistical questions, for example, "what percentage of the genome is required for anaerobic growth ?", however, it is desirable to have a pool of insertions that is as random as possible. Derivatives of Mu are thought to be the most random. Tn*5* and Tn*10* exhibit some degree of sequence specificity, and wild-type Tn*10* was shown to have hot spots in several genes (Kleckner *et al.*, 1977). The degree of specificity for these hot spots has been reduced in a mutant transposase derivative, ATS; see Section II,D,3,a (Bender and Kleckner, 1992).

2. Delivery Systems

a. P22 Transduction

Considerations in choosing a delivery system include not only which constructs are available for a particular transposon, but also the efficiency of a
delivery method, the probability of obtaining multiple transposon insertions,
and the degree to which independent insertions can be isolated.

P22 is a *Salmonella*-specific phage for generalized transduction. Isolation of
a high-frequency transducing derivative, P22 HT*int*, has allowed the generation
of lysates with a high proportion of transducing particles containing *Salmonella*
DNA (Schmieger, 1972). Delivery systems based on P22 transduction are therefore very efficient at introducing transposons. At high multiplicities of infection,
however, multiple transposon insertions and selection for LPS mutations are
possible; it is therefore preferable to use a low ratio of phage to bacteria.
Transduction allows isolation of independent insertions, because transductants
are not allowed to double before plating. Thus, each colony formed on the
transduction plate represents an independent transposition event, making isolation of siblings unlikely.

Working with P22.   When performing P22 transduction or using P22 as a
vehicle for transposon delivery, it is important to titer the donor lysate and
infect at a ratio of no more than 10 phages per bacterium. At higher multiplicities
of infection, we have found that the frequency of transductants containing a
rough, or incomplete, lipopolysaccharide (LPS) independent of the transposon
insertion increases. Since a complete LPS is required for P22 infection, a high
concentration of phage selects for mutants that are unable to adsorb (and
therefore are immune to) P22. Rough mutants are undesirable for two reasons:
an intact LPS is necessary for full virulence of *Salmonella* as well as for
adsorption of P22. If P22 cannot adsorb and infect the mutant, retransduction
into a fresh background is somewhat more labor-intensive, as it involves infection of the rough strain with P22 via conjugation. Strains have been constructed
in which P22 is carried as a lysogen on an F' plasmid (Elliott, 1989). Using
these strains, it is possible to generate a transducing lysate of a rough strain
by introduction of the F'::P22 via conjugation into the rough strain. Once it
enters the rough strain, the prophage is induced, generating a transducing lysate.
Although these lysates are of lower titer than those made by adsorption of P22,
they yield enough phage for transduction of the mutation.

b. Nonconjugative Plasmids

Several transposons are supplied on multicopy, nonconjugative plasmids.
Some, such as the mini-Tn*10*s, are supplied with an inducible transposase
outside the transposon. Since the transposase is expressed at high levels upon
induction (by IPTG), the frequency of transposition is also high. The high level
of transposase expression and the presence of the transposon in multiple copies
per bacterium, will, however, favor multiple transpositions. Induction of the

transposase during growth in liquid culture allows mutants several doublings, which will increase the probability of isolating siblings. When studying pathogenesis, it is desirable to get rid of the plasmid used to introduce the transposon. This is achieved by pooling mutants, growing P22 on the pool, and retransducing the mutations into a wild-type strain. Since the plasmids carrying the transposon may also be transduced by P22, recipients should be checked for loss of the plasmid's antibiotic resistance marker.

c. Conjugative Plasmids

Conjugative plasmids, such as F' or suicide plasmids, are also used to deliver transposons. Two features of wild-type *Salmonella* strains reduce the frequency of obtaining transconjugants. The first is lipopolysaccharide, which reduces the frequency of mating aggregation and transconjugant formation of wild-type strains approximately 20-fold in comparison to rough mutants (Sanderson *et al.*, 1981; Duke and Guiney, 1983). In addition, when the donor is a wild-type *Salmonella* strain, the expression of F-factor genes, including those for the F-pili necessary for mating aggregation, is repressed by the virulence plasmid. This repression reduces the frequency of transconjugants by 100- to1000-fold (Sanderson and Roth, 1988). Since conjugation is less efficient than P22 transduction at delivering transposons, a P22-based delivery system is preferable. For some transposons, such as Tn*pho*A, however, it is the most convenient system for delivery. An advantage of suicide vectors over F' plasmids is that they are unable to replicate in *Salmonella* and are lost, whereas F-factors are propagated in the recipient. Although the probability of multiple insertions is not high using this system of delivery, growth of the recipients during incubation of the conjugation mixture increases the likelihood of obtaining siblings. The most efficient method of conjugation differs according to the plasmid being transferred: F' plasmids transfer most efficiently in standing liquid culture, whereas suicide vectors containing the RP4 *mob* region are best mated on agar plates.

3. Some Frequently Used Transposons

a. Tn10 and Its Derivatives

Tn*10* and its derivatives have proven to be invaluable tools in *Salmonella* genetics. In addition to its use in random mutagenesis, Tn*10* can be used to create defined deletions of target DNA (Kleckner *et al.*, 1977). Analysis of mutants created using wild-type Tn*10* has in the past proven difficult because of the various genetic rearrangements caused by the transposase. A newer series of mini-Tn*10*s has, however, been constructed, eliminating the problems associated with transposase activity. Some of the constructs available are listed in Table I, and more detailed information can be found in Kleckner *et al.* (1991). Although Tn*10* has been shown to exhibit a preference for insertion at certain

Table I
Transposons for Mutagenesis of *Salmonella*

Transposon	Stable insertion[a]	Selectable marker	Fusion marker (fusion type)	Delivery system	Other features	Reference
Tn*10d*(Tc)	+	Tc	—	Conjugation via F' or transformation of multicopy plasmids	Transposase is on a separate plasmid	Elliott and Roth (1988)
Tn*10d*(Km)	+	Km	—			Kleckner et al. (1991)
Tn*10d*(Cm)	+	Cm	—			Castilho et al. (1984)
MudI1734 (Mud*J*)	+	Km	*lac* (operon)	Transitory *cis*-complementation		Hughes and Roth (1988)
MudII1734 (Mud*K*)	+	Km	LacZ (gene)			Castilho et al. (1984) Hughes and Roth (1988)
Mini-Mu*lux*	−	Km or Tc	*lux* (operon)	Helper phage Mu*cts*	Screens for envelope proteins	Engelbrecht et al. (1985)
Tn*phoA*	−	Km	PhoA (gene)	Conjugation of a suicide vector from a permissive (λ*pir*) E. coli host		Manoil and Beckwith (1985) Taylor et al. (1989)
Tn5-*rpsL*	−	Km	—	F' *lac* ts	Useful for curing *Salmonella* of virulence plasmids	Stojiljkovic et al. (1991)
Tn5-*oriT*	−	Km			Useful for transfer of large regions of the chromosome or virulence plasmid	Yakobson and Guiney (1984)
mini-Tn5 Sm/Sp	+	Sm/Sp	—	Conjugation of a suicide vector from a permissive (λ*pir*) E. coli host		de Lorenzo et al. (1990)
mini-Tn5 Tc	+	Tc	—			de Lorenzo et al. (1990)
mini-Tn5 Km	+	Km				de Lorenzo et al. (1990)
mini-Tn5 *lacZ1*	+	Km	*lac* (operon)			de Lorenzo et al. (1990)
mini-Tn5 *lacZ2*	+	Km	LacZ (gene)			de Lorenzo et al. (1990)
mini-Tn5 *phoA*	+	Km	PhoA (gene)		Screens for envelope proteins	de Lorenzo et al. (1990)
mini-Tn5 *luxAB*	+	Tc	*lux* (operon)			de Lorenzo et al. (1990)

[a] (+) indicates that the transposon does not carry its own transposase.

consensus sequences in the genome, insertion is considered to be sufficiently random to obtain mutants in almost any gene of interest. This preference for hot spots has been reduced in a mutant form of transposase (designated ATS for *a*ltered *t*arget *s*pecificity), which can be used with any of the mini-Tn*10*s (Kleckner *et al.*, 1991).

As indicated above, P22 transduction is an efficient system for delivering transposons. A protocol for using this procedure to perform a random mutagenesis with Tn*10d*(Tc) can be found below (Section II,D,4,d). The P22 lysate used for delivery of Tn*10d*(Tc)is obtained by growing P22 HT*int* on the *Salmonella* strain TT10423, which carries the transposon on a F' plasmid (Elliott and Roth, 1988). In this system, the recipient *Salmonella* strain must first be transformed with a plasmid, pNK2881, which carries the transposase gene under control of the *tac* promoter (Kleckner *et al.*, 1991). Since *Salmonella* lacks the *lac* repressor, which acts on P_{tac} to repress transposase expression, the transposase is expressed constitutively. Upon introduction of Tn*10d*(Tc) by P22, transposase acts in trans on its target sequences in Tn*10* to catalyze transposition into the chromosome. An advantage of this procedure is that the fragment containing Tn*10* cannot persist independently in the recipient, which means that the recipient can gain the transposon's antibiotic resistance marker only by transposition of Tn*10* into the genome. In order to get rid of the plasmid containing the transposase, the resulting transductants are pooled and used to prepare a second P22 lysate. This lysate is then used to infect a wild-type *Salmonella* strain.

b. Derivatives of Phage Mu

A second group of transposons commonly used in mutagenizing *Salmonella* are derivatives of the phage Mu. Mu and its derivatives have been found to insert with low specificity into their target DNA. Of the many derivatives available, we have found Mu*d*J, a transposition-defective Mu derivative, to be most useful. Mu*d*J contains a kanamycin resistance gene as well as a promoterless *lac*Z gene, permitting isolation of operon fusions between a chromosomal promoter and *lac*Z. For a random mutagenesis of the genome, a convenient delivery system for Mu*d*J is the one developed by Hughes and Roth (1988), which uses transitory cis complementation of a transposition-defective prophage with the functions of an intact prophage (Fig. 1). P22 is used to make a lysate of the donor strain TT10288, which contains Mu*d*J inserted into the *his*D (histidine biosynthesis) gene. This strain also contains a second copy of Mu, Mu*d*I, which is inserted 4 kb from Mu*d*J in the *his*A gene. Unlike Mu*d*J, Mu*d*I contains a functional transposase, which can be packaged together with Mu*d*J into a P22 phage. Because of the distance between Mu*d*J and Mu*d*I (4 kb) and the size of DNA packaged by P22 (44 kb), it is impossible to package and deliver both copies of Mu simultaneously. Thus, by infecting the recipient strain with P22 grown on TT10288 and selecting for resistance to kanamycin, one can select for transposition events into the *Salmonella* genome. Because Mu*d*J is inserted into the *his* locus, however, a proportion of kanamycin-resistant colo-

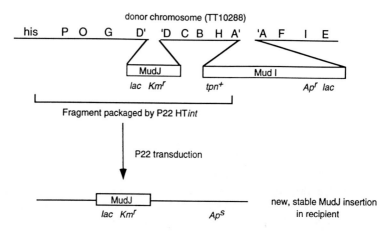

Fig. 1 The system for transitory *cis* complementation of the transposition-defective Mu*d*J. For explanation, see text. (Adapted from Hughes and Roth, 1988, with permission.)

nies will arise by homologous recombination between the flanking *his*D sequences of Mu*d*J from the donor lysate and the corresponding sequence in the recipient strain. Unlike mutants arising from transposition events, these recombinants will be phenotypically *his-* and can be eliminated either by plating the transductions on minimal medium (which will, however, eliminate all auxotrophs from the mutant bank) or by screening mutants for histidine auxotrophy on minimal medium with and without histidine. We have found the number of such *his-* recombinants to vary between 30 and 50% in different pools of mutants tested.

c. Tn5 and Derivatives

Since one would expect many virulence determinants to be located on the bacterial surface or secreted into the surrounding medium, it is useful to have a transposon that will provide an enrichment for such proteins. Tn*pho*A, a derivative of Tn*5,* is a probe that identifies exported proteins (Manoil and Beckwith, 1985). This approach is based on the finding that alkaline phosphatase (PhoA), which is a periplasmic protein, is inactive when its export to the periplasm is blocked (Derman and Beckwith, 1991). The copy of *pho*A that is inserted in Tn*pho*A has its export signal deleted; thus, only when the transposon inserts into a gene possessing an signal does the resulting fusion protein possess alkaline phosphatase activity. Signals sufficient for export of a *pho*A fusion protein have been found in proteins of the outer membrane, the periplasmic space, and the cytoplasmic membrane, as well as in secreted proteins.

Before mutagenizing *Salmonella* with Tn*pho*A, it is necessary to determine whether the strain to be used for the procedure possesses a high endogenous phosphatase activity, which could potentially mask that of a PhoA fusion protein. In *S. typhimurium,* the phosphatase encoded by the *pho*N gene causes a high background activity. One way to avoid this is to perform the Tn*pho*A mutagenesis in a *pho*N background, since PhoN is not required for virulence (Fields *et al.,* 1989).

A transposition defective derivative of Tn*pho*A has been constructed by de Lorenzo *et al.* (1990). Mini-Tn*pho*A can be introduced into a wild-type *Salmonella* strain by conjugation of a suicide vector, carrying the transposon and the Tn*5* transposase outside Tn*5* in cis, from its permissive *Escherichia coli* host (de Lorenzo *et al.,* 1990). Since the plasmid cannot replicate in the *Salmonella* recipient, the kanamycin resistance encoded by Tn*pho*A can be maintained only if the transposon hops into the genome. This procedure is described in Section II,D,4,f.

Several other useful mini-Tn*5* transposons are available, which include constructs for generation of gene and operon fusions (de Lorenzo *et al.,* 1990). In addition, constructs encoding a variety of antibiotic markers are included in this group. This set of mini-Tn*5*s is supplied on conjugable suicide plasmids and carry the Tn*5* transposase in cis, but external to the transposon.

4. Protocols for Transposon Mutagenesis

a. Generation of a P22 HT int Lysate

To obtain a high titer phage P22 lysate of a strain of interest, grow the strain to be lysed at 37°C with aeration in 0.5 ml of LB with appropriate antibiotics until the culture is saturated (OD_{600} is about 1; takes approx 5–8 hr). Add 2 ml of P22 broth [LB with 0.1 M glucose, 1X E-salts and ca. 10^9 plaque-forming units (pfu)/ml P22 HT *int* grown on wild-type *Salmonella*] and continue growth for 5 hr to overnight. (Cell debris is a sign that lysis has occurred, but unlysed cultures may also produce usable lysates.) Spin down cells and debris for 2 min in a microfuge. Carefully remove the supernatant to a fresh tube (screw-cap tubes are best) and add a few drops of chloroform to the lysate. Vortex thoroughly and invert the tube to distribute the chloroform, then place the tube on ice for ca. $\frac{1}{2}$ hr to allow the chloroform to settle before using. These lysates are stable for several months when stored at 4°C. (Note: since chloroform does not kill P22, contamination of chloroform bottles with phage may result in cross-contamination of lysates!) E-Salts are described in Vogel and Bonner (1956). Alternative methods for generating P22 HT *int* lysates, as well as a collection of recipes for specialized media used with protocols involving P22 transduction, can be found in Maloy *et al.* (1994). P22 HT*int* is available from K. Sanderson (Salmonella Genetic Stock Centre, Calgary, Canada).

Lysates can be titred by making serial 10-fold dilutions of 10^{-1} to 10^{-12} in LB and dropping 10 μl of each dilution onto an LB plate that has been spread with 3 ml of top agar (0.7% agar in LB), containing 0.1 ml of an overnight culture of a wild-type *Salmonella* strain (Add the bacteria to a small test tube containing the melted agar which has been cooled to 45°C. Mix and pour immediately onto an LB plate, tilting the plate to distribute the top agar evenly. Allow the agar to solidify before dropping the lysate dilutions onto the plate.) Incubate this plate until plaques (clear spots in the lawn of bacteria) are visible in the lysate drops. Count the number of plaques in a drop that contains well-separated plaques and calculate the number of phages per milliliter of the original lysate (No. of plaques x dilution factor counted/0.01 ml).

b. P22 Transduction

Culture your recipient strain overnight in LB. To 0.1 ml of this culture, add 10 μl of an appropriate dilution (in LB) of a P22 lysate grown on the strain carrying the marker you intend to transduce. The ratio of phage to bacteria should be between 1 and 10. Incubate at 37°C without shaking for 10–15 min to allow adsorption of P22 to the bacteria. Then add 1 ml of LB, 10 mM EGTA, pH 7.5, to prevent further phage adsorption. Incubate a further 15–20 min, then plate the transduction on appropriate antibiotic plates containing 10 mM EGTA, pH 7.5. Incubate these plates overnight. Pick transductant colonies carefully using a sterile toothpick or yellow pipette tip and streak each colony picked sequentially onto (1) LB + appropriate antibiotic and (2)EBU (Evans blue-uranine) agar (used to eliminate phage from the transductants; Bochner, 1984.). Incubate plates overnight. Compare LB/antibiotic and EBU plates to ensure that colonies picked from the transduction plates contain the appropriate resistance marker and are free of phage. Phage-contaminated colonies will appear dark blue on EBU plates, whereas uncontaminated ones are light blue. Purify by streaking light colonies from EBU plates (which have grown on the antibiotic plates) onto EBU plates again. If any dark blue colonies are still present in the streak, repeat the purification until no phage contamination is evident. As a final control, check purified transductants for sensitivity to P22 by streaking the colony across a P22 lysate on an EBU plate. After incubation, streaks should be light and turn dark where they cross the lysate. Phage-contaminated strains will be blue across the entire streak, whereas P22-resistant (LPS defective mutants or P22 lysogens) mutants will remain light-colored. The parent strain (which should be sensitive to P22) and a P22-resistant strain (a P22 lysogen or E. coli) should be included on each plate as controls. It is important to ensure that the transductants are free of phage and not lysogenic for P22, as P22 lysogeny will prevent further transduction of the marker.

c. Generation of a Transducing Lysate from Rough Strains Using F'::P22 HT int

Grow E. coli TE 1335 (Elliott, 1989) and the strain containing the marker to be retransduced to mid log-phase (OD$_{600}$ = 0.5–0.6) with antibiotic selection. Wash out antibiotics by spinning down cells and resuspending twice in the original culture volume of LB without antibiotics. Pipette 2 ml of each culture into a 50-ml flask; the culture should just cover the bottom of the flask. Allow the flask to stand without shaking at 37°C for 3 hr to overnight. At this point, cells should appear lysed, but unlysed cultures often yield usable lysates. Lyse the cells completely by adding a few drops of chloroform and incubate with rapid shaking for 5 min at 37°C. Spin the culture down for 15 sec in a microfuge to pellet cell debris. Transfer the supernatant to a new tube and add a few drops of chloroform to prevent bacterial growth. Allow the choroform to settle to the bottom of the tube before using the lysate. We generally obtain titers of 10^5–10^7 pfu/ml with this method. Store lysates at 4°C.

d. Mutagenesis with Tn10d(Tc)

1. Construct recipient strain. Transform your recipient wild-type *Salmonella* strain with the plasmid pNK2881, which carries the Tn*10* transposase under the control of the *tac* promoter (Kleckner *et al.,* 1991). Select for ampicillin-resistant transformants.

2. Generate P22 lysate for delivery of Tn*10d*(Tc). Make a P22 lysate (see Section II,D,4,a) of *S. typhimurium* TT10423, which carries Tn*10d*(Tc) on an F' plasmid (Elliott and Roth, 1988).

3. Mutagenize your recipient *Salmonella* strain. Introduce Tn*10d*(Tc) into recipient strain by transduction (see Section II,D,4,b) using P22 lysate generated in step 2, selecting for tetracycline-resistant transductants on LB plates containing 10 m*M* EGTA.

4. Eliminate transposase from Tn*10* mutants by retransducing the Tn*10* insertions. Pool transductants from step 3 by flooding Tc plates with 1.5 ml of LB, 10 m*M* EGTA and suspending bacteria with a spreader. Remove suspended bacteria to an Eppendorf tube and rinse plates with 1.5 ml LB, 10 m*M* EGTA. Add the wash to a second tube. Pellet cells and wash each tube twice with 1.5 ml of LB, 10 m*M* EGTA. Resuspend bacteria in LB. Grow P22 on this pool (see section II,D,4,a) and retransduce the *Salmonella* strain from which the recipient was constructed. Select for tetracycline-resistant retransductants and patch these onto LB Ap (50 μg/ml) plates to ensure that the plasmid carrying the transposase has been lost.

e. MudJ Mutagenesis

1. Test donor strain TT10288 to ensure that transposons are at the correct location (in the *his* locus) before beginning. Use P22 HT*int* grown on a wild-type strain to transduce TT10288 to histidine prototrophy. Plate transductions on E-plates + 10 m*M* EGTA. Patch transductants onto LB Ap100 and LB Km100. Some revertants to histidine prototrophy should be sensitive (i.e., they should not grow on the antibiotic plates) to both antibiotics, which will indicate that they have lost both Ap (from Mud*I*) and Km (from Mud*J*) markers from the *his* locus, where the two transposons are inserted (see Fig. 1). TT10288 should always be grown at 30°C, as it contains a temperature-sensitive Mu repressor. Growth at higher temperatures can result in induction of the Mud*I* prophage and transposition of Mud*I* to another location in the genome.

2. Construct donor lysate for MudJ. Make a P22 lysate of *S. typhimurium* TT10288 (Hughes and Roth, 1988; see Section II,D,4,a).

3. Mutagenize recipient strain. Transduce the *Salmonella* strain to be mutagenized by infecting it with the TT10288 lysate, as described in Section II,D,4,b. Plate on LB Km60, 10 m*M* EGTA or on E-Plates (Vogel and Bonner, 1956) with 100 μg/ml Km, 10 m*M* EGTA if auxotrophs are to be eliminated from the bank.

f. TnphoA Mutagenesis

1. Mutagenize recipient strain: Set up 10 parallel conjugations between *E. coli* SM10λ*pir* (pUT/mini Tn*phoA*) (de Lorenzo *et al.*, 1990) and the recipient *Salmonella* strain on LB plates by pipetting 50 μl of washed overnight cultures (to remove antibiotics) of donor and recipient and spreading together onto the plate. Incubate overnight at 37°C.

2. Select for mutants: Flood each plate with 1.5 ml of M9 glucose (Miller, 1972). Suspend bacteria with a spreader and remove liquid to a microfuge tube. Wash cells twice with 1.5 ml M9 glucose medium. Dilute suspension 1:10 in M9 glucose and spread varying amounts of diluted and undiluted bacteria onto M9 glucose plates containing 80 μg/ml D,L-valine (as a counterselection for the *E. coli* donor; Glover, 1962), 40 μg/ml XP (5-bromo-4-chloro-3-indolyl phosphate, an indicator substrate for PhoA) and 100 μg/ml kanamycin (to select for transposition of Tn*phoA*). Incubate over night at 37°C. Active PhoA fusions will be blue on these plates. The color of weakly blue colonies can be intensified by placing plates (after overnight incubation at 37°C) at 4°C. In order to confirm that mutants are not spontaneous valine-resistant mutants of *E. coli,* colonies can be picked onto Bismuth sulfite agar (Difco). *Salmonella* strains appear shiny black on this agar, whereas *E. coli* will be brown.

III. Screening of Variants

A. General Considerations

A bank consisting of mutants that have only a single defect must be relatively large if the entire genome, consisting of ~4000 genes, is to be examined. Thus, a compromise between the most efficient way of screening and the least (time-consuming and probably money-consuming) way must be found. The only way to determine the effect of a mutation on virulence is either by natural infection or by using an animal model. However, many *in vitro* assays simulating certain steps during infection have been developed and proven to be useful in enriching large banks of mutants for attenuated variants. Some stress conditions simulated in *in vitro* assays, including carbon starvation, phosphate starvation, osmotic stress, oxidative stress, and changes in pH may be too nonspecific and most likely occur during growth of *Salmonella* outside the host as well. In contrast, assays incorporating cell culture models seem to approximate *in vivo* conditions more closely and have been used successfully to screen mutant banks of various *Salmonella* serovars (Fields *et al.,* 1986; Stone *et. al.,* 1992). Results obtained using *in vitro* models of infection do not, however, always correlate with data obtained *in vivo.* For example, whereas plasmidless *S. typhimurium* behave exactly as their plasmid-containing parent strains *in vitro,* they are markedly attenuated *in vivo* (Jones *et al.,* 1982). The opposite was found for *S. typhimurium inv*A mutants, which exhibit decreased invasion of tissue culture cells, but are only slightly attenuated in the mouse (Galán and Curtiss, 1989).

In the mouse model of infection, *S. typhimurium* has been found to enter distinct cell types: epithelial and M-cells of the intestine during passage of the lining epithelium of the gut, and, during the systemic phase of infection, macrophages of the liver, spleen, and Peyer's patches. Therefore, cell culture models provide a tool for analyzing different steps of a *Salmonella* infection by using different cell types. Three distinct steps in the interaction of *Salmonella* with host cells can be addressed with different cell culture assays—adhesion, invasion, and intracellular persistence.

B. *In Vitro* Models of Infection

1. Methods for Studying Adhesion and Invasion

Various methods have been used to study adhesion to cells. One of these involves growing cells on coverslips, adding bacteria, and fixing and staining to count adherent bacteria. This method of counting can be tedious, but it yields information about whether the bacteria adhere singly or in clusters. Staining for adherent bacteria does not, however, allow for differentiation between live and dead bacteria. A second method entails addition of bacteria to cells and washing to remove bacteria that are not cell-associated. With this method, both adherent and intracellular bacteria are counted. This assay can be used to determine invasion rates by adding gentamicin or colistin, which do not penetrate tissue culture cells, to kill extracellular (adherent) bacteria. A third method allows determination of adherent bacteria only, by allowing microbes to adhere to fixed (i.e., intact but killed) cells, which do not allow penetration of the bacteria. In these last two methods, bacteria are counted by release from the cells and plating to determine colony-forming units (CFU).

2. Survival within Macrophages

Research in our laboratory has focused on the ability of *Salmonella* to survive within murine phagocytes. The assay described below can be performed using either primary macrophages (elicited peritoneal macrophages, or macrophages isolated from the bone marrow or spleen) or cell lines. We routinely use the murine macrophage-like cell line J774 for screening because it is derived from the Balb/c mouse, which we use as an animal model of infection (Ralph and

Nakoinz, 1975). In addition, J774 cells are easy to cultivate. Although some primary macrophages have a greater microbicidal effect on *Salmonella* than does J774 (Buchmeier and Heffron, 1989), we have found generally that inability of mutants to survive in J774 cells correlates with decreased virulence in the mouse (Fields *et al.*, 1986). As this cell line sometimes loses its ability to generate an oxidative burst after several passages in culture, we test cultures periodically for generation of an oxidative burst following stimulation with phorbol myristyl acetate (PMA), as described by Damiani *et al.*, (1990).

C. Parameters Affecting Tissue Culture Assays

1. Growing *Salmonella* for Assays

Adhesion and invasion assays have been described for many different pathogenic bacteria. One aspect of working with *Salmonella* to keep in mind when performing these assays is that, in comparison to other pathogens, it invades and grows more rapidly in cell cultures. For example, Salmonellae can be detected within epithelial cells within 30 min.

Most adhesion and invasion protocols use logarithmic phase bacterial cultures. The reason for this is that the highest levels of invasion can be achieved using logarithmically growing cultures for infection.

2. Multiplicity of Infection

There appear to be a limited number of binding sites for bacteria at the cell surface. Therefore, when cells are infected at a multiplicity of infection (MOI) that is higher than the number of binding sites, the proportion of bound bacteria in the inoculum will decrease as the MOI is increased. It is therefore necessary to infect at identical MOI with mutants and the wild type controls. In addition that means that only experiments with similar MOI can be compared.

3. Antibiotic-Resistant Bacteria

Many protocols call for addition of gentamicin to kill bacteria that have not invaded the tissue culture cells. This antibiotic is used because it does not accumulate within cells and therefore can be used to kill only extracellular bacteria. Our wild-type *S. typhimurium* strain, ATCC 14028, as well as SR11 and SL1344, is sensitive to gentamicin. Many field isolates of *Salmonella* are, however, resistant to gentamicin (J.G. Kusters, personal communication). For these strains, gentamicin can be substituted by colistin (150 μg/ml; Kusters *et al.*, 1993).

A further effect caused by antibiotic resistance can be seen while performing these assays. If cell lines are carried with antibiotics (usually penicillin/streptomycin), these may accumulate within cells, thereby conferring a survival advantage on strains that are resistant to these antibiotics (Leung *et al.*, 1992). For this reason, we carry cells for ca. 24 hr before an assay is performed without antibiotics, or culture cells without antibiotics altogether.

4. Spinning Bacteria onto Cells

One point about which differences in opinion exist is the practice of spinning bacteria onto cell monolayers. Centrifugation synchronizes the infection process, as many smooth strains may remain in suspension for hours, without contacting the cell monolayer. Synchronization of infection thus allows study of infection kinetics, as most of the bacteria will have been taken up within a short period of time. Others have argued that, since we do not know exactly how adhesion and invasion factors work, spinning bacteria onto cells may allow bacteria deficient in these factors to achieve intimate contact with, and entry into, cells. We therefore use centrifugation for kinetics studies in macrophages and eliminate the step from adhesion and invasion assays. We have found, using fixed macrophages, that the number of adherent bacteria decreases about 10-fold when bacteria are not spun onto the cells (R. Tsolis and F. Heffron, unpublished observations).

5. Sensitivity of Bacteria to Detergents

Since detergents are used in tissue culture assays to lyse tissue culture cells and recover bacteria, it is important to determine whether the bacteria are actually resistant to the detergents (usually 1% sodium deoxycholate or 1% Triton X-100) used. Some mutations, especially those in envelope proteins, may cause defects in the bacterial outer membrane, thus allowing detergents to penetrate and lyse the bacteria. This would lead, falsely, to the conclusion that the mutant is defective in adhesion, invasion, or intracellular survival. In order to test this possibility, mutants can be incubated for varying periods of time in the concentration of detergent used and plated before and after incubation to determine sensitivity. Alternatively, tissue culture cells can be lysed using distilled water, in order to avoid damaging the bacteria. Different cell lines will vary in the amount of time needed to lyse in water, so lysis should be followed under the microscope.

6. Choice of Cell Lines

Various cell lines have been used to study adhesion and invasion (Table II). Considerations in the choice of cell lines to use might include the species and organ from whence the cell line was derived, in order to ensure that the specificity matches that of the *Salmonella* serovar being studied. Historically, however, researchers have tended to use cell lines that are readily available and easily cultivated, and results of invasion studies have shown only minimal differences in invasion rates of different cell lines by Salmonellae.

Table II
Cell Lines Used to Study Interaction of *Salmonella* with Epithelial Cells and Macrophages

Cell line	Origin	Reference (originator)	Comments	Use with *Salmonella* (reference)
Epithelial-like lines				
Caco-2	Human colon adenocarcinoma	Fogh *et al.* (1977)		Gahring *et al.* (1990)
CHO	Chinese hamster ovary	Puck *et al.* (1958)		Stone *et al.* (1992)
HeLa	Human cervical carcinoma	Gey *et al.* (1952)		Kusters *et al.* (1993)
Hep-2	Human larynx carcinoma	Moore *et al.* (1955)	Easy to handle; withstands changes in temperature, nutrition, and environment without loss of viability.	Stone *et al.* (1992) Lee *et al.* (1992)
Int-407	Human embryonic jejunum and ileum	Henle and Deinhardt (1957)		Galán and Curtiss (1989) Kusters *et al.* (1993)
MDCK	Canine kidney	Madin and Darby (1958) American Type Culture Collection	Can be used to generate polarized monolayers	Finlay *et al.* (1988)
T-84	Human colon	Murakami and Masui (1980) Dharmsathaporn *et al.* (1984)	Grows slowly in culture; forms polarized monolayers	
Macrophage-like lines				
J774	Murine (Balb/c) reticulum cell sarcoma	Ralph and Nakoinz (1975)		Buchmeier and Heffron (1989)
U937	Human histiocytic lymphoma	Sundstrom and Nilsson (1976)	Can be induced to terminal differentiation by phorbol esters, vitamin D3, retinoic acid, interferon-γ, and TNF	

D. Assay Protocols

1. Adhesion to Glutaraldehyde-Fixed Cells

1. Wash confluent epithelial cell monolayers (approx 5×10^5 cells / well) in 24-well tissue culture plates twice with cold (4°C) PBS. Add 2 ml of 2% glutaraldehyde in PBS and incubate at 4°C for 1hr. Remove the glutaraldehyde and rinse four times with 2 ml of cold PBS, incubating in between washes 30 min at 4°C.

2. To obtain a logarithmic phase bacterial culture (approx 5×10^7CFU/ml), dilute a stationary (2×10^9 CFU/ml) culture of the strain to be tested 1:100

in prewarmed LB. Incubate 2hr at 37°C, then spin down cells (5 min at $3000 \times g$) and resuspend in prewarmed cell culture medium without supplements. Prepare 10-fold dilutions of this suspension to obtain 5×10^4–5×10^7 CFU/ml, which will result in a MOI of 0.1 to 100 bacteria per cell using 2 ml of these dilutions as inoculum. Rinse cells with 2 ml per well of prewarmed culture medium, then add 2 ml of each dilution to duplicate wells (once the adhesiveness of your wild-type strain to your particular cell line is established, it will be possible to perform the assays using the MOI that gives countable adhesive bacteria and omitting the other dilutions). Place inocula on ice to prevent growth before diluting and plating to determine CFU. Incubate plates for 30 min at 37°C to allow bacteria to adhere to the monolayer (a CO_2 incubator is preferable, but not critical, as these cells are already dead).

3. Remove the inoculum and wash five times with 2 ml of PBS to remove nonadherent bacteria. Lyse cells with 0.5 ml of 1% sodium deoxycholate (or 1% Triton X-100) and pipette vigorously to release bacteria from the cells and to break up the cellular DNA (this is important, as the DNA will otherwise become viscous, making the dilutions for determining adherent bacteria inaccurate). Remove the lysate to a test tube and rinse each well with 0.5 ml PBS, adding the rinse from each well to the cell lysate. Keep samples on ice. Make 10-fold serial dilutions of each sample, vortexing vigorously (20–30 sec) between dilution steps and spread 0.1 ml of each dilution on agar plates to determine adherent CFU. Incubate agar plates overnight and count CFU, then calculate proportion of adherent bacteria in the inoculum.

2. Invasion of Epithelial Cells

Prepare inoculum as described for adhesion assay above. Rinse confluent monolayers of epithelial cells (precultured for 24 hr without antibiotics!) in 24-well plates with 2 ml of cell culture medium without supplements, then incubate with 2 ml of this medium for 30 min in a CO_2 incubator to remove serum components. Add inoculum (2 ml per well) and incubate for 5 min to 3 hr in a CO_2 incubator at 37°C. Wash cells five times with 2 ml of PBS to remove nonadherent bacteria and add 1 ml of medium containing 150 μg/ml of gentamicin or 150 μg/ml of colistin to each well. Incubate 90 min to kill extracellular bacteria, then wash cells three times with 2 ml of PBS to remove the antibiotic. Lyse cells, dilute, and plate to determine intracellular CFU as described for adhesion assay above.

3. Macrophage Survival Assay

1. In order to obtain adherent monolayers of cells, 5×10^5 cells are aliquoted into each well of a 24-well plate and allowed to adhere and spread overnight in medium without antibiotics.

2. Directly before infection, fresh medium (0.5 ml/well) is added to the macrophages. To prepare the bacterial inoculum, a stationary phase (overnight) culture is washed in 10 vol of PBS. The following opsonization step is optional: after removing the PBS, the pellet is resuspended in 20 μl of normal mouse serum and incubated for 15 min at 37°C to opsonize the bacteria. This suspension is then diluted to 5×10^7 bacteria/ml with tissue culture medium and 0.1 ml of this suspension is added to each well. (The final ratio of bacteria to cells is 10:1; wild-type strains grow to a density of approx 5×10^9 overnight.) After infection, inocula are placed on ice to prevent growth before CFU determination by dilution and plating on the appropriate media.

3. To synchronize infection, bacteria are spun onto the cells for 5 min at 1200 rpm in a prewarmed (37°C) Sorvall centrifuge. Plates are then incubated for 20 min at 37°C to allow uptake of the bacteria.

4. Monolayers are washed three times with PBS to remove non-cell-associated bacteria and fresh tissue-culture medium containing 150 μg/ml of gentamicin is added to kill extracellular bacteria. This is considered to be the "zero time point" in the assay.

5. Samples are collected at 0, 1, 4, and 20 hr (after this point, our wild-type strain lyses the macrophages, making it impossible to determine the number of intracellular bacteria accurately). After 90 min, the wells for the remaining time points are washed and fresh medium supplemented with 20 μg/ml gentamicin is added to prevent bacterial replication. To collect samples, the supernatant is removed and discarded, then cells are washed three times with PBS to remove the antibiotic. One percent sodium deoxycholate (0.5 ml) is added to lyse the macrophages. This is collected and the well rinsed with 0.5 ml PBS. This wash is added to the lysate and samples are stored on ice before plating.

6. Inocula and samples are diluted and plated to determine viable counts. Results of a typical macrophage survival assay using wild-type and mutants defective in intracellular survival are shown in Fig. 2.

IV. Genetic Analysis

A. Cloning and Sequencing DNA Flanking the Transposon

The fastest way to determine whether a transposon has inserted into a known, sequenced gene is to sequence the flanking DNA. Since the sequences of most transposon ends are known, it is possible to amplify the flanking DNA at one end by inverse PCR and either to sequence the PCR product directly or to clone it in one of the available vectors (such as TA cloning kit from Invitrogen) for sequencing. This strategy has been used to identify genes carrying insertions of Tn10 and MudJ (Bäumler *et al.*, 1994; R. Tsolis and F. Heffron, unpublished results). Transposon ends contain palindromic DNA sequences and the sequence inside the transposon often contains additional repeats. Therefore, when

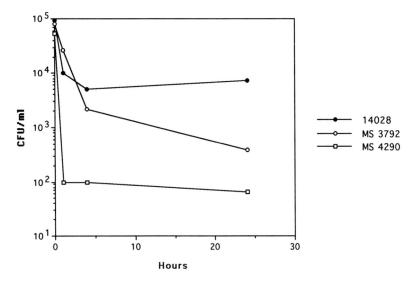

Fig. 2 Survival of *S. typhimurium* and mutants defective in intracellular persistence in splenic (primary) macrophages as determined by the macrophage survival assay described in Section III,D,3. Solid circles, *S. typhimurium* ATCC 14028 (wild-type parent); open circles and squares, two transposon mutants defective in macrophage survival. (Reproduced from Buchmeier and Heffron, 1989, with permission.)

designing PCR primers, it is helpful to compare the primer sequence to the entire transposon sequence in order to ensure single priming. One drawback of this method is that, because of the small size (~200 bp) of transposon flanking DNA fragments obtained by inverse PCR, only sequences that are identical or highly homologous to entries in the database searched can be identified.

B. Mapping Strategies

1. Genetic Mapping

a. Mud-P22 Prophages

Many more *Salmonella* loci have been characterized genetically than have been sequenced; therefore a genetic mapping approach is likely to yield more information about a particular mutant. A variety of tools have been developed for analysis of transposon insertions. If the transposon used for mutagenesis carries resistance to tetracycline, the insertion site can be mapped to within approximately 3 min of the chromosome using the method developed by Benson and Goldman (1992). This method employs a series of defective Mu-P22 hybrid prophages inserted at various points in the chromosome. Upon induction by mitomycin C, these phages package the adjacent 3 minutes (three P22 "heads full'') of the chromosome on one side of the phage insertion. Isolation of the phage thus provides an enrichment for a 3-min region of chromosomal DNA.

One method employing this set of Mu*d*-P22 strains for mapping is hybridization with cloned transposon-flanking DNA. Each of the phage DNAs is isolated and applied to a slot blot, which is then probed with the transposon-flanking DNA. For hybridization, it is necessary to eliminate all of the sequence from the probe that originates from the transposon ends (especially from Mu derivatives) in order to avoid nonspecific hybridization to the phage DNA present in each preparation.

A second approach that utilizes the set of Mu*d*-P22 strains is based on loss of tetracycline resistance by transduction. If a Mu*d*-P22 phage packages the region corresponding to a transposon insertion specifying tetracycline resistance, the phage DNA can recombine with that of the mutant, causing loss of tetracycline resistance. Since the antibiotic fusaric acid can be used to select for tetracycline-sensitive recombinants, mutants that have lost a transposon by recombination with the Mu*d*-P22 DNA are able to grow on these plates (Maloy and Nunn, 1981). Thus, by infecting a mutant with a series of these Mu*d*-P22 lysates, it is possible to determine the map location of the transposon insertion by the number of fusaric acid resistant colonies obtained with each lysate. Since this method is based on detection of reversion to tetracycline sensitivity, it is most convenient to have a tetracycline resistance marker on the transposon used for mutagenesis. Transposons containing other resistance markers can often be converted to tetracycline resistance by marker exchange, i.e., using a related transposon, containing the *tet* resistance gene, to replace the transposon present in the mutant by homologous recombination. For example, Mu*d*J, which contains a kanamycin resistance marker, can be exchanged with mini-Mu(Tet^r) to obtain a tetracycline resistant mutant for mapping (Belas *et al.*, 1984).

b. Mapping by Cotransduction with Known Markers

Once the insertion site has been narrowed down to a 3-min region of the chromosome, the mutation can then be mapped more precisely by cotransduction with known markers (Sanderson and Roth, 1988). Most of these known markers are Tn*10* insertions that are in or linked to known genes. A large collection of mapped Tn10 mutants has been assembled at the Salmonella Genetic Stock Centre—these are described in Sanderson and Roth (1988).

2. Physical Mapping

In addition to genetic mapping methods, physical maps of the *Salmonella* genome using the restriction enzymes *Xba*I and *Bln*I (*Avr*II) have been developed (Liu and Sanderson, 1992; Wong and McClelland, 1992). When used to digest chromosomal DNA of *S. typhimurium,* these enzymes produce a range of fragments that can be separated using pulsed-field gel electrophoresis (PFGE). Insertion of a transposon containing *Xba*I or *Bln*I sites introduces additional cleavage sites into the chromosome that can be used for restriction mapping of the transposon insertion. Protocols for preparation of chromosomal DNA and digestion and preparation of samples have been described by Liu and Sanderson (1992), Wong and McClelland (1992), and Bäumler *et al.* (1994).

An advantage of this method is that, using *Xba*I and *Bln*I, a precise physical map position can be determined for most transposon insertions. Physical maps have been published for other serovars, such as *S. enteritidis*, allowing comparison of transposon insertions found in different Salmonellae (Liu *et al.*, 1993). However, many transposons, including Mu*d*J and Tn*pho*A do not contain the necessary restriction sites for precise physical mapping of transposon insertion sites. Wild-type Tn*10* contains restriction sites for both *Bln*I and *Xba*I, but, of the newer derivatives, only Tn*10d*(Tc) contains an *Xba*I site. The sites for *Bln*I have also been deleted in the construction of the mini-Tn*10*s. Kleckner *et al.* (1991) have, however, made available mini-Tn*10* constructs carrying a polylinker containing rare-cutting restriction enzyme sites for physical mapping of transposon insertions.

In order to create new restriction sites for *Xba*I and *Bln*I in existing transposon mutants, our laboratory has used suicide vectors carrying a cloned fragment of the transposon and a polylinker containing *Xba*I and *Bln*I sites. The entire construct integrates into the transposon by homologous recombination between the transposon and the corresponding fragment cloned into the suicide vector, thus introducing the restriction sites into the transposon (R. Tsolis and F. Heffron, unpublished results). This approach has made it possible to obtain physical map positions for Mu*d*J insertions. In order to correlate the physical mapping data with the genetic information available on *Salmonella*, it is best to use the physical mapping methods in combination with the genetic ones outlined above for mapping a transposon insertion. Although physical mapping can yield a precise map location in kilobases, the physical and genetic maps of *Salmonella* are not fully colinear, and cotransduction of the transposon insertion with known markers (see section on genetic mapping) will correlate the physical map location with the body of available genetic data.

Finally, cloning of the gene carrying the insertion from the wild-type strain and complementation of the mutation with the cloned gene will show that the gene interrupted by the transposon insertion is necessary for virulence and rule out polar effects of the transposon insertion on downstream genes. If Benson–Goldman mapping has been used, the phage DNA from the region containing the transposon provides convenient starting material for cloning the intact gene.

V. Perspectives

Inactivating genes by transposon mutagenesis is just one strategy for finding virulence-associated genes. Use of operon fusions has given us information about which genes are expressed under *in vitro* conditions. Ideally, however, we would like to know which bacterial genes are activated *in vivo* during the course of an infection.

Mahan *et al.* (1993) have developed a system, called IVET (for *in vivo* expression technology), which is designed to detect genes which allow *Salmonella* to grow in the mouse. This novel strategy uses a *pur*D strain of *S. typhimurium,* which is incapable of growth in the mouse, because of the low availability of purines. The *pur*D mutant serves as the recipient for a bank of small *S.*

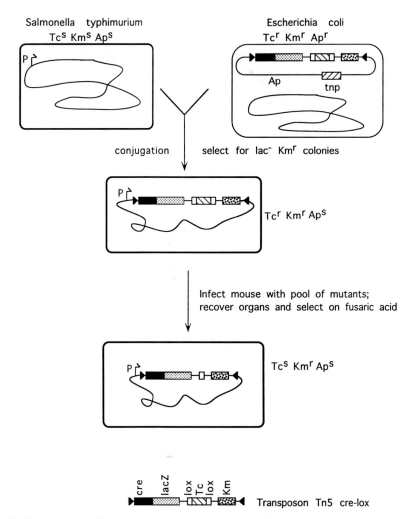

Fig. 3 Strategy for finding *Salmonella* genes expressed in the mouse. The mini-Tn5*cre lacZ* is introduced by conjugation from its permissive *E. coli* host into *S. typhimurium.* When the *cre* recombinase is expressed from a promoter (P) in the chromosome, mutants become sensitive to tetracycline. Tc, tetracycline; Km, kanamycin; Ap, ampicillin.

typhimurium genomic fragments cloned into a suicide vector in front of a promoterless *purD* gene fused to *lacZ*. Inclusion of *lacZ* in the construct allows elimination of clones containing promoters expressed *in vitro*. Strains that are Lac⁻ on laboratory media are then used to infect mice. Only strains containing a cloned promoter that is expressed in the mouse will be able to synthesize purines and, thus, grow and be recovered from the animal. IVET should therefore yield a fascinating set of *S. typhimurium* genes. One limitation of this method, however, is that it does not detect genes expressed only transiently (i.e., at the initiation stages) during infection.

Our laboratory is currently testing a second method for detection of genes expressed solely *in vivo* (Fig. 3; P. Valentine and F. Heffron, unpublished results). The system is carried on a mini-Tn5 derivative containing a promoterless *cre* recombinase gene from phage P1 as an operon fusion to *lacZ*. In addition, the transposon encodes a tetracycline resistance gene flanked by *lox*P sites (target sites for *cre* recombinase; Hoess *et al.*, 1982).When the transposon carrying *cre* inserts behind a promoter expressed *in vivo*, *cre* recombinase will act on the *lox*P sites, causing loss of tetracycline resistance. By infecting a mouse with a pool of these mutants, removing the organs colonized by *Salmonella*, and plating on fusaric acid, mutants will be recovered in genes that have been expressed in the mouse, and that have consequently lost the tetracycline gene (Maloy, and Nunn, 1981). An advantage of this system over IVET is that it allows us to identify genes that are expressed only transiently in the mouse. Other systems designed to detect genes expressed *in vivo* are currently being developed in several laboratories and this type of approach promises to take us a large step forward in defining the interaction of *Salmonella* with its host.

Acknowledgments

The authors thank J. G. Kusters, A. Bäumler, and P. Valentine for critical reading and discussion of the manuscript; N. Buchmeier for supplying figures; and S. Maloy and T. Elliott for helpful suggestions.

References

American Type Culture Collection, *Catalogue of Cell Lines & Hybridomas. 7th edition, (1992). Rockville, MD (USA).*

Bäumler, A. J., Kusters, J. G., Stojiljkovic, I., and Heffron, F. (1994). *Salmonella typhimurium* loci involved in macrophage survival. *Infect. Immun.* **62,** 1623–1630.

Belas, R., Mileham, A., Simon, M., and Silverman, M. (1984). Transposon mutagenesis of marine *Vibrio* spp. *J. Bacteriol.* **158,** 890–896.

Bender, J., Kleckner, N. (1992). IS*10* transposase mutations that specifically alter target site recognition. *EMBO J* **11:**741–750.

Benjamin, W. H., Turnbough, C. L., Posey, B. S., and Briles, D. E. (1986). *Salmonella typhimurium* virulence genes necessary to exploit the Ity^s/s genotype of the mouse. *Infect. Immun.* **51,** 872–878.

Benson, N. R., and Goldman, B. S. (1992). Rapid mapping in *Salmonella typhimurium* with Mud-P22 prophages. *J. Bacteriol.* **174,** 1673–1681.

Bochner, B. R. (1984). Curing bacterial cells of lysogenic viruses by using UCB indicator plates. *BioTechniques* **2,** 234–240.

Buchmeier, N. A., and Heffron, F. (1989). Intracellular survival of wild-type *Salmonella typhimurium* and macrophage-sensitive mutants in diverse populations of macrophages. *Infect. Immun.* **57,** 1–7.

Castilho, B. A., Olfson, P., and Casadaban, M. (1984). Plasmid insertion mutagenesis and *lac* gene fusion with mini-Mu bacteriophage transposons. *J. Bacteriol.* **158,** 488–495.

Cookson, B. T., Berg, D. E., Goldman, W. E. (1990). Mutagenesis of *Bordetella pertussis* with transposon Tn5tac1:conditional expression of virulence-associated genes. *J. Bacteriol.* **172:**1681–1687.

Damiani, G., Kiyotaki, C., Soeller, W., Sasada, M., Peisach, J., and Bloom, B. R. (1990). Macrophage variants in oxygen metabolism. *J. Exp. Med.* **152:** 808–822.

deLorenzo, V., Herrero, M., Jakubzlk, U., and Timmis, K. N. 199 Mini-Tn*5* transposon derivatives for insertion mutagenesis, promoter probing, and chromosomal insertion of cloned DNA in gram-negative eubacteria, *J. Bacteriol.* **172:**6568–6572.

Derman, A. I., Beckwith, J. (1991). *Escherichia coll* alkaline phosphatase fails to acquire disulfide bonds when retained in the cytoplasm. *J. Bacteriol.* **173:**7719–1722.

Dharmsathaphorn, K., McRoberts, J. A., Mandel, K. G., Tisdale, L. D., and Masui, H. (1984). A human colonic cell line that maintains vectorial electrolyte transport. *Am J Physiol* **246:**G204–208.

Duke, J., and Guiney, D. G. (1983). The role of lipopolysaccharide structure in the recipient cell during plasmid-mediated bacterial conjugation. *Plasmid* **9,** 222–226.

Elliott, T. (1989). Cloning, genetic characterization, and nucleotide sequence of the *hemA-prfA* operon of *Salmonella typhimurium*. *J. Bacteriol.* **171,** 3948–60.

Elliott, T., and Roth, J. R. (1988). Characterization of Tn10d-Cam: A transposition-defective Tn10 specifying chloramphenicol resistance. *Mol. Gen. Genet.* **213,** 332–337.

Engebrecht, J., Simon, M., and Silverman, M. 1985. Measuring gene expression with light. *Science* **227:**1345–1347.

Fields, P. I., Swanson, R. V., Haidaris, C. G., and Heffron, F. (1986). Mutants of *Salmonella typhimurium* that cannot survive within the macrophage are avirulent. *Proc. Natl. Acad. Sci. USA* **83:**5189–5193.

Fields, P. I., Groisman, E. G., and Heffron, F. (1989). A *Salmonella* locus that controls resistance to microbicidal proteins from phagocytic cells. *Science* **243,** 1059–1062.

Finlay, B. B., Starnbach, M. N., Francis, C. L., Stocker, B. A. D., Chatfield, S., Dougan, G., and Falkow, S. (1988). Identification and characterization of Tn*phoA* mutants of *Salmonella* that are unable to pass through a polarized MDCK epithelial cell monolayer. *Mol. Microbiol.* **2,** 757–766.

Fogh, J., Fogh, J. M., and Orfoe, T. (1977). One hundred and twenty seven cultured human tumor lines producing tumors in nude mice. *J. Natl. Cancer Inst.* **59,** 221–226.

Gahring, L. C., Heffron, F., Finlay, B. B., and Falkow, S. (1990). Invasion and replication of *Salmonella typhimurium* in animal cells. *Infect. Immun.* **58:**443–448.

Galán, J. E., and Curtiss, R. (1989). Cloning and molecular characterization of genes whose products allow *Salmonella typhimurium* to penetrate tissue culture cells. *Proc. Natl. Acad. Sci. U.S.A.* **86,** 6383–6387.

Gey, G. O., Coffman, W. D., and Kubicek, M. T. (1952). Tissue culture studies if the proliferative capacity of cervical carcinoma and normal epithelium. *Cancer Res.* **12,** 264–265.

Glover, S. W. (1962). Valine-resistant mutants of *Escherichia coli* K-12. *Genet. Res.* **3,** 448ff.

Gulig, P. A., Caldwell, A. L., and Chiodo, V. A. (1992). Identification, genetic analysis and DNA sequence of a 7.8-kb virulence region of the *Salmonella typhimurium* virulence plasmid. *Mol. Microbiol.* **6,** 1395–1411.

Henle, G., and Deinhardt, F. (1957). The establishment of strains of human cells in tissue culture. *J. Immunol.* **79,** 54.

Hoess, R. H., Ziese, M., and Sternberg, N. (1982). P1 site-specific recombination: Nucleotide sequence of the recombining sites. *Proc. Natl. Acad. Sci. U.S.A.* **79,** 3398–3402.

Hone, D. M., Attridge, S. R., Forrest, B., Morona, R., Daniels, D., LaBrooy, J. T., Bartholomeusz, R. C., Shearman, D. J., and Hackett, J. (1988). A *galE via* (Vi antigen-negative) mutant of *Salmonella typhi* Ty2 retains virulence in humans. *Infect. Immun.* **56**, 1326–1333.

Hughes, K., and Roth, J. (1988). Transitory Cis-complementation: A general method for providing transposase to defective transposons. *Genetics* **119**, 9–12.

Jones, G. W., Rabert, D. K., Svinarich, M., and Whitfield, H. J. (1982). Association of adhesive, invasive and virulent phenotypes of *Salmonella typhimurium* with autonomous 60-megadalton plasmids. *Infect. Immun.* **38**:476–486.

Kleckner, N., Roth, J., and Botstein, D. (1977). Genetic engineering in vivo using translocatable drug-resistance elements. New methods in bacterial genetics. *J. Mol. Biol.* **116**, 125–159.

Kleckner, N., Bender, J., and Gottesman, S. (1991). Uses of transposons with emphasis on Tn10. *In* "Methods in Enzymology" (J. Miller, ed.), Vol. 204, 2 pp. 139–180. Academic Press, San Diego.

Krause, M., Roudier, C., Fierer, J., Harwood, J., and Guiney, D. (1991). Molecular analysis of the virulence locus of the *Salmonella dublin* virulence plasmid pSDL2. *Mol. Microbiol.* **5**, 307–316.

Kusters, J. G., Mulders-Kremers, G.A.W.M., van Doornik, C. E. M., and van der Zeijst, B. A. M. (1993). Effects of multiplicity of infection, bacterial protein synthesis, and growth phase on adhesion to and invasion of human cell lines by *Salmonella typhimurium*. *Infect. Immun.* **61**, 5013–5020.

Lee, C. A., Jones, B. D., and Falkow, S. (1992). Identification of a *Salmonella typhimurium* invasion locus by selection for hyperinvasive mutants. *Proc. Natl. Acad. Sci. U.S.A.* **89**, 1847–1851.

Leung, K. Y., Ruschkowski, S. R. and B. B. Finlay. (1992). Isolation and characterization of the *aadA* aminoglycoside-resistance gene from *Salmonella choleraesius*. *Mol. Microbiol.* **6**:2453–2460.

Liu, S., Hessel, A., and Sanderson, K. E. (1993) The *XbaI-BlnI-CeuI* genomic cleavage map of *Salmonella enteritidis* shows an inversion relative to *Salmonella typhimurium* LT2. *Mol. Microbiol.* **10 (3)**, 655–664.

Liu, S. L., and Sanderson, K. E. (1992). A physical map of the *Salmonella typhimurium* LT2 genome made by using XbaI analysis. *J. Bacteriol.* **174**, 1662–1672.

Mahan, M. J., Slauch, J. M., and Mekalanos, J. J. (1993). Selection of bacterial virulence genes that are specifically induced in host tissues. *Science* **259**, 686–688.

Maloy, S. R., and Nunn, W. D. (1981). Selection for loss of tetracycline resistance by *Escherichia coli*. *J. Bacteriol.* **145**, 1110–1112; erratum: **146**, 831.

Maloy, S. R., Stewart, V. J., and Taylor, R. K. (1994). "Experiments in Bacterial Pathogenesis." Cold Spring Harbor Lab., Cold Spring Harbor, NY.

Manoil, C., and Beckwith, J. (1985). Tn*phoA*: A transposon probe for protein export signals. *Proc. Natl. Acad. Sci. U.S.A.* **82**, 8129–8133.

Miller, J. H. (1972). "Experiments in Molecular Genetics." Cold Spring Harbor Lab., Cold Spring Harbor, NY.

Moore, A. E., Sabaschewsky, L., and Toolan, H. W. (1955). Culture characteristics of four permanent lines of human cancer cells. *Cancer Res.* **15**, 598.

Puck, T. T., Cieciura, S. J., and Robinson, A. (1958). Genetics of somatic mammalian cells. III. Long-term cultivation of euploid cells from human and animal subjects. *J. Exp. Med.* **198**, 945.

Ralph, P., and Nakoinz, I. (1975). Phagocytosis and cytolysis by a macrophage tumour and its cloned cell line. *Nature (London)* **257**, 393–394.

Sanderson, K. E., and Roth, J. R. (1988). Linkage map of *Salmonella typhimurium*, edition VII. *Microbiol. Rev.* **52**, 485–532.

Sanderson, K. E., Janzer, J., and Head, J. (1981). Influence of lipopolysaccharide and protein in the cell envelope on recipient capacity in conjugation of *Salmonella typhimurium*. *J. Bacteriol.* **148**, 283–293.

Schmieger, H. (1972). Phage P22 mutants with increased or decreased transduction abilities. *Mol. Gen. Genet.* **119**, 75–88.

Silva, B., Gonzalez, C., Mora, G.C., and Cabello, F (1987). Genetic characteristics of the *Salmonella typhi* Ty21a vaccine. *J. Infect. Dis.* **155**, 1077–1078.

Slauch, J. J., and Silhavy, T. J. (1991). Genetic fusions as experimental tools. *In* "Methods in Enzymology", (J. Miller ed.), Vol. 204, pp. 213–248. Academic Press, San Diego.

Stojiljkovic, I., Trgovcevic, Z., and Salaj-Smic, E. (1991). Tn*5-rpsL:* A new derivative of transposon Tn5 useful in plasmid curing. *Gene* **99,** 101-104.

Stone, B. J., Garcia, C. M., Badger, J. L., Hassett, T., Smith, R. I. F., and Miller, V. (1992). Identification of novel loci affecting entry of *Salmonella enteritidis* into eukaryotic cells. *J. Bacteriol.* **174,** 3945–3952.

Sundstrom, C., and Nilsson, K. (1976). Establishment and characterization of a human histiocytic lymphoma cell line (U-937). *Int. J. Cancer* **17:**565–577.

Taylor, R. K., Manoil, C., and Mekalanos, J. J. (1989). Broad-host-range vectors for delivery of Tn*phoA:* Use in genetic analysis of secreted virulence determinants of *Vibrio cholerae. J. Bacteriol.* **171,** 1870–1878.

Tinge, S. A., and Curtiss, R. (1990). Conservation of *Salmonella typhimurium* virulence plasmid maintenance regions among *Salmonella* serovars as a basis for plasmid curing. *Infect. Immun.* **58,** 3084–3092.

Vogel, H. J., and Bonner, D. M. (1956). Acetyl ornithase of *Escherichia coli:* Partial purification and some properties. *J. Biol. Chem.* **218,** 97–106.

Wong, K. K., and McClelland, M. (1992). A *Bin*I restriction map of the *Salmonella typhimurium* LT2 genome. *J. Bacteriol.* **174:**1656–1661.

Yakobson, E. A., and Guiney, D. G. (1984). Conjugal transfer of bacterial chromosomes mediated by the RK2 plasmid transfer origin cloned into transposon Tn5. *J. Bacteriol.* **160:**451–453.

Zahrt, T. C., Mora, G. C., Maloy S. (1994). Inactivation of mismatch repair overcomes the barrier to transduction between *Salmonella typhimurium* and *Salmonella typhi. J. Bacteriol.* **176:**1527–1529.

CHAPTER 6

Mycobacterium: Isolation, Maintenance, Transformation, and Mutant Selection

Nancy D. Connell

Department of Microbiology and Molecular Genetics
University of Medicine and Dentistry of New Jersey
New Jersey Medical School
National Tuberculosis Center
Newark, New Jersey 07103

I. Introduction
II. Biosafety Considerations
III. Culture Media and Conditions
IV. Isolation of Mycobacteria
V. Maintenance of Stocks
VI. Genetic Techniques
    A. Mycobacteriophage Infection
    B. Conjugation
    C. Plasmid Transformation of DNA
VII. Mutagenesis
    A. Chemical Mutagenesis
    B. Transposon Mutagenesis
    C. Shuttle Mutagenesis
VIII. Strain Construction
    A. Homologous Recombination/Gene Replacement
    B. Nonhomologous Recombination
IX. Mutant Selection and Isolation
    A. Screening
    B. Selection
    C. Complementation
References

I. Introduction

A hallmark of the genus *Mycobacterium* is its intracellular habitat. However, genetic analysis of the mycobacterial pathogenesis has been greatly hampered by two salient features of mycobacterial physiology: their slow growth rates and their complex and highly hydrophobic cell walls. These features run as themes through all aspects of our understanding of the physiology, genetics, and pathogenesis of this genus and have played major roles in erecting impediments to our understanding. In this chapter, basic methods for culturing the bacteria, isolating genetic mutants, and moving DNA in and out of the cells will be described.

Experimental methods for culturing and genetically manipulating *M. avium*, *M. tuberculosis*, *M. smegmatis*, and BCG (bacille Calmette Guerin) will be discussed in this chapter. Unfortunately, the genetics of *M. avium* lags far behind the those of the other three species; differences among the bacteria will be delineated when appropriate.

II. Biosafety Considerations

The reemergence of tuberculosis as a public health crisis in the 1990s has fueled a considerable interest in basic mycobacterial reseach. Although many clinical and hospital mycobacteriology laboratories operate at biosafety level two (BSL-2) requirements, research activities involving *M. tuberculosis* and *M. bovis* require BSL-3 practices, facilities, and containment equipment. The airborne nature of tubercuolsis transmission, via infectious droplet nuclei of approximately 1–5 μm in size, requires a combination of biosafety approaches, which include laboratory techniques designed to minimize the production of aerosols. Many laboratoy manipulations yield aerosols comprising droplet nuclei in the 1- to 5 μm range. Therefore all manipulations of infectious material must be performed in properly installed and maintained biological safety cabinets that constitute a primary barrier to transmission. Directional airflow should ensure that air moves from clean areas of the laboratory to potentially contaminated areas, constituting the secondary barrier to transmission. Additional controls include training and monitoring of laboratory personnel, and supplemental controls such as germicidal ultraviolet lights and HEPA filtration. The reader is referred to the CDC/NIH Biosafety Guidelines and laboratory manuals [Centers for Disease Control (CDC), 1990], available upon request.

Introduction of mycobacterial DNA into any host (including *Escherichia coli*) requires approval form the investigator's Biosafety Committee; the National Institutes of Health Recombinant DNA Advisory Board recommends that any BSL-2-level organisms containing DNA from a BSL-3 level organism be maintained in a BSL-3-laboratory.

III. Culture Media and Conditions

There is a wide range of growth characteristics among the more than 50 species of the genus, from the fast-growing, nonpathogenic *M. smegmatis* to *M. leprae,* which has not yet been cultured in the laboratory. Nevertheless, the mycobacteria are auxotrophs and require no amino acids, purines, or pyrimidines for growth; salts and a source of both carbon and nitrogen will suffice for culture of the organisms. However, under minimal conditions, the rates of growth of the slow-growing species will be decreased to impractical levels. Early formulations of media with which to culture the mycobacteria reflect efforts to recreate the milieu of intracellular growth (host environment). Thus, modified Lowenstein–Jensen egg medium contains whole eggs, potato starch, and a number of simple components.

The current standard agar media are Middlebrook 7H10 and 7H11: inspection of the components reveals that these are simple media, composed of salts, buffers and pyridoxin, biotin and glutamic acid. Both formulations are designed for use with additions: bovine serum albumin fraction V, dextrose and sodium chloride (ADC) for most slow growing species, and ADC plus sodium oleate for *M. bovis* and *M. tuberculosis.* 7H10 is used for BCG and *M. smegmatis,* 7H11 for *M. avium* and *M. tuberculosis.* The oleate serves as a nutrient source. Oleic acid can be toxic, but albumin absorbs and neutralizes the free fatty acid.

MADCTW agar

19 g Difco (Detroit, MI) Middlebrook 7H10 or 7H11 agar
900 ml ddH$_2$O
Autoclave and cool to 55°C
Add 100 ml ADC enrichment and 2.5 ml Tween-80

ADC enrichment

Dissolve 50 g bovine serum albumin (Boehringer Mannheim, Indianapolis, IN) in 1 liter ddH$_2$O at room temperature (do not heat)
Add 20 g glucose and 8.5 g NaCl
Filter sterilize and store at 4°C.

A number of liquid media are currently in use and are convenient for defined media formulations. Middlebrook 7H9 is the most widely used. ADC is required for efficient culture of BCG, *M. avium,* and *M. tuberculosis,* but can be omitted for *M. smegmatis.* Proskauer–Beck medium can be used for *M. tuberculosis* and BCG when the (O)ADC additives are precluded for experimental reasons. Tween-80, a nonionic detergent, is included in liquid media to disperse the natural clumping of the cells. However, microscopic inspection of a mycobacterial culture grown in Tween-80 will reveal clumps of cells: a further precaution before plating single-cell suspensions is to pass the suspension through a 23-

gauge needle. It must be remembered that Tween-80 is a carbon source for mycobacteria and can interfere with the design of minimal media for mutant selection.

MADCTW liquid

4.7 g 7H9 broth base
2.0 ml glycerol
900 ml ddH2O
Autoclave, cool, and add ADC and 2.5 ml 20% Tween-80

Tween-80, 20%

Add 20 ml Tween-80 (polyoxyethylene sorbitan monooleate) to 80 ml ddH$_2$O, heat to 55°C to disolve, and filter sterilize.

Proskauer–Beck minimal medium

Dissolve, per liter:
5 g asparagine
5 g KH$_2$PO$_4$
0.5 gm K$_2$SO$_4$
20 ml glycerol
Adjust pH to 7 and autoclave
Add 2.5 ml 20% Tween-80 and 1.5 g Mg citrate

Another minimal media for mutant screens and selections for BCG and *M. smegmatis* is Sauton's minimal medium.

Sauton's minimal

Dissolve, per liter:
0.5 g KH$_2$PO$_4$
0.5 g MgSO$_4$
4.0 g L-asparagine
60 ml glycerol
0.05 g Ferric ammonium citrate
2.0 g citric acid
0.1 ml 1% ZnSO$_4$
Adjust pH to 7
Autoclave, cool, and add 2.5 ml 20% Tween-80.

Since early media formulations were designed for isolation of mycobacteria from clinical samples, the inclusion of dyes (malachite green and/or crystal violet) acts to inhibit the growth of other bacteria. In addition, many workers include antibiotics such as cycloheximide to control yeast contaminants.

Conditions of culture incubation vary from laboratory to laboratory. Liquid cultures of *M. smegmatis* are generally grown in beveled Ehrlenmeyer flasks. The slow-growing species should be grown either standing or in slowly turning roller bottles to ensure a uniform suspension. The mycobacteria are strict aerobes, and cultures should be well-aerated. Slow-growing species require weeks of incubation, which risks desiccation: a water-jacketed incubator or wrapping stacks of plates in aluminum foil or plastic bags can prevent their drying out. Five to ten percent CO_2 enhances the growth of many species on the Middlebrook media. With the exception of *M. marinum* and *M. ulcerans* (30°C), the optimum temperature for growth is 35–37°C.

IV. Isolation of Mycobacteria

The isolation of mycobacteria from contaminated sources (for example, soil or sputum) requires decontamination, since specimens usually contain mixed bacterial flora and some degree of organic debris such as body fluids, blood cells, and tissue fragments. There are a number of isolation protocols but all require digestion, decontamination, and concentration steps. The mycobacteria must be recovered undamaged while organic debris is liquified and contaminating micororganisms are killed. The following protocol is designed for treatment of sputum samples and is suitable for use in *clinical* BSL-2 and experimental BSL-3 laboratories for the treatment of clinical samples. For other types of samples, and for further information, the reader is referred to the *CDC Laboratory Manual* (Strong and Kubica, 1981).

Treatment of sputum samples

N-Acetyl-L-cysteine (NALC) is a mucolytic agent that destroys mucous by splitting sulfide bonds. It is used as a 0.5% solution in 2% NaOH/1.5% Na citrate. The 2% NaOH acts as a mild bacterial decontaminant and can be used at concentrations up to 6%.

1. Use 10 ml of the NaOH/NALC mixture in a 50-ml disposable centrifuge tube.

2. Add sample and vortex without forming froth until liquified (5–20 sec).

3. Let stand at room temperature for 15 min.

4. Increase volume to 50 ml with sterile ddH_2O, mix gently by hand, and centrifuge at 2000–3000 × g for 15 min.

5. Decant or aspirate supernatant into discard bottle containing disinfectant. Add 1–2 ml sterile ddH_2O or saline solution to sediment for immediate plating or inoculation.

6. If inoculation is delayed, use 0.2% BSA (Fraction V) in 0.85% NACl, and store at 4°C.

7. For inoculation of agar plates, dilute 1:10 in sterile ddH$_2$O and plate 0.1 ml, and spread with a flamed glass rod spreader.

V. Maintenance of Stocks

Mycobacteria are stable to freezing in 7H9 media, most likely due to the protective effects of glycerol and Tween-80. Short-term stocks (up to 3 months) can be stored in 10% glycerol/H$_2$O at $-20°C$. Longer (indefinite) storage at $-70°C$ requires 50% glycerol.

VI. Genetic Techniques

The genetics of mycobacteria have lagged far behind those of many bacteria, due, for the most part, to the difficulties discussed at the outset of this chapter: their slow growth rate and their tendency to clump. Indeed, the basis of genetic analysis is the single cell. Thus systems for genetic exchange have been slow in developing. Fortunately, the last few years have seen an explosion in the application of molecular genetic techniques, both contemporary and classical, to these recalcitrant bacteria. For an excellent review of the history of mycobacterial genetics, the reader is referred to Grange (1982). For other recent surveys of contemporary molecular genetic techniques, see Jacobs et al. (1991) and Hatfull (1993).

A. Mycobacteriophage Infection

The mycobacteriophages were initially isolated and developed for the purpose of strain typing. Here we will discuss their use in transfer of genetic material for the construction of strains.

1. Transduction

Phage transduction can be classified in two groups. (1) Generalized transducing phages, during synthesis and assembly, occasionally package genomic DNA into their phage heads. When such a phage is transferred to another host cell, the genomic DNA can recombine into the genome of the new host by homologous recombination. (2) Restricted transduction is the result of aberrant release of temperate phages from the genome. The genome of the temperate phage inserts into the host chromosome at a defined site by recombination at regions homologous between the phage and the host. Occasionally recombination may take place between the phage and another host gene(s); upon excision, the bacterial genes are packaged and carried to the new host. The recombinant phage may be defective and incapable of further replication. Nonetheless, at a certain frequency, these defective phages will transduce bacterial genes from one bacterial strain to another. Three transducing phages have been used to transduce markers from one strain to another: phage BO2 (Gelbart and Juhasz, 1973), D29 (Jones and David, 1972), and I3 (Sundar Raj and Ramakrishnan, 1970). A protocol for the generalized transduction of kanamycin resistance by the I3 phage in *M. smegmatis* is described below.

2. Phasmids

Phage-based gene transfer systems serve as an efficient means of gene transfer in mycobacteria, since whole untreated bacteria are used as target cells. Useful phasmid vectors for mycobacteria are the temperate shuttle phasmids constructed by Jacobs and co-workers (Snapper *et al.*, 1988). L1 and TM4 mycobacteriophages, which produce turbid plaques on *M. smegmatis*, were recombined with *E. coli* cosmids (containing an *E. coli* origin, antibiotic resistance markers, and *cos* sites for λ-phage packaging). The phasmids were further modified to express resistance to kanamycin, using the Tn*903* gene encoding aminoglycoside phosphotransferase. Thus, shuttle phasmid DNA can be packaged into bacteriophage-λ particles for propagation and manipulation as plasmids in *E. coli*. The DNA can then be introduced by electroporation into mycobacterial strains, where they will replicate and produce purified phage stocks for efficient mycobacterial infection. Protocols for preparation of mycobacteriophage stocks and infection by mycobacteriophage lysates are given here.

3. Phage Protocols

a. Mycobacteriophage Lysate Preparation

To prepare a high titer lysate of mycobacteriophage, the titer of the starting phage lysate must be in the range $1 \times 10^{6-8}$ plaque-forming units (pfu)/ml. To determine the titer of the original lysate:

1. The host (or donor) strain is grown to OD_{600} of 0.5–0.7, in MADCTW without Tween-80 (Tween-80 will inhibit phage binding).

2. 200 μl of cells is mixed with 100-μl dilutions of phage lysate, diluted in MP buffer, and allowed to adsorb for 20 min at 37°C without disturbing.

3. Add 3 ml of mycobacteriophage top agar to each mixture, mix gently, and pour over Dubos plates.

4. Incubate at 37°C until plaques appear (24 hr for 5–7 days for slow-growers).

MP (mycobacteriophage) buffer:

10 mM Tris, pH 7.6; 100 mM NaCl; 10 mM MgSO$_4$; 2 mM CaCl$_2$

Dubos plates: 20 g Dubos oleic aicd agar base (Difco), 1 g NaCl, 7.5 g glucose in 1 liter ddH$_2$O. Autoclave, cool, and add MgSO$_4$ to 10 mM, CaCl$_2$ to 2 m*M*.

Mycobacteriophage top agar: 0.5 g Middlebrook 7H9 broth base; 0.1 g NaCl; 0.75 g glucose; 0.7 g agar in 100 ml ddH$_2$O. For BCG, add 1.0 g proteose peptone 3 (Difco); 0.5 g yeast extract (Difco) and 5 ml 20% glycerol.

Preparation of bulk lysate:

1. Repeat the infection procedure with 2 ml host cells (at 3×10^8 cells/ml) plus 1 ml of phage diluted to 5×10^4 pfu/ml.

2. Adsorb as above.

3. Add 0.3 ml of phage cell mixture to 3 ml mycobacteriophage top agar and pour over plates.

4. When plaques appear, add 5 ml chilled MP buffer to each plate and incubate at 4°C overnnight.

5. Collect MP buffer, pool, and centrifuge at $3000 \times g$ for 15 min at 4°C.

6. Filter the supernatant through 0.45-μm filter unit.

7. Concentrate by spinning at $20,000 \times g$ for 5 hr at 4°C.

8. Resuspend pellet in 1 ml of phage buffer. Do not use chloroform, since mycobacteriophage coats contain lipids.

b. Transduction by I3 Mycobacteriophage

A high-titer lysate of I3 is made on the donor strain using the plate method, above. The titer of the lysate on the recipient strain is determined, and the lysate is used in the following protocol.

1. The recipient is then grown to OD$_{600}$ of 1.5.

2. The cells are spun down and resuspended at a concentration of 10^{10} cells/ml in MADCTW (a concentration of 1:20), and heat killed at 70°C for 30 min.

3. Mix together 1×10^8 recipient cells (200 μl) and 100 μl of I3 at 10^9 pfu/ml and allowed to adsorb for 30 min at 37°C. (This is a multiplicity of infection of 1.)

4. After adsorbtion, the cells are pelleted at $3000 \times g$ for 10 min and the supernatant is discarded. (Alternatively, the supernatant can be titered to determine the number of phage that were adsorbed.)

5. The pellet is resuspended in 200 μl MADCTW (or minimal selective media, if appropriate) with 100 μl heat-killed recipient cells.

6. Incubate at 37°C for 2 hr for expression of selectable marker and/or antibiotic resistance.

7. Add 3 ml of selective media with 1% Na citrate to each tube, mix, and pour onto selective agar plates.

B. Conjugation

Transfer of plasmids by conjugation is not restricted to closely related bacterial species or genera. In fact, conjugation is possible between Gram-negative and Gram-positive bacteria. Lazraq *et al.* (1990) describe the transfer of kanamycin resistance from *E. coli* to *M. smegmatis* using a chimaeric plasmid composed of the pAL5000 replication origin (Rauzier *et al.*, 1988) and the origin of transfer of pRK2, with additionally required genes supplied by a helper plasmid carried by the donor cell. Gormley and Davies (1991) demonstrated transfer of of the conjugative plasmid RSF1010 from *E. coli* to *M. smegmatis,* with stable inheritance of streptomycin and sulfonamide resistance.

Conjugation protocol

1. Donor *E. coli* cells are grown to early stationary phase in Luria broth [per liter: Bacto-tryptone (Difco) 10 g; Bacto-yeast extract (Difco), 5 g; NaCl 5 g)].

2. Recipient *M. smegmatis* are grown to stationary phase in 7H9 (MADCTW), washed, and resuspended in LB at a concentration of 10^9 cells/ml (OD_{600} of approximately 2)

3. The cells are washed and resuspended at a concentration of 5×10^8 cells/ml (OD_{600} of approximately 1).

4. 0.1 ml of each are mixed and spread on sterile 0.45-μm filters (Millipore) and incubated overnight on LB plates at 37°C.

5. The filters are then washed and suspended in sterile ddH$_2$0 and dilutions are plated on 7H10 with antibiotic (streptomycin 20 μg/ml).

C. Plasmid Transformation of DNA

Early techniques used protoplast formation for introduction of plasmid DNA into mycobacteria (Jacobs *et al.*,1987). However, highly efficient plasmid transformation is acheived by electroporation (Snapper *et al.*, 1990). The frequency of transformation varies among species: *M. smegmatis* wild-type strain ATCC 607, for example, has a transformation rate of >10 colonies/μg DNA, using a pAL5000-based plasmid (Rauzier *et al.*, 1988) (see below); a high-frequency transforming derivative of that strain, mc^{2}155, exhibits transformation frequencies of >10^5colonies/μg DNA (Snapper *et al.*, 1990). On the other hand, *M. tuberculosis* and *M. bovis* BCG appear to have consistently lower but still useful transformation frequencies without requiring the isolation of a high-frequency transforming strain.

There are a number of plasmids available for expression of heterologous genes or other purposes in mycobacteria. A widely used mycobacterial plasmid system relies on the mycobacterial origin of replication (ori) first identified by Gigel-Sanzy and co-workers on the plasmid pAL5000 of *M. fortuitum* (Rauzier *et al.*, 1988). This original plasmid carries no identified antibiotic marker. A number of groups have developed recombinant shuttle plasmids bearing the pAL5000 origin, an *E. coli* origin, and the kanamycin resistance marker from Tn*903* (Rauzier *et al.*, 1988; Aldovani and Young, 1991; Snapper *et al.*, 1988). Stover and co-workers (1991) further modified their vectors by replacing the mycobacterial origin with phage (L5) integration sites, permitting the stable integration of the vector into the mycobacterial chromosome. The copy number of the pAL5000-based shuttle plasmds is estimated to be 3–10 copies per cell.

Additional plasmids have been developed from the *M. scrofulaceum* plasmid pMSC262 and from the broad host range plasmids RSF1010 and pNG2 (reviewed in Hatfull, 1993). A general electroporation protocol for introducing DNA is described here.

Electroporation of Plasmid DNA into Mycobacteria

1. Inoculate 1 liter of MADCTW with 10 ml of a mycobacterial culture at approximately $OD_{600} = 0.7$–1.

2. Grow at 37°C until $OD_{600} = 0.7$–1.0 (for BCG, approximately 1 week; for *M. smegmatis*, 48 hr). [Note: Aldovini *et al.* have increased the transformation efficiency of BCG by (1) adding 1.5% glycine to the culture medium for the last 24 hr before preparation for electroporation and (2) maintaining the cells in logarithmic phase for 2 months by diluting the cells 1:4 every 2 days (Aldovini *et al.*, 1993).]

3. Place cultures on ice for 20–40 min, but not longer than 1 hr. All subsequent steps should be performed on ice.

4. Centrifuge in chilled 250-ml bottles at 5000 × *g* for 15 min at 4°C.

5. Pour off supernatant and resuspend pellets in 200 ml chilled 10% glycerol.

6. Spin and wash in same volume again. Remove supernatant carefully: the absence of Tween-80 results in a very loosely packed pellet.

7. Resuspend in 100 ml 10% glycerol, spin, and remove supernatant.

8. Resuspend in 3–5 ml 10% glycerol. The final concentration factor should be approximately 1:20 for electrocompetent cells.

9. 500 μl of cells are used per transformation. (Note: the cells can be aliquoted at this point and quick-frozen on dry ice–ethanol to be stored at −70°C for later use. The transformation frequency will drop by a factor of 10–100 if the cells are frozen and stored.)

10. Up to 5 mg of DNA is mixed with 500 μl of cells and placed into a 0.4-cm electrode gap cuvette (Bio-Rad).

11. The pulser (Bio-Rad) is set at 2500 V, 25 mF, and 1000 Ω. After the voltage is applied, the time constant should be in the range of 18–22. (Note: If "arcing" should occur, add 100 μl of chilled 10% glycerol to the cuvette and repeat voltage application. If arcing still takes place, there is excess salt in either the cell suspension or the DNA preparation (a control cuvette without

DNA can easily distinguish between these alternatives). The cells should be washed again in 10% glycerol, or the DNA should be precipitated in ammonium acetate (3.5 M) and 70% ethanol and resuspended in ddH$_2$0.)

12. After electroporation, the cells should be removed immediately from the cuvette and placed in 5 ml of rich medium (MADCTW) without selective agent for expression, 2 hr at 37°C.

13. The cells are then spun down and plated on selective agar plates [for example, kanamycin (50 μg/ml) MADCTW plates].

An important step after transformation of mycobacteria is to streak out transformed colonies on selective media to single colonies. The extra precaution is neccessary since the cells tend to clump together, and it is possible that a single colony on a plate may have arisen from two or more cells stuck together. To determine whether the transformed DNA is stably maintained in the mycobacterial cell, the following modification of the alkaline lysis protocol is used (Birnboim and Doly, 1979).

Plasmid Preparation from Mycobacteria

1. Grow a 5 to 10-ml culture of the plasmid-containing strain in selective media.

2. Spin down the culture and decant the supernatant.

3. Resuspend the pellet in 100 μl Solution A (25 mM Tris–HCl, pH 8; 10 mM EDTA; 50 mM glucose) and transfer to Eppendorf tube.

4. Add fresh lysozyme at final concentration of 10 mg/ml and incubate overnight at 37°C.

5. Add 200 μl Solution B (fresh 0.2 mM NaOH; 1% SDS), mix well, and incubate on ice for 10 min.

6. Add 150 μl 5 M potassium acetate (KAc) and incubate on ice for 10 min.

7. Centrifuge at high speed in a microfuge at 4°C for 30 min.

8. Remove supernatant and extract with 500 μl [chloroform:phenol:isoamyl alcohol (CPI) at 24:25:1].

9. Transfer aqueous phase to clean centrifuge tube and add 1 ml chilled ethanol; incubate on ice for 20 to 60 min.

10. Spin down DNA pellet at 4°C for at least 15 min.

11. Remove supernatant and rinse with chilled 70% ethanol.

12. Resuspend pellet in 10–25 μl TE buffer (10 mM Tris–HCl, pH 8; 5 mM EDTA).

The yield from this plasmid preparation is small: it is suggested that the total DNA obtained in this manner be transformed directly into *E. coli* for amplification and analysis.

VII. Mutagenesis

There are a number of methods for mutagenizing the mycobacterial genome. Techniques for the generation of both random and targeted mutations will be discussed here. Random mutations can be induced by chemical agents (mutagens) and transposons; targeted mutagenic techniques include shuttle mutagenesis and gene replacement.

The DNA of the mycobacterial genome is subject to both spontaneous and induced mutation. Agents used to induce mutations in mycobacteria include acriflavin, ethyl methane sufonate (EMS), hydroxylamine, nitrous acid, N-methly-N-nitroso-N-nitroguanidine (MNNG), and gamma and ultraviolet radiation (reviewed by Konicakova-Radochova et al., 1970).

Other techniques for random mutagenesis depend on the phenomenon of homologous recombination. Homologous recombination has been efficiently demonstrated in M. smegmatis but not in BCG or M. tuberculosis.

A. Chemical Mutagenesis

1. Mid-log cells [$OD_{600} = 0.8$, or $2–3 \times 10^8$ colony-forming units (cfu) ml^{-1}] are grown in minimal medium plus 0.1% Tween-80.

2. Aliquots concentrated 1:10 in basal media are treated with MNNG in the range 100–1000 mg ml^{-1}.

3. The mutagen DEO has proved useful for mutant induction: it causes small deletions of 2–7 base pairs. This mutagen is used in the range 1–10%, in the absence of Tween-80.

4. The incubation conditions for both mutagens are 1 hr at 37°C, with shaking.

5. Cells are then washed twice in MADCTW and resuspended at the original density for expression of the mutants.

6. The extent of killing by each mutagen that is appropriate for increased mutagenic frequencies varies from 50 to 99%. For each mutagenic treatment, a survival curve should be performed.

The classical technique of penicillin enrichment for auxotrophs has been modified and applied to mycobacteria (Holland and Ratledge, 1971; Hinshelwood and Stoker, 1992). The technique relies on the cyclic exposure of a mixed population of wild-type and mutants to conditions under which wild-type cells grow and mutant cells cannot vs conditions under which both cell types can grow. Inhibition of growth of wild-type cells is obtained by exposure to nonlethal levels of inhibitors of cell wall synthesis such as penicillin. However, since the genus is resistant to β-lactams (due to the production of β-lactamases), the procedure has been modified. Isoniazid (INH), which blocks cell wall synthesis in mycobacteria (Winder and Brennan, 1964; Banerjee et al., 1994), is used to inhibit the growth of wild-type bacteria. Amino acid auxotrophic frequencies of $10^{-3}–10^{-4}$ have been obtained by this method.

INH Enrichment of Auxotrophs

1. After mutagenesis and expression in basal media plus Tween-80 and casamino acids (%), the treated cells are exposed to INH (40 μg/ml) in minimal media for 15 hr at 37°C. During this time actively growing cells should be killed, and mutant cells requiring an amino acid for growth should survive.

2. The cells are then washed free of INH in minimal media with Tween-80 and screened for auxotrophy (see mutant screening, below).

3. Alternatively, the cells can be resuspended in basal media plus Tween-80 and casamino acids, to allow the mutant cells to grow for 3–5 hr at 37°C.

4. The cycle of INH treatment and expression can be repeated twice for enrichment of mutant cells.

B. Transposon Mutagenesis

Transposons are mobile genetic elements that have been widely exploited as genetic tools on the basis of two properies: (1) transposons can be used to insert a marker into a target gene and (2) since many transposons can insert at random sites in the genome, they can be used to create panels of insertion mutants, which are by definition usually mutants lacking a gene function. In addition, some transposons have been modified for use as promoter probes, by the inclusion of a truncated structural gene whose expression can be monitored (for example, β-galactosidase and *pho*A).

One transposon (Martin *et al.*, 1990) and several insertion sequences (Green *et al.*, 1989; McAdam *et al.*, 1990; Thierry *et al.*, 1990; Cirillo *et al.*, 1991; Kunze *et al.*, 1991) have been described in mycobacteria. Attempts to use existing transposons for the construction of insertion mutants in mycobacteria have met with limited success (see discussion of shuttle mutagenesis below). Tn*610* (Martin *et al.*, 1990) appears to insert at random in the genome of *M. smegmatis*. Tn*1096* (Cirillo *et al.*, 1991), recovered from *M. smegmatis* but not present in *M. tuberculosis* or BCG, is the most reliable candidate for insertional mutagenesis.

There are a variety of transposon delivery systems that are currently under development for use in mycobacteria: suicide vectors (conditionally replicating plasmids or phages); minitransposon derivatives with inducible transposase; temperature-sensitive plasmids (Guilhot *et al.*, 1992).

C. Shuttle Mutagenesis

In the absence of an efficient transposon system for insertional mutagenesis, a system of shuttle mutagenesis was developed for mycobacteria (Kalpana *et al.*, 1991). A library of genomic mycobacterial DNA is generated and subjected to transposon mutagenesis in *E. coli*. The resulting transposon-containing plasmids are introduced into the *M. smegmatis* genome by homologous recombination (see below) (Husson *et al.*, 1990). Interestingly, in both BCG and *M.*

tuberculosis, the same technique results in a large degree of illegitimate recombination (Kalpana *et al.,* 1991) (see nonhomologous recombination, below).

There are a number of systems for random insertional mutagenesis in *E. coli* (Kleckner *et al.,* 1991). The basic steps of the technique are outlined here; modifications may be neccessary according to the particular transposon system used.

Shuttle Mutagenesis

1. A genomic library of *M. smegmatis* DNA (plasmid or cosmid) is transformed into an *E. coli* strain carrying an active transposon or insertion sequence capable of random hopping and bearing a drug resistance marker.

2. Transformants are divided into pools and grown overnight to allow transposition.

3. Aliquots of overnight cultures are plated onto agar plates containing high levels of antibiotic (thereby enriching for hops onto a multicopy plasmid) ("hyperresistance selection").

4. Colonies from each plate are collected and maintained in pools. This will decrease somewhat the occurrence of sister plasmids accumulated during the overnight transposition events.

5. Extrachromosomal DNA is prepared from the individual pools of *E. coli* (see Plasmid Preparation, above).

6. Pools of plasmid DNA are then transformed into the mycobacterial recipient (see Plasmid Transformation, above) and submitted to appropriate selection or screening.

VIII. Strain Construction

A. Homologous Recombination/Gene Replacement

Homologous recombination is reasonably efficient in *M. smegmatis,* but rare in *M. tuberculosis* or BCG. Husson *et al.* (1990) devised a gene replacement system using an interrupted copy of the *pyr*F gene of *M. smegmatis* carried on a plasmid lacking a mycobacterial origin. *pyr*⁻ mutants fell into two classes: one class is the result of a double crossover in which either (1) sequences at the *pyr* locus are replaced by sequences inserted in the homologous copy carried on the plasmid or (2) the entire plasmid is integrated followed by deletion of duplicated sequences. The second class is the result of a single crossover, in which the entire plamid is integrated into the *pyr*F gene. These workers were able to derive between 10 and 500 transformants/ug plasmid DNA. A prerequisite for using this technique is the availability of positive selection for the phenotype required. Homologous recombination was also studied in BCG by Aldovini *et al.* Using a protocol for increased transformation efficiency, 20%

of the transformants exhibited recombination of transformed *ura*A sequences into the homologous site (Aldovini *et al.,* 1993). Thus, it is poosible to construct a strain by homologous recombination, although the process is inefficient in BCG.

B. Nonhomologous Recombination

1. Site-specific: Site-specific nonhomologous recombination, based on phage integration systems, has distinct advantages in the design and construction of mycobacterial strains. In some cases, single-copy expression of a gene might be required. In addition, such recombinants are stable in the absence of selection, for use in animal infection models or vaccine strain construction, when selection for plasmid maintenance is not feasible. A plasmid is constructed in *E. coli,* which lacks a mycobacterial origin but contains the attachment sequences from a lysogenic mycobacteriophage. The plasmid must also bear the gene encoding the phage integrase, or *int* gene. Once transformed into the recipient cell, the plasmid will integrate at the phage attachment site (*att*B) in the mycobacterial genome. A number of such plasmids have been constructed for the purpose of site-specific integration of sequences from such phages as L5 (Lee *et al.,* 1991) and FRAT1 (Haeseleer *et al.,* 1993).

2. Mutational: Kalpana *et al.* (1991), in designing a shuttle mutagenesis system (see Shuttle Mutagenesis, above), found a significant degree of illegitimate recombination in BCG and *M. tuberculosis.* In fact, linear DNA fragments containing a selectable marker were incorporated at a frequency of 10^4–10^5 relative to the number of transformants obtained with autonomously replicating vectors. The authors proposed the exploitation of this system for generating random insertional mutants in the slow-growing species of mycobacteria.

IX. Mutant Selection and Isolation

A. Screening

Searching for a mutant by screening a population of mutagenized mycobacteria for a specific phenotype requires careful attention to the phenomenon of clumping. For example, in a screen for mutants unable to utilize a specific nutrient, a wild-type cell stuck to a mutant cell can mask the mutant phenotype. As mentioned above, populations plated on agar for screening by colony lifts or patching must be grown in Tween-80 and dispersed by passing through a 23-gauge needle. If replica plating is used, the master plates should include Tween-80 (at the same concentration used in liquid media); colonies of mycobacteria grown on agar/Tween-80 are smooth and domed (as opposed to flat and dry) and are easily transferred by sterile velvet or filter paper.

B. Selection

Direct selection for mutant phenotpye is a powerful tool that solves the problems of bacterial clumping: only cells with the required phenotype will survive the selection. However, the mycobacteria are notoriously resistant to many selective agents, such as antibiotics. In fact, there are few antibiotics available as genetic markers: kanamycin (Ranes *et al.*, 1990; Snapper *et al.*, 1988), chloramphenicol (Snapper *et al.*, 1988), streptomycin (Hermans *et al.*, 1990), and sulfonamide (Gormley and Davies, 1991). Tetracycline kills mycobacteria but its half-life in media is short and is therefore only useful for *M. smegmatis* (G. V. Kalpana and W. R. Jacobs, Jr., unpublished observations). Gentamycin can be used with *M. smegmatis* only (Gormley and Davies, 1991) and hygromycin (Radford and Hodgson, 1991) can be used with both *M. smegmatis* and BCG. Care should be taken when considering introducing antibiotic resistance determinants into potentially infectious species such as *M. avium*.

Many toxic analogs that have been used to select mutants in *E. coli* are ineffective with mycobacteria. For example, canavanine, a toxic analog of arginine, has been used to select *E. coli* mutants in arginine transport and metabolism: it is effective at concentrations of 1–5 μM. *M. smegmatis,* however, can survive quite well in concentrations of 100 mM cn agar plates (N. D. Connell, unpublished observation). An obvious explanation for this result is that *M. smegmatis* secretes free arginine into the medium, which inhibits the uptake of arginine. Another, not mutually exclusive, is that the local concentrations of canavanine near the colony are lowered to subtoxic levels, since the cells are strict aerobes and sit well on the surface of the agar medium. Growing the cells in liquid media containing high concentrations of canavanine for several hours before plating for single colonies on selective agar plates lowered the requirement for canavanine to 20 mM. As always, care should be taken to ensure that the cells are as close to single-cell suspension as possible.

C. Complementation

Complementation of mutants, either inter- or intra-species, has been a fruitful method of identifying specific genes of mycobacteria. Genes of *M. tuberculosis* (Garbe *et al.*, 1991) and *M. leprae* (Jacobs *et al.*, 1986) have been cloned by intraspecies complementation of *E. coli*. Interspecies complementation of *M. smegmatis* mutants have been used to clone the autologous *his*D gene (Hinshelwood and Stoker, 1992) and the gene for triose isomerase (Tuckman and Connell, personal communication). A gene cluster of *M. avium* encoding a highly antigenic glycopeptidolipid was isolated by screening an *M. avium* cosmid library expressed in *M. smegmatis* (Belisle *et al.*, 1991). Here we present a method for isolation of chromosomal DNA from which appropriate genomic libraries can be constructed.

Genomic DNA Isolation

1. Inoculate 5 ml of MADCTW and grow at 37°C.

2. If *M. tuberculosis*, heat to 80°C for 20 min.

3. Spin down culture, resuspend pellet in 500 μl TE, and transfer to 1.5-ml Eppendorf tube.

4. Add 50 μl 10 mg/ml lysozyme (Serva), mix, and incubate 1 hr at 37°C.

5. Add 70 μl 10% SDS and 6 μl 10 mg/ml proteinase K (Boehringer Mannhein), mix, and incubate 10 min at 65°C.

6. Add 100 μl of 5 M NaCl and mix thoroughly.

7. Add 80 μl CTAB/NaCl, mix thoroughly, and incubate 10 min at 65°C.

8. Add an equal volume chloroform/isoamyl alcohol (24:1), mix thoroughly, and spin in microfuge for 5 min at room temperature.

9. Remove aqueous supernatant to clean tube and add 0.6 vol isopropanol, incubate at -20°C for 30 min, and spin 20 min in microfuge at 4°C.

10. Pour off supernatant and wash pellet with chilled 70% EtOH. Remove wash and spin pellet. Dry at room temperature and redissolve pellet in TE.

TE: 10 mM Tris, pH 8; 1 mM EDTA.

CTAB/NaCl: 4.1 g NaCl in 80 ml ddH$_2$O. While stirring, add 10 g CTAB (*N*-cetyl-*N,N,N*,-trimethyl ammonium bromide, Merck). If neccessary, heat solution to 65°C. Adjust volume to 100 ml with ddH$_2$O.

This protocol can be scaled up to 10- or 20-fold with no significant reduction in yield. High-speed Eppendorf spins should be replaced by 12,000 $\times$ g.

Acknowledgments

I thank the following people for their generous contributions of results before publication and information: JoAnne Flynn, John Chan, Ruth McAdam, Amanda Brown, Bill Jacobs. Special thanks go to Mitchell and Eloise Gayer for contributions to the biosafety section.

References

Aldovini, A., and Young, R. A. (1991). Humoral and cell-mediated immune response to live recombinant BCG-HIV vaccine. *Nature (London)* **351**, 479–482.

Aldovini, A., Husson, R. H., and Young, R. A. (1993). The *ura*A locus and homologous recombination in *Mycobacterium bovis* BCG. *J. Bacteriol.* **175**, 7282–7289.

Banerjee, A., Dubnau, E., Quemard, A., Balasubramanian, V., Um, K. S., Wilson, T., Collins, P., deLisk, G., and Jacobs, W. R., Jr. (1994). *inh*A, a gene encoding a target for isoniazid and ethionamide in *Mycobacterium tuberculosis*. *Science* **263**, 227–229.

Belisle, J. T., Pascopella, L., Inamine, J. M., Brennan, P. J., and Jacobs, W. R., Jr. (1991). Isolation and expression of a gene cluster responsible for biosynthesis of the glycopeptidolipid antigens of *Mycobacterium avium*. *J. Bacteriol.* **173**, 6991–6997.

Birnboim, H., and Doly, J. (1979). A rapid alkaline extraction procedure for screening recombinant plasmid DNA. *Nucleic Acids Res.* **7**, 1513–1523.

Centers for Disease Control (CDC). (1990). Guidelines for preventing the transmission of tuberculosis in the health care settings, with special focus on HIV-related issues. *Morbid. Mortal. Wkly. Rep.* **39**, 2–3.

Cirillo, J. D., Barletta, R. G., Bloom, B. R., and Jacobs, W. R., Jr. (1991). A novel transposon trap for mycobacteria: Isolation and characterization of IS*1096*. *J. Bacteriol.* **173**, 7772–7780.

Garbe, T., Servos, S., Hawkins, A., Dimitriadis, G., Young, D., Dougan, G., and Charles, I. (1991). The *Mycobacterium tuberculosis* shikimate pathway genes-evolutionary relationship between biosynthetic and catabolic 3-dehydrogenases. *Mol. Gen. Genet.* **228**, 385–392.

Gelbart, S. M., and Juhasz, S. E. (1973). Transduction in *M. phlei. Antonie van Leeuwenhoek* **39**, 1–10.

Gormley, E. P., and Davies, J. (1991). Transfer of Plasmid RSF1010 by conjugation from *Escherichia coli* to *Streptomyces lividans* and *Mycobacterium smegmatis. J. Bacteriol.* **173**, 6705–6708.

Grange, J. M. (1982). The genetics of mycobacteria and mycobacteriophages. *In* "The Biology of the Mycobacteria" (C. Ratledge and J. Stanford, eds.), Vol. 1, pp. 309-353. Academic Press, San Diego.

Green, E. P., Tizard, M. L. V., Moss, M. T., Thompson, J., Winterbourne, D. J., McFadden, J., and Heron-Taylor, J. (1989). Sequence and characteristics of IS*900*, an insertion element identified in a human Crohn's disease isolate of *Mycobacterium paratuberculosis. Nucleic Acids Res.* **17**, 9063–9073.

Guilhot, C., Gicquel, B., and Martin, C. (1992). Temperature-sensitive mutants of the *Mycobacterium* plasmid pAL5000. *FEMS Microbiol Lett.* **98**, 181–186.

Haeseleer, F., Polet, J. F., Bollen, A., and Jacobs, P. (1993). Molecular cloning and sequencing of the attchment site and intergrase gene of the temperate mycobacteriophage FRAT1. *Mol. Biochem. Parasitol.* **57**, 117–126.

Hatfull, G. F. (1993). Genetic transformation of mycobacteria. *Trends Microbiol.* **1**, 310–314.

Hermans, P. W., Schuitema, A. R. J., van Soolingen, D., Verstynen, C. P. H. J., Bik, E. M., Thole, J. E. R., Kolk, A. H. J., and vanEmbden, J. D. A. (1990). Specific detection of *Mycobacterium tuberculosis* complex strains by polymerase chain reaction. *J. Clin. Microbiol.* **28**, 1204–1213.

Hinshelwood, S., and Stoker, N. G. (1992). Cloning of mycobacterial histidine synthesis genes by complementation of a *Mycobacterium smegmatis* auxotroph. *Mol. Microbiol.* **6**, 2887–2895.

Holland, K. T., and Ratledge, C. (1971). A procedure for selecting and isolating specific auxotrophic mutants of *Mycobacterium smegmatis. J. Gen. Microbiol.* **66**, 115–118.

Husson, R. N., James, B. E., and Young, R. A. (1990). Gene replacement and expression of foreign DNA in Mycobacteria. *J. Bacteriol.* **172**, 519–524.

Jacobs, W. R., Jr., Docherty, M. A., Curtiss, R., III, and Clark-Curtiss, J. E. (1986). Expression of *Mycobacterium leprae* genes from a *Streptococcus mutans* promoter in *Escherichia coli* K-12. *Proc. Natl. Acad. Sci. U.S.A.* **83**, 1926–1930.

Jacobs, W. R., Jr., Tuckman, M., and Bloom, B. R. (1987). Introduction of foreign DNA into mycobacteria using a shuttle phasmid. *Nature (London)* **327**, 532–535.

Jacobs, W. R., Jr., Kalpana, G. V., Cirillo, J. D., Pascopella, L., Snapper, S. B., Udani, R. A., Jones, W., Barletta, R. G., and Bloom, B. R. (1991). Genetic systems for mycobacteria. *In* "Methods in Enzymology" (J. Miller, ed.), Vol. 204, pp. 537–555. Academic Press, San Diego.

Jones, W. D., and David, H. L. (1972). Preliminary observations on the occurrence of streptomycin R-factor in *Mycobacterium smegmatis* ATCC 607. *Tubercle* **53**, 35–42.

Kalpana, G. V., Bloom, B. R. , and Jacobs, W. R., Jr. (1991). Insertional mutagenesis and illegitimate recombination in mycobacteria. *Proc. Natl Acad. Sci. U.S.A.* **88**, 5433–5437.

Kleckner, N., J. Bender, and Gottesman, S. (1991). Uses of transposons with emphasis on Tn*10*. *In* "Methods in Enzymology" (J. Miller, ed.), Vol. 204, pp, 139–180. Academic Press, San Diego.

Konicakova-Radochova, M., Konicek, J., and Malek, I. (1970). The study of mutagenesis in *Mycobacterium phlei. Folia Microbiol.* **15**, 88–102.

Kunze, Z. M., Wall, S., Appelberg, R., Silva, M. T., Portaels, F., and McFadden, J. J. (1991).

IS *901,* a new member of a widespread class of atypical insertion sequences, is associated with pathogenicity in *Mycobacterium avium. Mol. Microbiol.* **5,** 2265–2272.

Lazraq, R., Clavel-Seres, S., David, H. L., and Roulland-Dussoix, D. (1990). Conjugative transfer of a shuttle plasmid for *Escherichia coli* to *Mycobacterium smegmatis. FEMS Microbiol Lett.* **69,** 135–138.

Lee, M. H., Pascopella, L., Jacobs, W. R., Jr., and Hatfull, G. F. (1991). Site-specific integration of mycobacteriophage L5: Integration-proficient vectors for *Mycobacterium smegmatis., Mycobacterium tubercuolsis,* and bacille Calmette-Guérin. *Proc. Natl. Acad. Sci. U.S.A.* **88,** 3111–3115.

Martin, C., Timm, J., Rauzier, J., Gomez-Lus, R., Davies, J., and Giquel, B. (1990). Transposition of an antibiotic resistance element in mycobacteria. *Nature (London)* **345,** 739–743.

McAdam, R., Hermans, P. W. H., van Soolingen, D., Zainuddin, Z. F., Caty, D., van Embden, J. D. A., and Dale, J. W. (1990). Characterization of a *Mycobacterium tuberculosis* insertion sequence belonging to the IS*3* family. *Mol. Microbiol.* **25,** 796–801.

Radford, A. J., and Hodgson, A. L. M. (1991). Construction and characterization of a mycobacterium-*Escherichia coli* shuttle vector. *Plasmid* **25,** 149–153.

Ranes, M., Rauzier, J., LaGranderie, M., Gheorghiu, M., Gicquel, B. (1990). Functional analysis of pAL5000, a plasmid from *Mycobacterium fortuitum:* Construction of a "mini" mycobacterium-*Escherichia coli* shuttle vector. *J. Bacteriol.* **172,** 2793–2797.

Rauzier, J., Moniz-Pereira, J., and Giquel-Sanzy, B. (1988). Complete nucleotide sequence of pAL5000, a plasmid from *Mycobacterium fortuitum. Gene* **71,** 315–321.

Snapper, S. B., Lugosi, L., Jekkel, A., Melton, R., Kieser, T., Bloom, B. R., and Jacobs, W. R., Jr. (1988). Lysogeny and transformation in mycobacteria: Stable expression of foreign genes. *Proc. Natl. Acad. Sci. U.S.A.* **85,** 6987–6991.

Snapper, S. B., Melton, R. E., Mustafa, S., Kieser, T., and Jacobs, W. R., Jr. (1990). Isolation and characterization of efficient plasamid transformation mutants of *Mycobacterium smegmatis. Mol. Microbiol.* **4,** 1911–1919.

Stover C. K., De La Cruz, V. F., Fuerst, T. R., Burlein, J. E., Benson, L. A., Bennet, L. T., Bansal, G. P., Young, J. F., Lee, M. H., Hatfull, G. F., Snapper, S. B., Barletta, R. G., Jacobs, W. R., Jr., and Bloom, B. R. (1991). New use of BCG as a vector for vaccines. *Nature (London)* **351,** 456–460.

Strong, B. E., and Kubica, G. P. (1981). "Isolation and Identification of *Mycobacterium tuberculosis,*" HHS Pub. No. (CDC) 81-8390. U. S. Govt. Printing Office, Washington, DC.

Sundar Raj, C. V., and Ramakrishnan, T. (1970). Transduction in *Mycobacterium smegmatis. Nature (London)* **228,** 280–281.

Thierry, D., Cave, M. D., Eisenach, K. D., Crawford, J. T., Bates, J. H., Gicquel, B., and Guesdon, J. L. (1990). IS*6110,* an IS-like element of *Mycobacterium tuberculosis. Nucleic Acids Res.* **18,** 188.

Winder, F., and Brennan, P. (1964). The accumulation of free trehalase by mycobacteria exposed to isoniazid. *Biochim. Biophys. Acta* **90,** 442–444.

PART II

Microbial Adherence and
Invasion Assays

CHAPTER 7

Modulation of Murine Macrophage Behavior *in Vivo* and *in Vitro*

Gregory J. Bancroft, Helen L. Collins, Lynette B. Sigola, and Caroline E. Cross

Department of Clinical Sciences
London School of Hygiene and Tropical Medicine
London WC1E 7HT, England

I. Introduction
II. Reagents and Solutions
    A. Culture Media
    B. Eliciting Agents for Obtaining Peritoneal Macrophages
    C. Culture Media for Growth of Bone Marrow-Derived Macrophages
    D. Commonly Used Targets for Phagocytosis
    E. Buffer for Fluorescent Labeling of Microorganisms
    F. Anti-fade Mounting Fluids
III. Animal Husbandry and Maintenance of Immunocompromised Mice
    A. General Guidelines
    B. Maintenance of Immunodeficient Strains
IV. Eliciting and Harvesting Macrophages *in Vivo*
    A. General Guidelines
    B. Eliciting and Obtaining Peritoneal Macrophages
    C. Harvest of Splenic Macrophages
    D. Harvest of Kupffer Cells
    E. Harvest of Alveolar Macrophages
    F. Preparation of Bone Marrow-Derived Macrophages
V. Adherence of Macrophages to Solid Substrates
VI. Assays for Macrophage Phagocytic Activity
    A. General Guidelines
    B. Direct Labeling of Particles
    C. Analysis of Phagocytic Activity by Indirect Immunofluorescence
    D. Differentiation of Bound versus Ingested Particles
VII. Analysis of MHC Class II Antigen Expression
    A. General Guidelines

METHODS IN CELL BIOLOGY, VOL. 45

B. Visual Assessment by Indirect Immunofluorescence
C. Quantitation by Flow Cytometry
References

I. Introduction

Mononuclear phagocytes play a central role in the regulation of immune responses and in defense against infection. These cells originate in the bone marrow as monoblasts and promonocytes, migrate through the bloodstream as monocytes, and can then be found as mature cells in virtually all tissues of the body. Resident mononuclear phagocytes are often uniquely differentiated according to the specific tissue site such as Kupffer cells in the liver, microglial cells in the nervous tissue, and osteoclasts in bone. During episodes of inflammation or infection these can be supplemented by freshly derived monocyte/macrophages recruited from the bone marrow. Monocytes and macrophages secrete a broad range of bioactive molecules including components of the coagulation and complement cascades as well as immunoregulatory cytokines including IL-1, IL-6, TNFα, and IL-10. They also express a variety of cell surface receptors and glycoproteins, allowing physical interaction with components of the extracellular matrix, other cell types within the immune system, and pathogenic microorganisms. It is therefore not surprising that their activity is strictly regulated and influenced by the tissue of origin, state of maturation, and immunological environment.

A common theme is that macrophages can be activated by host-derived and/or microbial stimuli resulting in dramatic changes in their cell surface phenotype and function. In general, macrophages respond to four different forms of immune stimulation: first, they can be activated by cytokines secreted from natural killer (NK) cells and T cells such as IFNγ (Bancroft *et al.*, 1991; Paulnock, 1992). Second, T cells have been suggested to provide cognate signals for macrophage activation possibly via adhesion molecule-mediated cell contact or expression of membrane bound TNF (Stout, 1993). Third, interaction with extracellular matrix proteins such as laminin and fibronectin can regulate cell migration and function (Brown and Lindberg, 1993). Finally, components of microorganisms such as endotoxin (LPS) from Gram-negative bacteria or muramyl dipeptide from *Mycobacteria* can directly stimulate macrophage responses or act as ''second signals'' following priming by host cytokines.

More recently there has been an increased interest in the mechanisms of macrophage deactivation. Thus cytokines such as IL-4, IL-10, and macrophage deactivating factor (MDF) can inhibit macrophage cytokine synthesis and antimicrobial activity and can antagonize the activating effect of IFNγ (Stout, 1993). It is also becoming less appropriate to broadly characterize individual cytokines as either activating or deactivating since the response is often dependent upon the specific macrophage function being assayed. For example, IFNγ activates macrophages for expression of Ia antigens or microbicidal activity, but inhibits phagocytosis of complement opsonized particles. Conversely, IL-4 inhibits secretion of proinflammatory cytokines but is a potent activator of mannose receptor expression and function (Stein *et al.*, 1992). These observations reinforce the idea that macrophage phenotype in the host reflects the combined outcome of multiple influences, which will vary according to the type of antigenic or infectious challenge.

The purpose of this chapter is to provide a general introduction to obtaining and manipulating macrophages *in vitro* and *in vivo*. Clearly, the multiple influences on macrophage function summarized above must be taken into account in any experiment involving primary macrophage populations. This is less problematic when using macrophage cell lines that offer the advantage of supplying large numbers of cells with a uniform phenotype under controlled conditions. These have been used successfully to study aspects of macrophage–pathogen interactions (e.g., Tilney *et al.*, 1992) and can respond to exogenous cytokines such as IFNγ in some circumstances. However, other functions may either be absent or not appropriately regulated in these cell lines and care must be taken before extrapolating these results to normal cells. Here, methods are presented for harvesting primary macrophage populations from the common tissue sites, their cultivation *in vitro*, and performing assays of macrophage activity, namely phagocytosis and expression of class II MHC antigens. Further details of procedures on microbicidal and antigen presenting functions using these cells will be found in other chapters in this volume.

II. Reagents and Solutions

A. Culture Media

The majority of *in vitro* manipulations of macrophages do not involve proliferation of the cells and can be performed in standard culture media such as RPMI 1640 supplemented with 10% heat-inactivated FCS plus 10 mM HEPES, 100 IU/ml of penicillin, and 100 μg/ml of streptomycin (RPMI/10). For example, adherent peritoneal macrophages can be maintained in culture for 5–7 days, allowing analysis of responses to various cytokines and microbial "second signals" such as LPS or heat-killed *Listeria monocytogenes*. An exception to this rule is the growth of purified macrophages from bone marrow precursors. Here, DMEM supplemented with glucose and a source of colony-stimulating factors must be used to allow proliferation and differentiation into mature macrophages, with the cells being returned to RPMI/10 media for subsequent activation steps (see below).

Note

It is essential that levels of LPS be kept to a minimum in any procedure involving the manipulation of macrophages *in vitro*, since significant changes in cell phenotype and function can occur in the presence of picogram quantities of this material. Furthermore, although LPS is a potent second signal for macrophage activation if added after priming of the cells with cytokines such as IFNγ, addition before the cytokine can actually inhibit some activation events *in vitro*.

Liquid culture media prepared by the manufacturer is essentially endotoxin free, but if powdered stocks are used to prepare bulk volumes of media "in house" then strict attention must be placed on the quality of the water used and the possibility of contamination from recycled glass bottles. Batches of FCS differ in their content of LPS and should be screened for suitability in the specific macrophage assay being investigated. Endotoxin levels in culture media can be readily determined by ELISA-based kits or using the original amebocyte lysate gel assay (Whittaker Bioproducts or Sigma).

B. Eliciting Agents for Obtaining Peritoneal Macrophages

1. Proteose peptone No. 3: (Supplier: Difco). Prepare a 10% solution in distilled water and heat to 56°C for 1 hr to dissolve. Centrifuge the solution at 3000 rpm/25 min to remove undissolved material, autoclave to sterilize, and store at 4°C. Prewarm to 37°C and inject 1–5 ml ip per mouse.

2. Brewers' thioglycollate medium: (Supplier: OXOID). Prepare a 10% solution as above, autoclave, and store at room temperature for several months before use. Inject 1–5 ml ip prewarmed to 37°C.

3. Concanavalin-A (Type IV, Sigma). Prepare at 1 mg/ml in pyrogen-free saline, filter sterilize, and store in aliquots at $-20°C$. Inject 0.2 ml per mouse ip (200 μg).

Note

Any diluent used to administer substances to mice *in vivo* must be free of bacterial endotoxin (LPS) to avoid artifactual changes in macrophage phenotype. This is easily achieved by using either serum-free culture medium or saline suitable for clinical administration.

C. Culture Media for the Growth of Bone Marrow-Derived Macrophages

Bone marrow macrophages are obtained by culturing murine bone marrow cells in DMEM containing 4.5 g/liter L-glucose supplemented with 20% FCS, 10 mM HEPES, 1 mM sodium pyruvate, 2 mM glutamine, 100 μg/ml each of penicillin and streptomycin, 5×10^{-5} M β2 – mercaptoethanol and 20–30% v/v L cell conditioned medium (LCM). L cell conditioned medium is prepared by growing L929 murine fibroblasts in 150-cm² flasks in DMEM/10 until confluent and the culture medium changes from red to yellow-orange in color. The supernatant is decanted, centrifuged to remove cell debris, and stored in aliquots at $-20°C$ until needed (stable for 12 months).

Note

The source of L cells for this step will influence the concentration of LCM needed to generate bone marrow macrophages in culture and the number of times the cultures must be supplemented. L cell stocks known to be high secretors of colony-stimulating factors should be used where possible and large batches of LCM prepared, aliquotted, and titrated for their effective concentrations in order to avoid variation between experiments.

D. Commonly Used Targets for Phagocytosis

1. Latex beads: (Supplier: Sigma). 3-μm diam. 10% solids = approx 7×10^9 beads/ml.

2. Zymosan: (Supplier: Sigma). 1 mg/ml = approx 3×10^7 particles/ml.

3. SRBC-Ig (EA). SRBC: (Suppliers: Whittaker Bioproducts, GIBCO, UK). Use within 14 days of date of packaging. Take 5 ml of SRBC and wash three times at 2000 rpm/5 min/4°C in $1\times$ Veronal buffered saline (VBS; $5\times$ VBS stock solution, Whittaker Bioproducts). Resuspend in 20 ml of VBS to give approximately 10^9 SRBC/ml (OD approx 0.4 at 541 nm). Dilute the anti-SRBC antibody (anti-sheep RBC antisera IgG, Cordis Laboratories) to the required concentration (suggested range 1/500–1/5000) and add dropwise to 1 ml of SRBC while gently vortexing. Incubate for 30 min at room temperature or 15 min at 37°C, wash twice with VBS, and store at 4°C for up to 1 week.

E. Buffer for Fluorescent Labeling of Microorganisms

Mix 0.5 M NaHCO$_3$ (8 vol) plus 0.5 M Na$_2$CO$_3$ (1 vol) plus 0.9% NaCl (27 vol). Prepare FITC or TRITC (Sigma) in this carbonate buffer to 0.5 mg/ml.

F. Anti-fade Mounting Fluids

Final mounting of macrophage monolayers for immunofluorescence analysis should be done in the presence of an anti-fading compound, particularly if high magnification objectives lenses are being used (e.g., counting fluorescent phagocytosis assays of bacteria within macrophages), where bleaching of the sample can occur during counting. We routinely use either:

1. 0.4 g propylgallate (Sigma) in 10 ml PBS plus 10 ml glycerol OR

2. Mowiol 4-88 (Calbiochem). Add 2.4 g Mowiol to 6 ml glycerol and vortex. Add 6 ml water and leave at room temperature for 2 hr. Add 12 ml 0.2 M Tris, pH 8.5, and incubate at 55°C until dissolved. Centrifuge at 4000–5000 rpm for 20 min and aliquot. Store at room temperature; stable for 12 months.

III. Animal Husbandry and Maintenance of Immunocompromised Mice

A. General Guidelines

The microbiological quality of animals used to harvest primary macrophage populations is the single most important influence on the characteristics of the cells obtained. Underlying infections with any of a number of different murine pathogens can interfere with baseline immunological responses and have the potential to activate or inhibit macrophage activity. Animals should be purchased from accredited suppliers and certified free of pathogens including: pneumonia virus of mice (PVM), mouse hepatitis virus (MHV), Sendai virus, Ectromelia (mouse Pox), Reo 3, GDVII (Theilers), *Mycoplasma pulmonis,* and *Bacillus piliformis* (Tyzzers disease) as well as helminths and intestinal protozoa. Most animals facilities will routinely monitor their unit for the major murine pathogens as part of an ongoing health surveillance program. This is achieved by screening pathogen-free sentinel animals by direct microbiological culture of specimens from autopsy and by serological monitoring for exposure. Where this is not possible, used bedding from key mouse cages can be provided to sentinel animals in another room and seroconversion to specified pathogens measured 2–3 weeks later. Two additional checks that can be made by the investigator are first to monitor the number of peritoneal cells obtained from unstimulated mice and second to measure the frequency of macrophages expressing class II MHC antigens (above a usual background of 10–20% Ia$^+$; see Section VII for method). Increases in either parameter in otherwise untreated animals are often an early indicator of infection and can occur in the absence of clinical signs.

B. Maintenance of Immunodeficient Strains

The use of mice with severe immunodeficiencies such as *scid* and *nude* mutants or RAG-1/2 knock-outs allows analysis of macrophage function in the absence of T or B cells and provides a focus on components of the innate immune system. However, additional precautions are needed to prevent overwhelming infection with pathogens that may be subclinical in immunocompetent mice. The greatest threats are MHV and murine *Pneumocystis carinii* pneumonia. There are several methods to maintain mice under aseptic conditions, which vary in their cost and level of security. A common requirement is the provision of presterilized food, water, and housing achieved by either autoclaving, irradiation (2.5 MRad), or liquid disinfection (e.g., using either Alcide or Virkon disinfectants) depending on the article.

The most commonly used barrier systems are:

1. Flexible Film Isolators

These units are based on a "plastic bubble" design with glove ports to allow animal handling, HEPA filters for protection, and a sealable air lock chamber (preferably with solid rather than flexible plastic doors) to allow transfer of materials. These provide the most stringent microbiological containment and can be used under positive pressure to provide a sterile environment for the animals or under negative pressure for containment of pathogens introduced during the experiment (e.g., SCID mice infected with *Mycobacterium tuberculosis*). The microbiological status of the unit is dependent upon the decontamination procedures for entry and removal of material and any breach of this barrier will affect all the animals. Delicate procedures such as intravenous dosing are extremely cumbersome through the glove ports and increase the risk of needle stick injury to the investigator. (Supplier: Harlan Olac, UK; MDH Ltd., UK; Banton and Kingman, Hull, UK.)

2. Laminar Flow Rack Housing

Cages are placed in open racks within a laminar flow, HEPA filtered air system under positive pressure (i.e., blowing sterile air over the cage top and into the room). This has the disadvantages of increasing allergen exposure to the investigator and is not practical for experiments involving airborne pathogens. Nevertheless, this system does reduce the risk of cross-contamination from cage to cage. (Suppliers: Tecniplast Gazzada, Italy.)

3. Microisolator Systems

This is the simplest and most cost-effective method and has revolutionized the maintenance of animals in an aseptic environment. Each cage is covered with an individual bonnet incorporating a filter mesh to prevent influx of contaminated bedding or dust. A sterile work area (such as a class 100 tissue culture hood or a class II microbiological safety cabinet) is needed to prevent contamination during routine cage changing or experimental manipulation of the animals. (Suppliers: Lab Products; Tecniplast Gazzada, Italy.)

For long-term breeding and use of severely immunocompromised strains (e.g., SCID), we maintain a core breeding stock within a flexible film isolator and transfer animals into microisolator cages after weaning. This combination provides added security for the breeding nucleus and easy experimental manipulation of stock animals. Regardless of the containment system used, strict attention must be paid to the "standard operating procedures" untilized by both animal technicians and investigators to prevent a breakdown of the barrier. Also, transformed cell lines are potential sources of viral pathogens and can contaminate animal stocks following *in vivo* passage. Transfer of normal lymphocytes from subclinically infected mice into immunocompromised recipients can also contaminate these mutants. It is possible to provide prophylactic antibiotics in the drinking water on a routine basis. However, this is not a universally adopted technique and is not appropriate for studies of bacterial and parasitic infections. If required, prepare a stock suspension of 40 mg Trimethoprim plus 200 mg sulfamethoxazole/5 ml of water. Add 0.125 ml of this to every 4 ml of drinking water. Provide for 3 days per week, turning the bottles daily and supply normal water for the remaining time.

In the case of severely immunocompromised animals such as SCID mice, serological monitoring of health status is not an option and emphasis is placed on direct culture of the pathogen. Except for routine monitoring of "resting" cell yields (an increase in cell numbers in the spleen or PEC is a good marker of infection) or liver histology, reliable tests for immunological exposure are not available. Presumably it may be possible in the future to use polymerase chain reaction based techniques to test for the presence of microbial nucleic acids (e.g., for MHV) directly from tissue samples.

IV. Eliciting and Harvesting Macrophages *in Vivo*

A. General Guidelines

Resident mononuclear phagocytes can be identified histologically in a variety of anatomical sites including the lymphoid organs, liver, lungs, central nervous system, skin, and kidneys. These mature cells are relatively long-lived and show intrinsic differences in surface phenotype and metabolic activity according to their tissue location (Gordon *et al.*, 1992a,b; see also Table I for a selected list of commonly used surface markers). In terms of harvesting cells for experimentation, the peritoneal cavity of mice provides the most commonly utilized site. Macrophages can also be routinely obtained from other sites such as the lung, liver, and spleen, but this often requires additional manipulation to obtain cells in sufficient number and purity. Nevertheless, this is important for specific models of infection, such as investigating the alveolar macrophage response to pulmonary pathogens. It is also possible to induce selective growth and differentiation of macrophage bone marrow precursors *in vitro* using culture media supplemented with hemopoietic growth factors. This provides large numbers of mature macrophages at high purity, with little phenotypic variation from batch to batch. These cells are often used to investigate the effects of T-cell-derived cytokines such as IFNγ and IL-4 on microbicidal or tumoricidal activity. However, bone marrow macrophages have obviously been cultured in a mixture of colony stimulating factors, which can influence some responses such as the regulation of CR3-dependent phagocytosis.

Table I
Membrane Antigens Used to Characterize Macrophage Populations from Various Tissues

Marker	Antibody	Distribution	Function and Regulation
F4/80	F4/80	Mature macrophages not dendritic cells decreased expression in T cell-dependent areas	Decreased expression on *in vitro* activation with IFN; decreased expression *in vivo* following inflammatory stimuli
Macrosialin (CD68)	FA.11	Macrophages and dendritic cells	Predominantly intracellular on endosomal membranes; different isoforms in resident versus elicited cells
CR3 CD11b/CD18	5C6 M1/70	Granulocytes, macrophages, monocytes, NK cells	Heterogeneous expression on macrophages; involved in adhesion to endothelium; phagocytic receptor recognizing C3bi/C3b
Sialoadhesin	SER-4	Stromal macrophages, marginal zone cells	Recognition of sialylated structures; sheep red cell receptor involved in binding hemopoietic cells

B. Eliciting and Obtaining Peritoneal Macrophages

Peritoneal cells can be obtained from untreated animals to provide a source of resident macrophages with low expression of class II MHC antigens and little or no microbicidal or tumoricidal activity. Alternatively, to obtain a greater number of cells and/or to acquire an activated phenotype, mice can be treated with eliciting agents for 3–7 days prior to cell harvest. Harvesting before this time usually increases contamination with granulocytes, which represent the initial cellular infiltrate, and then decline in numbers after several days. Agents that can elicit macrophages in this manner are usually classified into two broad categories. One category contains sterile irritants such as proteose peptone and thioglycollate broth, which nonspecifically elicit inflammatory cells with an activated morphology and increased phagocytic activity for complement opsonized particles. In contrast, the other category comprises "immunological activators" such as Con-A or microorganisms such as *L. monocytogenes* or Bacille Calmette Guerin (BCG), which also trigger secretion of IFNγ *in vivo* and can promote increased expression of class II MHC antigens and induction or priming for microbicidal or tumoricidal activity.

Mice are killed by cervical dislocation and the fur over the peritoneum is prepared with 70% ethanol. A 3 to 5-mm incision is made in the midline, level with the top of the pelvis and the skin reflected back to reveal the peritoneal wall. Ten milliliters of cold RPMI/1 is injected via an 18-gauge needle into the lower left or right abdominal quadrant, preferably near the inguinal fat pads, which will block the injection site once the needle is withdrawn and prevent leakage of fluid. Fluid is withdrawn from the upper right quadrant, taking care not to lacerate the liver, which will contaminate the preparation with erythrocytes. The harvested cells are centrifuged at 1100 rpm/7 min/4°C, washed once, resuspended in cold RPMI/10, counted, and plated as required. [Expected yields: naive adult mice will provide $1–3 \times 10^6$ cells per mouse (of which 20–30% will be adherent), elicited populations $3–15 \times 10^6$ per mouse (60–>80% adherent). Naive SCID mice will yield $0.5–1 \times 10^6$ cells per mouse (>80% adherent).]

C. Harvest of Splenic Macrophages

Preparation of splenic macrophages can be achieved by direct adherence of unfractionated spleen cell suspensions. However, to obtain a high density of macrophages usually requires prior enrichment due to the low frequency of these cells in the spleen suspensions. Sterile cell suspensions prepared by gentle disruption of the organ with forceps or passage through a stainless-steel mesh can be layered onto a four-step discontinuous Percoll gradient, final densities 1.06, 1.075, 1.085, and 1.11 g/ml, and centrifuged at $1100 \times g$ for 20 min/4°C. Cells harvested from the 1.06/1.075 and 1.075–1.085 interfaces (enriched for macrophages) are then washed three times in cold RPMI/10, resuspended in the same medium, and plated as required.

D. Harvest of Kupffer Cells

Kill the mouse by cervical dislocation and expose the liver. Cannulate the inferior vena cava with a 25-gauge butterfly needle and perfuse with 10 ml of prewarmed HBSS. After the liver begins to blanch and the portal vein distends, cut the portal vein and complete the injection. Continue perfusing with 10 ml of HBSS/0.05% collagenase (Type IV Sigma) prewarmed to 37°C.
Note
If possible screen different sources and batches of collagenase for the best results, as these can vary in efficiency and toxicity.
Excise the liver, chop into pieces with a sterile scalpel, and incubate in 20 ml per liver of HBSS/HEPES/0.05% collagenase plus 100 μg/ml DNAase (Sigma) for 10 min at 37°C. Pour through a metal sieve (250 μm, No. 60) to remove undigested material. Wash the cell suspension three times in HBSS for 10 min/4°C/1100 rpm and resuspend the pellet in 1 ml of HBSS per liver and pass through a Nytex filter to remove clumps. Layer one liver equivalent (1 ml) onto 9 ml of 1.037 g/ml Percoll and centrifuge at 1800 rpm/30 min/4°C. Remove the interface between the medium and the Percoll layers (containing debris, dead cells, and hepatocytes) and harvest the cell pellet (enriched for macrophages). Wash the pellet three times in medium to remove excess Percoll, resuspend in RPMI/10, and adhere to glass or plastic for final enrichment of the macrophages (modified from Crocker *et al.*, 1984).

E. Harvest of Alveolar Macrophages

The animal is anesthetized, the sternum is opened, and the lungs are cleared of blood by inserting a needle into the right ventricle and flushing with medium after cutting the vena cava and aorta. Identify the trachea taking care not to sever the major blood vessels, insert a ligature under it, and tie loosely. Make a small hole between the cartilage rings using a needle or scissors and insert a nylon catheter (i.d. 1 mm, o.d. 1.3 mm, attached to a 21-gauge needle and two 10-ml syringes via a three-way Luer lock stopcock) approximately 3–4 mm into the trachea. Tighten the ligature to hold the catheter firmly and infuse approximately 0.7–1.0 ml of prewarmed lavage fluid (PBS/0.1% BSA ±12 mM Lidocaine (Sigma), 25 mM HEPES). Change the stopcock position, aspirate the fluid using the second syringe, and repeat the procedure until approximately 10 ml of fluid has been flushed through. The cells obtained are washed once or twice and adhered to enrich for alveolar macrophages. Approximate yield in the presence of Lidocaine: $0.9–2.4 \times 10^6$ cells per mouse (for further details, see Holt, 1979).

F. Preparation of Bone Marrow-Derived Macrophages

Mouse femurs are harvested into 15 ml of DMEM/10 on ice, taking care to make clean, sharp incisions so not to splinter the bone and decrease the cell harvest. After removing excess muscle tissue, a cut is made at the end of the femur and the bone marrow cell plug is eluted by inserting a 26-gauge needle into the marrow cavity and flushing with cold DMEM/10. The cells are aspirated gently to create a single-cell suspension and washed by centrifugation at 1100 rpm/7 min/4°C. Following washing, the cells are resuspended in DMEM/10 and diluted to a concentration of 4×10^6/ml. One milliliter of cells and 9 ml of DMEM/10 supplemented with 20–30% LCM are added to 100-mm-diameter bacteriological grade (not tissue-culture) petri dishes and incubated at 37°C. After 3 days, an additional 10 ml of DMEM/10/LCM is added to each plate. A visible monolayer of macrophages will be apparent after 4–8 days according to the potency of the LCM used. To harvest these cells for experimentation, the medium is removed, 10 ml of ice cold sterile PBS is added, and the cells are incubated on a cold surface at 4°C for 1 hr. The cell monolayers are then resuspended using this fluid and the resulting cell suspension is washed once and plated as required. As an estimate, 10^7 bone marrow derived macrophages can be obtained per 100-mm dish.

V. Adherence of Macrophages to Solid Substrates

According to the different requirements of the assay system, macrophages can be adhered to a variety of different culture surfaces.

Note

The plating densities stated below are for initial guidance only and should be modified according to the particular assay used.

Flat-bottom 96-well tissue culture plates (Falcon):
Used for multiple replicates of adherent macrophages in antigen presentation assays, some microbicidal assays, and for binding of radiolabeled antibodies. Allows removal of nonadherent cells by washing with 8- or 12-channel pipettors for increased efficiency. Suggested plating density: 0.2 ml per well at 2×10^6 cells/ml.

24-well, (16-mm-diam) plates (Costar):
Used for adherence onto glass coverslips or when increased numbers of cells are required such as in preparing cell lysates for SDS–PAGE Western blotting or extracting mRNA for cytokine expression. Suggested plating density 0.5 ml per well at 2×10^6/ml.

Labtek 8-well slide chambers:
Glass (or plastic) slides with a plastic multiwelled chamber attached via a silicone gasket. Labtek slides allow sterile culture and manipulation of macrophage monolayers, which can then be directly visualized by light microscopy. We have found these to be excellent for studies of macrophage phagocytic activity. Suggested plating density 0.3 ml per well at 10^6–2×10^6/ml.

Covalent modification of glass substrates:
Covalent modification of the culture surface allows binding of macrophages to insoluble extracellular matrix proteins and other specific ligands. This usually involves coating the glass surface with poly-L-lysine and crosslinking the ligand with either glutaraldehyde or l-ethyl-3-(3-dimethylaminopropyl) carbodiimide (Sung *et al.*, 1983). This method can also be used to study receptor migration and mobility within the macrophage membrane. In this case, macrophages are plated onto ligand-coated surfaces, resulting in migration of the associated receptors to the basal surface of the cell. Quantitation of the number and function of the remaining receptors on the apical surface can give clues to the migratory potential of the molecules or the ability of ligation of one receptor to co-cap a second cell surface molecule.

In many of the examples given above it is difficult to remove adherent macrophages from tissue culture treated surfaces without significant loss of viability. Culture of macrophages in suspension can be achieved for 1–3 days by using polypropylene tubes (Falcon) rather than the standard polystyrene materials.

VI. Assays for Macrophage Phagocytic Activity

A. General Guidelines

Macrophages are capable of ingesting a variety of different particles and ligands, primarily through cell surface receptors such as CR3, mannose receptor, and Fc receptors. These allow either direct recognition of pathogens (e.g., mannose receptor binding to *Candida* or *Pneumocystis*) and/or indirect binding to host-derived opsonins attached to the microbial surface (e.g., binding of CR3 to C3bi deposited on the capsule surface of *Cryptococcus neoformans*). *In vitro* assays of phagocytic activity are used to confirm the purity of macrophage populations harvested from mice or for more specific investigations into the cell biology of the receptors per se.

In the case of latex beads or sheep erythrocytes, it is possible to count the ingested particles directly by light microscopy and express this information in terms of the percentage of cells which are phagocytic and/or the average number of particles per 100 macrophages. An alternative approach is to attach a fluorescent label to the particle, allowing easy identification under a fluorescence microscope. In this case the label can be either directly incorporated prior to phagocytosis or indirectly incorporated via specific antibodies after ingestion has taken place.

B. Direct Labeling of Particles

Direct conjugation of fluorescein isothiocyanate (FITC) or tetramethylrhodamine isothiocyanate (TRITC) to microorganisms such as yeasts and bacteria can be readily performed using a carbonate buffer system. Prewash the organisms once in carbonate buffer (see Reagents) and add 2 ml of FITC/TRITC solution to the pellet of bacteria (e.g., 2×10^9 *Listeria*) or yeasts (e.g., 10^8 zymosan or *Cryptococcus neoformans*). Leave for 2 hr at room temperature with constant gentle shaking—a roller apparatus is optimal. Wash the organisms five times in PBS or until the supernatant is clear of obvious fluorescent material.
Note

This method provides phagocytic targets labeled to extremely high efficiency (see Color Plate 1) before they have contact with the macrophage. It is therefore important to confirm initially that the labeling does not artifactually alter the characteristics of binding and ingestion.

C. Analysis of Phagocytic Activity by Indirect Immunofluorescence

In the case of microorganisms that are difficult to label directly, an indirect immunofluorescence assay can be used to tag microbes within macrophages for easier counting. This method requires either polyclonal antisera or a mAb specific for the organism under investigation and labels the organism or particle after ingestion by the macrophage. The example given is an analysis of phagocytosis of the AIDS-related fungal pathogen *C. neoformans* (Collins and Bancroft, 1992).

Prepare the macrophage population of interest in RPMI/10 and plate in 300-μl vol into 8-chamber glass Labtek slides (GIBCO). For analysis of phagocytic activity by resting peritoneal macrophages plate cells at $1-2 \times 10^6$/ml.

Allow the cells to adhere for 2 hr at 37°C and wash in warm RPMI/10 to remove non adherent cells.

Note

At this point it is possible to insert an *in vitro* activation period into the assay to monitor the effects of exogenous cytokines on macrophage function. In this example, recombinant murine cytokines such as TNFα and GM-CSF are diluted and added for 1–3 days prior to measurement of phagocytic activity.

The cell monolayers are washed twice to remove exogenous cytokines and the test particles are added to the monolayers at ratios of from 1 to 100 particles per cell. Binding and ingestion are allowed to take place from 30 to 120 min at 37°C and the monolayers are washed four times to remove unbound organisms. The plastic chamber is gently removed from the slide, leaving the gasket intact to provide a well for the staining reagents. The monolayers are then fixed in 100% methanol for 10 min at room temperature to permeabilize the cells and rabbit anti-*C. neoformans* capsule anti-sera diluted 1:500 in PBS/10% donor calf serum is added (35–50 μl per well) for 45 min at 4°C. After washing in three changes of cold PBS, TRITC conjugated goat anti-rabbit Ig diluted in PBS/DCS is added and the staining repeated. The organisms are visualized using a fluorescence microscope (phase contrast objectives allow easy viewing of adherent macrophages) and the number of organisms bound/ingested is calculated by: (total number of fluorescent organisms per field/total number of macrophages per field) × 100.

D. Differentiation of Bound versus Ingested Particles

Depending upon the particle and receptors involved, binding and ingestion by the macrophage may not be identical. For example, resident macrophages will bind complement opsonized SRBC but require exogenous activation by cytokines or components of the extracellular matrix for ingestion. Under these conditions it is important to determine whether the particle is extracellular or actually internalized by the cell. There are three common methods to visualize whether a particle is either bound or ingested. In the case of SRBC, treatment of the monolayers with 0.1 × PBS for 1–3 min will lyse extracellular erythrocytes without damaging either macrophages or intracellular SRBC, allowing a differential count to be made. For directly fluoresceinated particles, quenching of extracellular, but not intracellular, fluorescence can be achieved by addition of 0.5 mg/ml crystal violet to the medium. In the example of indirect, antibody-mediated identification described in the previous section, incubation with rabbit anti-*C. neoformans* antibody followed by a second layer goat anti-rabbit Ig FITC antiserum is first performed PRIOR to methanol fixation to label only extracellular organisms. Following methanol permeabilization of the cell membrane, the first-layer staining is repeated but then followed with TRITC-labeled anti-Ig antibody. The number of organisms ingested can then be calculated by subtracting the number of yeasts stained green (bound only) from the total number stained red (bound and ingested).

═══ VII. Analysis of MHC Class II Antigen Expression

A. General Guidelines

The regulated expression of class II MHC (Ia) antigens is a convenient assay of macrophage activation. The frequency of macrophages expressing Ia molecules in a resting state varies according to the site of origin (peritoneal macrophages: 10–20% Ia$^+$; splenic, 40–50%; liver, approx 50%). Because of their low baseline levels of expression, peritoneal macrophages are commonly used to assay the ability of different pathogens to augment Ia expression *in vivo*. This response predominantly reflects the action of IFNγ. For example, infection of SCID mice with *L. monocytogenes* raises Ia expression from 7% to >95% Ia$^+$ cells at 12 days postinfection and this can be abolished by prior administration of neutralising mAb specific for IFNγ. Ia expression can also be enhanced in immunocompetent mice by administration of T cell mitogens such as Con-A *in vivo* (see Fig. 1). Alternatively, peritoneal cells can be harvested from naive mice and cultured with recombinant cytokines for 24–72 hr to induce Ia expression *in vitro*. It is important to note that even the basal Ia level observed upon initial cell harvest will decay to zero by 48 hr in culture in the absence of any other stimuli. This is a useful trick to allow low starting levels of expression prior to the addition of specific activators.

Quantitation of Ia expression is performed by binding of haplotype specific mAb (e.g., MKD6 anti-Ia^d or 10.36.2 anti-Ia^k) followed by radiolabeled or fluorescent anti-Ig reagents.

B. Visual Assessment by Indirect Immunofluorescence

Macrophages are plated onto glass coverslips placed in 24-well cluster plates (Costar) at $1–2 \times 10^6$/ml in 0.5 ml of medium. The cells are allowed to adhere for 1.5–2 hr at 37°C, washed with warm medium to remove nonadherent cells, and either stained immediately to reflect *in vivo* Ia expression or cultured for a further 24–72 hr in the presence of cytokines. For Ia staining, 50 μl of anti-Ia mAb is placed on the bottom of each well of a 24-well plate, the coverslips are washed once in medium, the excess fluid is removed by touching against the edge of an absorbent tissue, and they are placed "cells down" onto the bottom of the well. After incubation at 4°C for 30 min, 1 ml of cold PBS is added per well, the coverslip is removed and washed by dipping five times into each of three beakers of cold PBS, and the staining procedure is repeated with fluoresceinated F(ab)′2 rabbit anti-mouse Ig. After further washing in PBS the coverslips are mounted onto glass slides with an anti-fade mounting fluid (see Reagents), sealed with nail varnish, and kept in the dark at 4°C until viewing. Initially, binding of a species and isotype matched mAb against an irrelevant MHC haplotype (e.g., MKD6 on Ia^k cells) should be included to confirm the specificity of the response. Addition of 10% normal rabbit serum to all dilutions of antibody reagents will reduce nonspecific binding problems by blocking

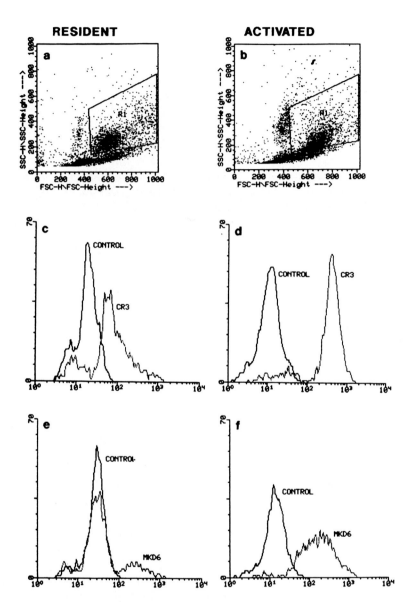

Fig. 1 FACScan analysis of CR3 and Ia expression by murine peritoneal macrophages. Peritoneal exudate cells were harvested from BALB/c mice treated 4 days previously with either pyrogen free saline (RESIDENT: a,c,e) or 200 μg Con-A ip (ACTIVATED: b,d,f). (a, b) Forward and side scatter profiles. (c, d) Expression of CR3 using 5C6 mAb plus FITC OX12 versus FITC OX12 alone (control). (e, f) Ia antigen expression using MKD6 anti Ia^d mAb plus FITC F(ab)′2 anti mouse Ig versus anti mouse Ig alone (control). See section VII,B.

macrophage Fc receptors. This method measures the frequency of Ia-bearing macrophages as well as allowing observations of cellular morphology that can also reflect macrophage activation.

C. Quantitation by Flow Cytometry

This is based on a similar procedure as described above, but using cells in suspension, and provides the additional advantage of greater sensitivity, quantitation, and through-put of samples (see Fig. 1). However, macrophages have a higher autofluorescent background than lymphocytes, which becomes particularly apparent with the sensitivity of the FACS. Expression of Fc receptors also causes raised levels of nonspecific staining and can be overcome by use of F(ab)$'_2$ fluorescent antibodies, microfuging reagents for 10 min at 13,000 rpm before use, and prior blocking with 100 μl of undiluted serum of the same species used as the detecting layer (e.g., normal rabbit serum) for 10 min at 4°C before staining.

Acknowledgments

The research in our laboratory that uses these methods is supported by the Wellcome Trust and the GLAXO Action TB Initiative.

References

Bancroft, G. J., Schreiber, R. D., and Unanue, E. R. (1991). Natural immunity: A T cell independent pathway of macrophage activation, defined in the scid mouse. *Immunol. Rev.* **124,** 5.

Brown, E. J., and Lindberg, F. P. (1993). Matrix receptors of myeloid cells. In "Blood Cell Biochemistry" (M. A. Horton, ed.), Vol. 5, p. 279. Plenum, New York.

Collins, H. L., and Bancroft, G. J. (1992). Cytokine enhancement of complement dependent phagocytosis by macrophages: Synergy of tumour necrosis factor α and granulocyte-macrophage colony stimulating factor for phagocytosis of *Cryptococcus neoformans. Eur. J. Immunol.* **22,** 1447.

Crocker, P. R., Blackwell, J. M., and Bradley, D. J. (1984). Expression of the natural resistance gene Lsh in resident liver macrophages. *Infect. Immun.* **43,** 1033.

Gordon, S., Lawson, L., Rabinowitz, S., Crocker, P. R., Morris, L., and Perry, V. H. (1992a). Antigen markers of macrophage differentiation in murine tissues. *Curr. Top. Microbiol. Immunol.* **181,** 1.

Gordon, S., Fraser, I., Nath, D., Hughes, D., and Clarke, S. (1992b). Macrophages in tissues and *in vitro. Curr. Opin. Immunol.* **45,** 25.

Holt, P. G. (1979). Alveolar macrophages. I. A simple technique for the preparation of high numbers of viable alveolar macrophages from small laboratory animals. *J. Immunol. Methods* **27,** 189.

Paulnock, D. M. (1992). Macrophage activation by T cells. *Curr. Opin. Immunol.* **4,** 344.

Stein, M., Keshav, S., Harris, N., and Gordon, S. (1992). Interleukin 4 potently enhances murine macrophage mannose receptor activity: Marker of alternative immunologic macrophage activation. *J. Exp. Med.* **176,** 287.

Stout, R. D. (1993). Macrophage activation by T cells: Cognate and non-cognate signals. *Curr. Opin. Immunol.* **5,** 398.

Sung, S.-S. J., Nelson, R. S., and Silverstein, S. C. (1983). Yeast mannans inhibit binding and phagocytosis of zymosan by mouse peritoneal macrophages. *J. Cell Biol.* **96,** 160.

Tilney, L. G., DeRosier, D. J., and Tilney, M. S. (1992). How Listeria exploits host cell actin to form its own cytoskeleton. I. Formation of a tail and how the tail might be involved in movement. *J. Cell Biol.* **118,** 71.

CHAPTER 8

In Vitro Assays of Phagocytic Function of Human Peripheral Blood Leukocytes: Receptor Modulation and Signal Transduction

Eric J. Brown

Departments of Medicine, Molecular Microbiology, and Cell Biology and Physiology
Washington University School of Medicine
St. Louis, Missouri 63110

I. Introduction
II. Phagocytosis Assays
    A. Phagocytes
    B. Opsonized Particles
    C. Ingestion Assays
III. Phagocytic Receptors
IV. Stimulation of Phagocytosis
V. Phagosome Isolation
    References

I. Introduction

Ingestion and destruction of microorganisms are the major mechanisms of host defense against bacterial infection. This essential function is largely performed by the "professional phagocytes:" the bone marrow-derived leukocytes called neutrophils (polymorphonuclear leukocytes, PMN), monocytes, and macrophages. Twenty years ago the classic studies of Griffin and Silverstein showed that phagocytosis required sequential interaction of phagocyte receptors with the microbe, and that the cytoskeletal rearrangements that led to successful ingestion were limited to a small region around the recently ligated receptor (Griffin *et al.*, 1975, 1976; Griffin and Silverstein, 1974). It has been long established that phagocytosis also requires cytoskeletal rearrangements induced by phagocyte recognition of the invading microbe (Greenberg and Silverstein, 1993). These findings implied that phagocytosis required repeated receptor-initiated signal transduction. Unlike mitogenic signals, which travel through the cell from receptor to nucleus, signal transduction for phagocytosis was local, apparently confined to the plasma membrane and cytoskeleton in the immediate region of the occupied receptor. However, the biochemical nature of the signals involved remained, and to a large extent remain, unknown. The 20 years since Silverstein's experiments have seen an explosion in understanding of the nature and structure of phagocyte receptors and of the biochemistry of phagocyte signal transduction (see Greenberg and Silverstein, 1993, for a recent review). These new developments have led a number of investigators to reexamine the nature of the signals generated during receptor ligation that leads to successful phagocytosis. This chapter will describe techniques that have been used to enhance understanding of the phagocytic process, especially the nature of the receptors involved and the signal transduction events relevant to phagocytosis.

II. Phagocytosis Assays

There are a number of phagocytosis assays currently used by investigators that are really not in principle different from those used 20 years ago. Studies with human cells rely on recent techniques for cell purification that leave the cells as close as possible to the resting, unactivated state in which they exist in the circulation. This allows study of the process of activation of these cells to their full inflammatory potential as must occur *in vivo* as the phagocytes emigrate from the blood to sites of infection. The protocols for the preparation of assay components follow.

A. Phagocytes

1. Neutrophils (PMN)

PMN are freshly purified from human peripheral blood. Although a two-step purification of neutrophils involving dextran sedimentation and centrifugation through a single density of a mixture Ficoll and Hypaque has traditionally been used (Boyum, 1968a, b),we have recently switched to a step gradient purification using two densities of Histopaque (Zhou and Brown, 1993). This procedure separates PMN from erythrocytes better than the traditional method, although the yield is somewhat decreased. This abrogates the need to remove erythrocytes from the PMN by hypotonic lysis. Hypotonic lysis activates the PMN respiratory burst, induces fusion of some cytoplasmic vessicles with the plasma membrane, with consequent enhanced expression of many receptors including CD11b/CD18, and activates PMN adhesion. Thus, PMN purified with the step gradient of Histopaque are less activated than cells that have been exposed to hypotonic media to lyse erythrocytes. Our PMN purification scheme is shown in Protocol 1.

Protocol 1

1. To a 15-ml conical tube add

   0.2 ml heparin (1000 unit/ml)

   2 ml 6% of dextran T500 (Pharmacia Cat. No. 17-0320-01) in PBS

   10 ml of fresh whole blood

Mix well and let set at RT for 40–60 min, until almost all the erythrocytes have settled into the pellet at the bottom of the tube. The plasma layer contains platelets and leukocytes.

2. To fresh 15-ml tubes, add 3 ml of Histopaque-1119 (Sigma) and carefully overlay 3 ml of Histopaque-1077.

3. Take the plasma layer from the dextran sedimentation and layer it on top of above step gradient.

4. Centrifuge at 700 rpm (~80 × g) at RT for 30 min with brake off. PMN should be seen as a separate opaque layer between medium 1077 and 1119. Mononuclear cells should be seen between medium 1077 and plasma.

5. Carefully aspirate the PMN from the 1.077/1.119 interface with Pasteur pipette. Add 10 ml. of Hanks' balanced salts solution (without Ca^{2+}, Mg^{2+}, or phenol red), containing 4.3% $NaHCO_3$ and 20 mM HEPES (HBSS), and centrifuge at 1500 rpm for 10 min at room temperature. Wash twice more with HBSS and resuspend cells in HBSS containing 1% human serum albumin (HSA) at 2×10^6/ml.

We have found that HSA or other protein is necessary to preserve PMN function for even a few hours *in vitro*. Cells are used within 4 hr after purification for phagocytosis assays.

2. Monocytes

Human peripheral blood monocytes are purified by counterflow elutriation from leukophoresis packs (Lionetti *et al.*, 1980). We start with approximately 5×10^9 cells obtained from our blood bank as a byproduct of preparation of platelet packs for therapeutic use. The leukophoresed cells are largely mononuclear cells, but a further removal of platelets and contaminating granulocytes is obtained before elutriation. Typically, we get $1.5–2 \times 10^8$ cells from a single elutriation run, of which >90% stain with nonspecific esterase and express CD14. Our elutriation protocol is shown in Protocol 2.

Protocol 2

1. 10 ml of leukophoresed cells are combined with 25 ml of HBSS containing 10 mM EDTA in a 50-ml centrifuge tube. Ten millimeters of sterile Ficoll–Hypaque (Pharmacia) are pipetted under the mixture, and the tube is then centrifuged at 600 × g for 15 min at room temperature. The interface (containing monocytes, lymphocytes, and some platelets) is harvested, washed once in the HBSS/EDTA buffer. Cells are collected by centrifugation at 250 × g for 5 min at 4°C, then washed again and collected by centrifugation at 4°C for 10 min at 55 × g. Cells are resuspended in elutriation buffer (0.104 M NaCl; 13 mM PO$_4^{4-}$, pH 7.4; 1 mM MgSO$_4$; 5 mM KCl; 0.2% dextrose; 0.05% HSA) at 2 x 10^8/ml.

2. We use a standard Beckman elutriation chamber and a JE-6B elutriation rotor in a J2-21 Beckman centrifuge. Centrifuge speed is 2500 rpm (875 × g) and buffer flow rate is 21 ml/min; 10^9 cells in elutriation buffer are injected into the chamber and allowed to separate under these centrifuge and flow conditions for 20 min at room temperature. Platelets and lymphocytes flow out of the chamber, while the majority of monocytes are retained. After 20 min, purified monocytes are harvested by shutting off the centrifuge and collecting the expelled cells in a volume of ~40 ml.

Monocyte purification using a Sanderson chamber is also satisfactory, but requires different flow rates and centrifuge speeds. Monocytes may be stored in elutriation buffer overnight on ice without affecting phagocytic function.

3. Monocyte-Derived Macrophages

Normal human tissue macrophages are difficult to obtain, so we have frequently used as macrophages monocytes that have been allowed to mature for 5–7 days *in vitro*. These cells enlarge, lose myeloperoxidase, become highly adherent, and express several differentiation markers that distinguish macrophages from monocytes (Gresham *et al.,* 1989; Siffert *et al.,* 1993; Bohnsack *et al.,* 1985, 1986). These are called culture-derived macrophages, to distinguish them from cells which have matured to macrophages *in vivo*. Although all macrophages arise from circulating monocytes, these *in vitro* differentiated cells are sometimes called monocyte-derived macrohages as well.

There are three ways to obtain culture-derived macrophages from elutriated monocytes. One way is to allow monocytes to adhere to glass or to tissue culture dishes and then culture them in the adherent state. To do this, monocytes are suspended at 2.5 × 10^6/ml in RPMI containing 1% human AB serum. They are plated into the appropriate dish. We use 8 ml for a 100-mm tissue culture dish and 100 μl for a well of an eight-well LabTek slide. The cells are incubated at 37°C in an incubator with 5% CO$_2$ and a humidified atmosphere. After 1 hr, the adherent monocytes are washed gently with 37°C RPMI, and then the original volume of RPMI, now containing 15% fetal calf serum, 10 mM HEPES, and antibiotics, is added, and the cells are returned to the incubator. Ten percent human AB serum or donor serum may be substituted for the fetal calf serum. We have found no difference in phagocytosis between cells cultured with these two serum sources. There is no need to change media or feed the cells during the 5 to 7-day differentiation period. The recovery of these cells after 1 week in culture is excellent (generally >75%), and the cells are well differentiated, large, and adherent. The disadvantage of this technique for *in vitro* differentiation of monocytes to macrophages is that the cells cannot easily be removed from the adherent surface. Therefore, all assays must be done in the dishes or wells in which the cells are cultured.

A second method for monocyte differentiation involves culturing them in a way that prevents adhesion. These cells may then be transferred to any surface or vessel of the investigator's choice for phagocytosis assays or for biochemical measurements. Elutriated monocytes are suspended in RPMI-1640 containing 15% fetal calf serum or 10% human AB serum at 2×10^6/ml and 25 ml is added to Teflon beakers. The cells are then transferred to the CO_2 incubator and cultured at 37°C for the required 5–7 days. These cells are smaller than those cultured adherent and are less highly phagocytic. A major disadvantage is that for unclear reasons, the recovery of cells after the incubation period is quite low, often <40% of input cells.

A third method involves culturing cells on bacteriologic plastic. For this cells are plated exactly as described above for tissue culture plastic. Monocytes will adhere to the bacteriologic plastic sufficiently so that the low serum medium can be removed and replaced with high serum medium, as described above. However, unlike the case with tissue culture plastic or glass, the monocyte-derived MΦ can be released from bacteriologic plastic. To do this, we replace the medium with PBS containing 5 mM EDTA, pH 7.4, and place the dishes at 4°C for 2 hr. Following this incubation, cells release from the dish with vigorous pipetting. Seventy to ninety-five percent of the originally plated monocytes can be obtained in this fashion. The released cells may be studed in suspension or will readhere to a wide variety of surfaces when incubated in divalent cation-containing buffers.

B. Opsonized Particles

1. Opsonization of Sheep Erythrocytes and Glass or Latex Beads

The classic phagocytosis target is the opsonized sheep erythrocyte, coated with IgG antibody or complement. The reason this target is so commonly used is that it is quite simple to distinguish intracellular (ingested) particles from extracellular (bound) particles by lysing extracellular erythrocytes (E) with hypotonic buffer. Furthermore, sheep E are only 4 μm in diameter, which makes them easier to ingest than the somewhat larger E of other species. References for IgG and complement opsonization of E abound (see, e.g., Healy *et al.*, 1992; Fallman *et al.*, 1989; Ehlenberger and Nussenzweig, 1977; Pommier *et al.,* 1983). We use anti-E IgG and IgM from Diamedix (Miami, FL) and E which are ≤2 weeks old. To prepare C3b-coated E (EC3b), we first absorb the commercial anti-E IgM with protein A–agarose or fixed *Staphylococcus aureus,* since some IgM preparations contain traces of IgG that are sufficient to cause phagocytosis of complement-coated erythrocytes, even though not sufficient to cause phagocytosis without complement (Ehlenberger and Nussenzweig, 1977). C1, C4, C2, and C3 are added exactly as previously described (Pommier *et al.,* 1983, 1984; Graham *et al.,* 1989) in isotonic, low-ionic-strength buffer containing 0.15 mM Ca^{2+}, 1 mM Mg^{2+}, and 0.1% gelatin (Brown *et al.,* 1982). Following opsonization, EC3b are washed and stored in HBSS at 1.5 × 10^8/ml.

To convert the C3b on EC3b to iC3b (Brown, 1991), we use serum as a source of Factors H and I. EC3b are washed and resuspended at 1.5 × 10^8 in EDTA–DVBS, pH 6 (Gaither *et al.,* 1979). This buffer is 3.75% dextrose, 0.0225 M NaCl, 1 mM veronal, 10 mM EDTA, 0.1% gelatin, pH 6. Cells are then incubated with an equal volume of 1% human serum as a source of Factors H and I in the same buffer at 37°C for 2 hr. The resulting erythrocytes have primarily iC3b on their surfaces (Brown *et al.,* 1982).

For some studies, it is preferable to use other inert particles for phagocytosis. E may interfere in an unpredictable way with measurements of phospholipids and fatty acids. In addition, isolating phagosomes containing E is more difficult than for phagosomes containing target particles of density or composition markedly different from that of phagocyte membranes. For such experiments we have used opsonized glass or latex beads (Lennartz and Brown, 1991; Lennartz *et al.,* 1993; Zheleznyak and Brown, 1992). Protocol 3 details opsonization of glass and latex beads:

Protocol 3

1. 2 to 3 μm-diameter glass beads (Duke Scientific, Palo Alto, CA) are washed three times with PBS and coated with poly-L-lysine (50 μg/ml) in PBS for 30 min at room temperature.

2. Following washing × 2 in PBS, the beads are resuspended in 2.5% glutaraldehyde for 15 min to activate the lysines. Excess glutaraldehyde is removed by washing and the cells are incubated with the coating ligand overnight. We have used fibronectin (50 μg/ml), vitronectin (10 μg/ml), and BSA (20 mg/ml) successfully. To prepare IgG-coated beads, BSA-coated glass beads are incubated with IgG anti-BSA (Sigma) for 1 hr at room temperature.

2. Fluorescent Labeling of Ligand–Coated Particles

It is more difficult to distinguish bound from phagocytosed glass or latex particles than sheep erythrocytes, since hypotonic lysis does not remove bound but uningested glass or latex. Protocol 4 details the use of fluorescent ligands on these particles.

Protocol 4
Proteins

1. 1 part dry FITC is mixed with 10 parts dry celite (diatomaceous earth).

2. BSA (10 mg/ml in PBS) is brought to alkaline pH by addition of 0.1 vol of 5% $NaHCO_3$. The final pH should be ~9, and is checked with pH paper.

3. Add 2 mg of FITC/celite mixture per 10 mg of BSA and shake for 5 min at room temperature in the dark. Centrifuge at 1000 rpm to remove celite.

4. Separate derivitized protein from free FITC by chromatography on Sephadex G25. A very similar protocol is used with Lucifer yellow, using the vinyl sulfone (VS) derivative (Bailey *et al.,* 1983).

Microorganisms (We have used this protocol with *S. pneumoniae* and *Mycobacterium avium.*)

1. Bacteria are suspended at $10^8–10^9$/ml in 0.1 *M* $NaHCO_3$, pH 9.5, (Mycobacteria are used at 10^7/ml), containing 1 mg of Lucifer yellow–VS(Sigma). Incubation is continued for 2 hr, with occasional mixing.

2. Bacteria are washed $3\times$ with HBSS containing 1% BSA.

They can be directly labeled with fluorescein or Lucifer yellow. The fluorescence of extracellular particles may be quenched with trypan blue or gentian violet, leaving only intracellular particles fluorescent. We have used two different fluorescent dyes for this purpose, fluorescein and Lucifer yellow. The advantage of Lucifer yellow is that it has a pH-insensitive quantum yield (Stewart, 1978, 1981), so that fluorescence is independent of the pH of the intracellular compartment into which the organism is ingested, and quantitative studies on uptake may be automated. However, some fluorescence plate readers and fluorescence flow cytometers are not optimized for reading Lucifer yellow fluorescence, so it may be difficult to automate assays using this dye. Furthermore, quenching green fluorescence (either fluorescein or Lucifer yellow) with trypan blue or gentian violet causes a marked increase in autofluorescence at red wavelengths, so it is hard to use the quenching technique in combination with counterstaining with rhodamine, Texas red, or phycoerythrin-labeled antibodies. Microorganisms are frequently used, convenient, and biologically relevant phagocytic targets, which can also be labeled with Lucifer yellow VS or FITC (Protocol 4).

C. Ingestion Assays

The process of ingestion occurs when appropriately opsonized targets are incubated with phagocytes. The appropriate assay depends on whether the phagocytic target is an erythrocyte or a fluorescently tagged bead or microorganism. Moreover, phagocytosis can be performed by phagocytes adherent to glass, plastic, or biologically relevant matrices or by phagocytes in suspension.

Phagocytosis assays using opsonized E or fluorescent targets are detailed in Protocols 5 and 6.

Protocol 5: Phagocytosis of E
Phagocytes in suspension

1. Mix 2×10^5 phagocytes with 7.5×10^6 opsonized erythrocytes in a volume of 75 μl in a 12×75 mm sterile culture tube in phagocytosis buffer (HBSS containing 2 mM Ca^{2+}, 1.5 mM Mg^{2+}, 1% HSA).

2. Incubate at 37°C for 30 min without CO_2.

3. Transfer tube to ice bath. At this point an aliquot can be removed to determine the number of E bound to the phagocytes, if desired. Lyse extracellular E by adding 1.5 ml of 0.83% NH_4Cl for about 1 min.

4. Centrifuge at 2000 rpm for 5 min. Decant supernatant; add 50 μl of PBS and resuspend cells. We have found it easiest to count ingested E by staining with Giemsa. To do this, the resuspended cell pellet is incubated on an adhesion slide (MM Development Corp., Ontario, Canada) for 10 min. The supernatant is removed and 1% glutaraldehyde in PBS is added for 5 min to fix the cells. Thereafter, cells are stained with Giemsa stain and air-dried. Phagocytosis is determined by microscopy using $800\times$ magnification. We routinely count the percentage of phagocytes ingesting E and the number of E ingested per phagocyte. It is most useful to report these data as a phagocytic index, the number of E ingested/100 phagocytes.

Adherent phagocytes

1. If phagocytes have not been cultured in the assay chamber, 10^5 phagocytes are allowed to adhere to each well of an 8-well LabTek chamber (Nunc) by incubation in 300 μl of phagocytosis buffer for 30 min at 37°C. The surface of the chamber can be precoated with antibodies or other molecules of interest, as desired. Adherent cells are washed gently, and 7.5×10^6 opsonized E is added as above. To enhance interaction between phagocyte and target, the LabTek chambers are centrifuged at $\sim 50 \times g$ for 2 min. The assay then proceeds at 37°C as above.

2. The assay is stopped by removal of the LabTek chamber to ice. Unattached E are removed by gentle washing with phagocytosis buffer. E are lysed, cells and fixed and stained, and phagocytosis is scored exactly as above.

Protocol 6: Fluorescence assay for phagocytosis

1. Add fluorescent beads or bacteria to phagocytes as described for erythrocytes in Protocol 5. Incubate at 37°C. If phagocytes are adherent, wash gently with HBSS with 1% BSA to remove unbound targets. Score phagocytes with adherent/ingested targets by determining the number of fluorescent particles

associated with the phagocytes by fluorescence microscopy, using fluorescein filters.

2. Determine the number of ingested particles by adding crystal violet to a final concentration of 0.5% and recounting phagocyte-associated fluorescence.

This assay has been adapted to quantitation on a flow cytometer (Andoh et al., 1991; Sveum et al., 1986). Our procedure is as follows:

1. The flow cytometer (EPICS, Coulter Diagnostics) is set to measure forward and side scatter as well as fluorescein fluorescence (FL1). Forward and side scatter are used to set the gates on live, unaggregated phagocytes. The analysis is set to run for 10,000 events in the gated region. Populations of each cell type in the absence of beads are analyzed, and dead cells and cell aggregates are gated out of fluorescence measurements. The endogenous fluorescence of each cell population is determined, and an analysis region is placed outside the upper 95% interval of cellular fluorescence. When fluorescent beads are added, this permits the analysis of only those cells that have increased their fluorescence due to the binding of fluorescent ligands.

2. A 0.1% solution of trypan blue (Sigma) in assay buffer is then added at a 1:1 v:v ratio and the cells are read again to determine fluorescence under quenching conditions and a second analysis region is set for quenched cells.

3. Next, each preparation of fluoresceinated beads or bacteria is analyzed to determine the fluorescence of a single bead. This readout of FL1 should be a single, sharp peak; otherwise aggregates are present that may yield skewed results. The mean fluorescence of each population is determined for later calculations.

4. Samples for determination of adhesion and phagocytosis are run sequentially with all set gates and analysis regions in place. Equal volumes of trypan blue quenching solution are added to all samples and after a 5-min room temperature incubation the samples are run again.

5. Data collected from each run (i) are the mean fluorescence of the analysis region; (ii) the mean fluorescence of a single bead (from step 3); and (iii) the number of events (out of 10,000) that fall in the analysis region. The phagocytic index is calculated from the results of trypan blue quenched samples by multiplying the mean fluorescence of the analysis region (i) by the number of events in the region (iii), dividing the product by the fluorescence of a single bead (ii) to yield the total number of targets associated with all the cells and is finally multiplied by 0.01 to yield the attachment index (AI) (the number of beads attached to 100 cells) for fluorescence in the absence of trypan blue. Fluorescence in the presence of trypan blue, using the appropriate quenched fluorescence parameters from steps 2 and 3 yields the phagocytic index (PI).

The use of fluorescence quenching agents to distinguish ingested from attached phagocytic targets has become quite popular (Drevets and Campbell, 1991; Andoh et al., 1991; Hed and Stendahl, 1982).

III. Phagocytic Receptors

One of the most fundamental questions in studying the phagocytosis of any target is the nature of the initial interactions with the phagocytic cell. Methods commonly used to understand these interactions are (i) use of monoclonal antibodies to inhibit receptor–ligand interaction; (ii) use of monoclonal antibodies or ligands to deplete the apical cell surface of potential receptors; (iii) purification of receptors on ligand affinity columns; (iv) reconstitution of binding by transfecting receptor into a cell without constitutive binding ability. The last two options presuppose a detailed knowledge of receptor and ligand that often does not exist during the initial phases of examination of phagocytosis. Thus, the first two techniques are frequently used in initial screening. All phagocytes express receptors for the Fc domain of IgG (Rosales and Brown, 1993). Possible interaction of the Fc domain of monoclonal antibodies with these receptors can complicate interpretation of the effects of these monoclonals on phagocytic function. Frequently, isotype-matched, nonbinding monoclonals are used as controls for this possible effect. However, this is not an adequate control, because monoclonals that interact with the cell surface of the phagocyte can interfere with Fc receptor function much more readily than nonbinding antibodies (Kurlander, 1983). Therefore, we routinely use Fab or F(ab')$_2$ of potential antireceptor antibodies in these assays. As discussed below, this is not necessary when antibodies are attached to surfaces by protein A.

Using antibodies in solution to inhibit ligand binding and phagocytosis is the most direct way to determine a potential role for a receptor in phagocytic function. Generally, we preincubate effector cells with the monoclonals for 10–15 min prior to addition of the phagocytic targets, and we do not wash out the antibody before addition of the targets. These steps are necessary because particulate targets for phagocytosis are multivalent and their avidity for cells is very high. Thus, preincubation with antibodies and maintainance of a high concentration of antibody through the assay is necessary to achieve effective competition with the phagocytic target. Of course, other molecules in solution, such as peptides, proteins, and carbohydrates may be used in place of antibodies in these experiments to determine their effects on phagocytosis.

There are two distinct phases of ingestion at which a potential inhibitor may affect ingestion. The inhibitor may block initial interaction of the phagocytic target with its receptor(s) and thus interfere with binding of the phagocytic target. Alternatively, the inhibitor may not prevent binding, but may interfere with ingestion of the particle. There are several examples of cooperation of plasma membrane receptors for ingestion after the step of recognition and binding of the particulate target by the phagocyte (Wickham *et al.*, 1993; Graham *et al.*, 1989; Gresham *et al.*, 1991). To determine at which step a particular inhibitor acts, binding of the targets is compared directly to ingestion in the presence and absence of the inhibitor. This can be done for erythrocytes simply by not adding the lysis step; for flourescent particles by not quenching fluorescence.

Occasionally, antibodies can inhibit phagocytosis not because of direct competition with opsonized targets for binding or for other molecules involved in phagocytosis, but because the antibody activates signal transduction pathways that are inhibitory to some step of the phagocytic process (Gresham *et al.*, 1991). Thus, antibody inhibition of binding or ingestion cannot be definitive evidence that the antibody recognizes the phagocytic receptor. Rather, these experiments should be considered a mechanism for identifying receptor candidates, which can then be confirmed or rejected by other, more definitive techniques, such as expression of the receptors from cDNAs in other cell types (Mosser *et al.*, 1992).

A shortcoming of this experimental method is that to inhibit binding requires an antibody that interferes with ligand recognition. Frequently, such antibodies are not available, or the functional activity of available antibodies is unknown. In that case, receptor modulation becomes the screening assay of choice. In this circumstance, cells are adhered to surfaces coated with specific antibodies. Receptors recognized by the antibodies are trapped on the adherent surface and therefore not available to interact with phagocytic targets on the apical cell surface. The protocols we use for this way of studying receptor function employ either direct binding of monoclonal antibodies to surfaces or indirect binding via protein A and are detailed in Protocol 7.

Protocol 7: Protocols for receptor modulation
Poly-L-lysine to adhere antibody to dishes

1. 8-well glass or plastic LabTek slides (Nunc, Inc., Naperville, IL) are incubated with 300 μl of poly-L-lysine (PLL, 70–150 kDa, Sigma No. P1274) at 100 μg/ml in PBS for 10 min at room temperature. PLL can be stored as a stock solution of 10 mg/ml at $-20°C$ for several months.

2. Following several washes with PBS, the PLL-coated wells are incubated with 1% glutaraldehyde in PBS for 30 min at room temperature. The wells are then again washed extensively with PBS.

3. Monoclonal antibody is added at 10 μg/ml (300 μl) and incubated with the glutaraldehyde-activated surfaces for 1–2 hr. Monoclonal antibodies may be used as intact IgG, Fab, or F(ab')$_2$. If intact IgG is used, Fc receptor ligation will certainly occur. This will lead, for example, to nonspecific inhibition of IgG-mediated phagocytosis by any monoclonal antibody attached to the plate. Alternatively, protein A (see below), specific receptor ligands (such as mannan), or other proteins may be added to the glutaraldehyde-activated surfaces at this point.

4. Following several further washes, residual glutaraldehyde sites are blocked with 1 mM glycine or 2 mg/ml casein in PBS by incubation for 1 hr. Antibody-coated Lab-Tek chambers may be stored at 4°C for 1 week in PBS or HBSS with antibiotics or azide. If azide is used, the wells must be thoroughly washed before cells are added.

Protein A - dependent adhesion of monoclonal antibodies may be used as an alternative to direct binding of antibody to glutaraldehyde-activated wells.

1. We incubate glutaraldehyde-activated wells (prepared as in steps 1 and 2) with 100 μg/ml protein A, using the protocol described above.

2. After blocking residual glutaraldehyde sites with with glycine or casein, monoclonal antibodies are added at 20 μg/ml for ≥4 hr. At 20 μg/ml, all IgG subclasses seem to bind equivalently to the coated wells, no matter what their intrinsic ability to bind protein A in solution (Zhou and Brown, 1993). Since the IgG Fc is blocked by protein A, we have found no significant inhibition of Fc receptor function on cells adherent to these antibody-coated surfaces when antibodies directed at antigens other than Fc receptor are used (M. J. Zhou, and E. J. Brown, unpublished).

Coating surfaces with complement. We have used a variation of this protocol to coat glass surfaces with complement.

1. Human IgM is coated onto the glutaraldehyde-activated glass surfaces at 50 μg/ml.

2. Following blocking of residual glutaraldehyde sites with glycine or casein, whole human serum is added at 1:10–1:100 dilution in veronal-buffered saline with 0.1% gelatin.

3. Complement activation is initiated by incubating the wells at 37°C for 30 min.

4. Wells are washed 3× with buffer appropriate for incubation of cells.

Receptor modulation

1. Cells are added to antibody- or complement-coated wells for 1–4 hr at 37°C in RPMI-1640 medium containing 1% HSA. For receptor modulation experiments, PMN and monocytes are added at 5×10^4/well and monocyte-derived macrophages are added at 2×10^4/well. Maximal receptor modulation may take as long as 4 hr at 37°C.

As with direct inhibition by antibodies, there are several concerns that need to be addressed in interpretation of the results of receptor modulation experiments. First, monoclonal antibodies may not completely modulate receptor off of the apical surface of the cells. Experiments with monoclonal antibodies have demonstrated that 10–40% of the apical surface receptor may remain after maximal receptor modulation (Brown *et al.,* 1988; Michl *et al.,* 1979, 1983; Graham *et al.,* 1989). This remaining receptor may be equivalent to the "immobile fraction" of receptors seen in fluorescence photobleaching recovery experiments. Whether this residual receptor on the apical surface is sufficient to mediate function apparently depends on which functions are measured. A second concern in receptor modulation experiments is that receptors may comodulate. This has been shown for mannose/fucose receptors and Fcγ receptors on murine macrophages (Sung *et al.*, 1985). Finally, adherent ligands and antibodies, even more than those in the fluid phase, are able to activate signal transduction cascades in the adherent phagocytes. Thus, the ability of an antibody or a ligand to modulate a phagocytic function off the apical surface of a phagocyte does not necessarily imply direct interaction of the opsonized particle with the receptor recognized by the modulating ligand. Again, these experiments must be regarded as identifying receptor candidates, rather than directly identifying receptors involved in phagocytic function.

IV. Stimulation of Phagocytosis

Circulating monocytes and PMN are in a "resting" state. They are poorly phagocytic, not very adhesive, and minimally metabolic. Gene transcription is also minimal. At sites of inflammation, these cells change phenotype. They become much more phagocytic, highly adhesive, consume much more O_2, and generate highly reactive products of O_2 metabolism through the hexose monophosphate shunt pathway, including H_2O_2, superoxide, and HOCl. These toxic products are important in killing of pathogens and are also responsible for the nonspecific host tissue destruction that accompanies clearance of infection as well as idiopathic inflammation. The question of how circulating phagocytes achieve this activated phenotype at sites of inflammation is a problem in signal transduction. How do activators at sites of inflammation, such as bacterial chemotactic peptides, complement fragments, platelet products, cytokines, and extracellular matrix proteins, cause this phenotypic change from the resting to the activated state? Current evidence suggests that multiple signal transduction pathways can be involved, including phospholipases A, C, and D and both serine/threonine and tyrosine kinases (Greenberg and Silverstein, 1993). Exactly which cascades are involved most likely depends on the cell type, the receptor involved in the activation event, and the receptor involved in the effector event. For example, signal transduction pathways involving FcγRIII and its associated γ chain obviously cannot take place in peripheral blood monocytes that do not express FcγRIII or in PMN that express a phosphoinositol glycan-associated form of FcγRIII rather than the transmembrane form found in NK cells and macrophages. There are two methods we have found particularly useful in studying signal transduction from specific receptors for leukocyte activation. The first is to study the effects of adhesion to antireceptor antibodies. The second is to purify phagosomes and examine them for the presence of enzymes or substrates involved in signal transduction.

Signal transduction from adhesion to antireceptor antibodies: For these experiments, we generally examine generation of the respiratory burst, using the method of de la Harpe and Nathan (1985), as modified by Berton *et al.* (1992). Production of H_2O_2 in this assay appears to be as good an indicator of phagocyte activation as is stimulated ingestion. The method is simple and can be performed kinetically on a small numbers of cells (Protocol 8).

*Protocol 8: Respiratory burst production from
antireceptor monoclonal antibodies*

1. Wells of a 96-well flat-bottom ELISA plate are coated with monoclonal antibodies using the PLL–glutaraldehyde–Protein A technique described above. All antibodies are used at 20 μg/ml, which appears to lead to equivalent binding of all IgG suclasses (Zhou and Brown, 1993).

2. After antibody coating, wells are given a final incubation with 50 μl fetal calf serum, and then three washes with PBS. This additional step is important to block nonspecific H_2O_2 production by PMN or monocytes in contact with the ELISA wells.

3. 80 μl of an H_2O_2 detection mixture is added to each well. The mixture consists of 37.5 μM scopoletin, 1.25 mM NaN$_3$, and 1.25 U horseradish peroxidase in Krebs–Ringer phosphate glucose buffer (KRPG: 145 mM NaCl, 4.86 mM KCl, 1.22 mM MgSO$_4$, 5.7 mM Na$_2$HPO$_4$, 0.54 mM CaCl$_2$, 5.5. mM glucose, pH 7.4).

4. 5×10^4 PMN or monocytes are added to each well in 20 μl KRPG to initiate the reaction. The production of H_2O_2 in the wells is quantitated by determining the decrease in fluorescence using a fluorescence microplate reader. We use the Cytofluor 2300 (Millipore, Bedford, MA) with an excitation wavelength of 360 nm and an emission wavelength of 460 nm. H_2O_2 production is quantitated by comparison to a standard curve of H_2O_2 in the detection mixture (de la Harpe and Nathan, 1985). To determine the effect of pharmacologic inhibitors on generation of the respiratory burst, phagocytes are incubated with the inhibitors for an appropriate period of time before addition to the antibody-coated wells.

We have used a similar assay system to examine for colocalization of receptors and specific signal transduction enzymes as well as for tyrosine phosphorylation of specific targets on the adherent cell surface. Adherent cells can be solubilized in appropriate buffers for SDS–PAGE, immunoprecipitation, or immunofluorescence. We have found that a 5-min extraction of adherent phagocytes in 0.5% Triton X-100 in extraction buffer (10 mM Hepes, pH 7.2, 300 mM sucrose, 100 mM KCl, 3 mM MgCl$_2$ 10 mM EGTA, 1 mM PMSF, and 2 mM DFP)leaves the adherent cytoskeleton and associated proteins intact while removing granules and nucleii, for experiments directed specifically to molecules on the antibody-adherent membrane (Zhou and Brown, 1994). For these studies, we use 6- or 24-well rather than 96-well plates and 2.5×10^5–5×10^6 cells/well in order to enhance detection of specific proteins.

V. Phagosome Isolation

We have used the low density of opsonized latex beads as the basis for our phagosome isolation (Protocol 9).

Protocol 9: Isolation of phagosomes from human peripheral blood monocytes

1. Freshly isolated human peripheral blood monocytes ($0.5-2 \times 10^8$)are incubated at 4°C with a 20-fold excess of IgG- opsonized 3-μ-diameter latex beads in HBSS with 1 mM CaCl, 1 mM MgSO$_4$, 1% HSA containing 100 μg/ml each of leupeptin, pepstatin A, and aprotinin. We have found that inclusion of these protease inhibitors does not affect phagocytosis and does improve yields of several proteins (including intact protein kinase C) after phagosome isolation.

2. After incubation of the opsonized latex targets with the monocytes for 15 min, excess beads are removed by centrifugation at $55 \times g$ for 10 min at 4°C.

3. Phagocytosis is initiated by incubating the cell pellet resuspended in the same buffer at 37°C. The efficiency of both the adhesion and the ingestion steps is enhanced by rotation of the cell–target mixture during incubation.

4. Phagocytes are chilled on ice, pelleted by centrifugation as above, and resuspended in 10 ml of homogenization buffer (10 mM HEPES, pH 7.4, 5% sucrose, 1 mM DFP, leupeptin, aprotinin, and pepstatin as above). Cells are disrupted in a Dounce homogenizer, and disruption is confirmed by light microscopy.

5. The suspension is loaded onto a continuous 30 ml 16–44% sucrose gradient in HBSS and centrifuged at 100,000xg for 1 hr at 4°C in an SW28 rotor. Phagosomes collect near the interface between the homogenization buffer and the 16% sucrose and are generally visible as an opaque band.

6. The phagosomes are collected as a 4-ml fraction, diluted to 20 ml in HBSS with protease inhibitors, and collected by centrifugation at 2500xg for 10 min. The pelleted phagosomes may then be resuspended in a buffer appropriate for further experiments.

Many other methods exist (for example, see Chapter 1, this volume). We have found that inclusion of protease inhibitors from very early in the isolation is essential for good recovery of protein kinase C activity (Zheleznyak and Brown, 1992), and we assume that this is the case as well for many other proteolytically sensitive molecules present in the phagosome. Because destruction of the ingested particle is an important aspect of phagocytosis, internalization via phagocytic receptors results in rapid fusion of phagosomes with lysosomes containing many hydrolases. Thus, these organelles are particularly rich in degradative enzymes, which must be inhibited for biochemical studies to be undertaken.

Acknowledgments

I thank the many members of my laboratory, who over the years have developed the protocols reported here. Ming-jie Zhou, Scott Blystone, Jeff Schorey, and Irene Graham were extremely helpful in the gathering and writing of these protocols. Work in the development of these protocols has been supported by NIH Grants GM38330, AI24674, AI33348, and AI35811.

References

Andoh, A., Fujiyama, Y., Kitoh, K., Hodohara, K., Bamba, T., and Hosoda, S.(1991). Flow cytometric assay for phagocytosis of human monocytes mediated via Fcgamma-receptors and complement receptor CR1 (CD35). *Cytometry* **12,** 677–686.

Bailey, M. P., Rocks, B. F., and Riley, C.(1983). Use of Lucifer yellow VS as a label in fluorescent immunoassays illustrated by the determination of albumin in serum. *Ann. Clin. Biochem.* **20,** 213–216.

Berton, G., Laudanna, C., Sorio, C., and Rossi, F.(1992). Generation of signals activating neutrophil functions by leukocyte integrins: LFA-1 and gp150/95, but not CR3, are able to stimulate the respiratory burst of human neutrophils. *J. Cell Biol.* **116,** 1007–1017.

Bohnsack, J. F., Kleinman, H., Takahashi, T., O'Shea, J. J., and Brown, E. J.(1985). Connective tissue proteins and phagocytic cell function: Laminin enhances complement and Fc- mediated phagocytosis by cultured human macrophages. *J. Exp. Med.* **161,** 912–923.

Bohnsack, J. F., Takahashi, T., and Brown, E. J.(1986). The cell binding domain of human fibronectin is necessary but inefficient for enhancement of CR1 mediated phagocytosis by human monocyte-derived macrophages. *J. Immunol.* **136,** 3793–3798.

Boyum, A.(1968a). Isolation of leucocytes from human blood. A two-phase system for removal of red cells with methylcellulose as erythrocyte-aggregating agent. *Scand. J. Clin. Lab. Invest., Suppl.* **97,** 9–29.

Boyum, A.(1968b). Isolation of mononuclear cells and granulocytes from human blood. Isolation of monuclear cells by one centrifugation, and of granulocytes by combining centrifugation and sedimentation at 1 g. *Scand. J. Clin. Lab. Invest., Suppl.* **97,** 77–89.

Brown, E. J.(1991). Complement receptors and phagocytosis. *Curr. Opin. Immunol.* **3,** 76–82.

Brown, E. J., Gaither, T. A., Hammer, C. H., Hosea, S. W., and Frank, M. M.(1982). The use of conglutinin in a quantitative assay for the presence of bound C3bi and evidence that a single molecule of C3bi is capable of binding conglutinin. *J. Immunol.* **128,** 860–865.

Brown, E. J., Bohnsack, J. F., and Gresham, H. D.(1988). Mechanism of inhibition of immunoglobulin G- mediated phagocytosis by monoclonal antibodies that recognize the Mac-1 antigen. *J. Clin. Invest.* **81,** 365–375.

de la Harpe, J., and Nathan, C. F.(1985). A semi-automated micro-assay for H2O2 release by human blood monocytes and mouse peritoneal macrophages. *J. Immunol. Methods* **78,** 323–336.

Drevets, D. A., and Campbell, P. A.(1991). Macrophage phagocytosis: Use of fluorescence microscopy to distinguish between extracellular and intracellular bacteria. *J. Immunol. Methods* **142,** 31–38.

Ehlenberger, A. G., and Nussenzweig, V.(1977). The role of membrane receptors for C3b and C3d in phagocytosis. *J. Exp. Med.* **145,** 357–371.

Fallman, M., Lew, D. P., Stendahl, O., and Andersson, T.(1989). Receptor-mediated phagocytosis in human neutrophils is associated with increased formation of inositol phosphates and diacylglycerol: Elevation in cytosolic free calcium and formation of inositol phosphates can be dissociated from accumulation of diacylglycerol. *J. Clin. Invest.* **84,** 886–891.

Gaither, T. A., Hammer, C. H., and Frank, M. M.(1979). Studies of the molecular mechanisms of C3b inactivation and a simplified assay of beta 1H and the C3b inactivator (C3bINA). *J. Immunol.* **123,** 1195–1204.

Graham, I. L., Gresham, H. D., and Brown, E. J.(1989). An immobile subset of plasma membrane CD11b/CD18 (Mac-1) is involved in phagocytosis of targets recognized by multiple receptors. *J. Immunol.* **142,** 2352–2358.

Greenberg, S., and Silverstein, S. C. (1993). Phagocytosis. *In* "Fundamental Immunology" (W. E. Paul, ed.), pp. 941–964. Raven Press, New York.

Gresham, H. D., Goodwin, J. L., Anderson, D. C., and Brown, E. J.(1989). A novel member of the integrin receptor family mediates Arg-Gly-Asp-stimulated neutrophil phagocytosis. *J. Cell Biol.* **108,** 1935–1943.

Gresham, H. D., Graham, I. L., Anderson, D. C., and Brown, E. J.(1991). Leukocyte adhesion deficient (LAD) neutrophils fail to amplify phagocytic function in response to stimulation: Evidence for CD11b/CD18-dependent and -independent mechanisms of phagocytosis. *J. Clin. Invest.* **88,** 588–597.

Griffin, F. M., Jr., and Silverstein, S. C.(1974). Segmental response of the macrophage plasma membrane to a phagocytic stimulus. *J. Exp. Med.* **139,** 323–336.

Griffin, F. M., Jr., Griffin, J. A., Leider, J. E., and Silverstein, S. C.(1975). Studies on the mechanism of phagocytosis. I. Requirements for circumferential attachment of particle-bound ligands to specific receptors on the macrophage plasma membrane. *J. Exp. Med.* **142,** 1263–1282.

Griffin, F. M., Jr., Griffin, J. A., and Silverstein, S. C.(1976). Studies on the mechanism of phagocytosis.II. The interaction of macrophages with anti-immunoglobulin IgG-coated bone marrow-derived-lymphocytes. *J. Exp. Med.* **144,** 788–809.

Healy, A. M., Mariethoz, E., Pizurki, L., and Polla, B. S.(1992). Heat shock proteins in cellular defense mechanisms and immunity. *Ann. N. Y. Acad. Sci.* **663,** 319–330.

Hed, J., and Stendahl, O.(1982). Differences in the ingestion mechanisms of IgG and C3b particles in phagocytosis by neutrophils. *Immunology* **45,** 727–736.

Kurlander, R. J.(1983). Blockade of Fc receptor-mediated binding to U-937 cells by murine monoclonal antibodies directed against a variety of surface antigens. *J. Immunol.* **131,** 140–147.

Lennartz, M. R., and Brown, E. J.(1991). Arachidonic acid is essential for Fc-receptor-mediated phagocytosis by human monocytes. *J. Immunol.* **147,** 621–626.

Lennartz, M. R., Lefkowith, J. B., Bromley, F. A., and Brown, E. J.(1993). Immunoglobulin G-mediated phagocytosis activates a calcium-independent, phosphatidylethanolamine-specific phospholipase. *J. Leukocyte Biol.* **54,** 389–398.

Lionetti, F. J., Hunt, S. M., and Valeri, C. R. (1980). Isolation of human blood phagocytes by counterflow centrifugation elutriation. *Methods Cell Sep.* **3,** 141–155.

Michl, J., Pieczonka, M. M., Unkeless, J. C., and Silverstein, S. C.(1979). Effects of immobilized immune complexes on Fc- and complement-receptor function in resident and thioglycollate-elicited mouse peritoneal macrophages. *J. Exp. Med.* **150,** 607–621.

Michl, J., Pieczonka, M. M., Unkeless, J. C., Bell, G. I., and Silverstein, S. C.(1983). Fc receptor modulation in mononuclear phagocytes maintained on immobilized immune complexes occurs by diffusion of the receptor molecule. *J. Exp. Med.* **157,** 2121–2139.

Mosser, D. M., Springer, T. A., and Diamond, M. S.(1992). Leishmania promastigotes require opsonic complement to bind to the human leukocyte integrin Mac-1 (CD11b/CD18). *J. Cell Biol.* **116,** 511–520.

Pommier, C. G., Inada, S., Fries, L. F., Takahashi, T., Frank, M. M., and Brown, E. J.(1983). Plasma fibronectin enhances phagocytosis of opsonized particles by human peripheral blood monocytes. *J. Exp. Med.* **157,** 1844–1854.

Pommier, C. G., O'Shea, J. J., Chused, T., Yancey, K., Frank, M. M., Takahashi, T., and Brown, E. J.(1984). Studies of the fibronectin receptors of human peripheral blood leukocytes: Morphologic and functional characterization. *J. Exp. Med.* **159,** 137–151.

Rosales, C., and Brown, E. J. (1993). Neutrophil receptors and modulation of the immune response. *In* ''The Neutrophil'' (J. S. Abramson and J. G. Wheeler, eds.), pp. 23–62. Oxford University Press, Oxford.

Siffert, J. C., Baldacini, O., Kuhry, J. G., Wachsmann, D., Benabdelmoumene, S., Faradji, A., Monteil, H., and Poindron, P.(1993). Effects of *Clostridium difficile* toxin B on human monocytes and macrophages: Possible relationship with cytoskeletal rearrangement. *Infect. Immun.* **61,** 1082–1090.

Stewart, W. W.(1978). Functional connections between cells as revealed by dye-coupling with a highly fluorescent naphthalimide tracer. *Cell (Cambridge, Mass.)* **14,** 741–759.

Stewart, W. W.(1981). Lucifer dyes—highly fluorescent dyes for biological tracing. *Nature (London)* **292,** 17–21.

Sung, S.-S. J., Nelson, R. S., and Silverstein, S. C.(1985). Mouse peritoneal macrophages plated on mannan- and horseradish peroxidase-coated substrates lose the ability to phagocytose by their Fc receptors. *J. Immunol.* **134**, 3712–3717.

Sveum, R. J., Chused, T. M., Frank, M. M., and Brown, E. J.(1986). A quantitative fluorescent method for measurement of bacterial adherence and phagocytosis. *J. Immunol. Methods* **90**, 257–264.

Wickham, T. J., Mathias, P., Cheresh, D. A., and Nemerow, G. R.(1993). Integrins $\alpha_v\beta_3$ and $\alpha_v\beta_5$ promote adenovirus internalization but not virus attachment. *Cell (Cambridge, Mass.)* **73**, 309–319.

Zheleznyak, A., and Brown, E. J.(1992). IgG-mediated phagocytosis by human monocytes requires protein kinase C activation: Evidence for protein kinase C translocation to phagosomes. *J. Biol. Chem.* **267**, 12042–12048.

Zhou, M.-J., and Brown, E. J.(1993). Leukocyte response integrin and integrin associated protein act as a signal transduction unit in generation of a phagocyte respiratory burst. *J. Exp. Med.* **178**, 1165–1174.

Zhou, M.-J., and Brown, E. J.(1994). CR3 (Mac-1, $\alpha_M\beta_2$, CD11b/CD18) and Fc(gamma)RIII cooperate in generation of a neutrophil respiratory burst: Requirement for Fc(gamma)RII and tyrosine phosphorylation. *J. Cell Biol.* **125**, 1407–1416.

CHAPTER 9

Bacterial Adhesion and Colonization Assays

Per Falk,★ Thomas Borén,† David Haslam,† and Michael Caparon†

★Department of Molecular Biology and Pharmacology
†Department of Molecular Microbiology
Washington University School of Medicine
St. Louis, Missouri, 63110

  I. Introduction
 II. *In Situ* Screening of Host Receptor Distribution
     A. Bacterial Labeling Procedure
     B. Tissue Section Overlay Assay
III. Bacterial Adherence to Cells in Culture
     A. Assay for Bacterial Adherence to Cultured Cells
 IV. Biochemical Characterization of the Molecular Nature of Receptors *in Situ*
     A. Periodate Oxidation
  V. Bacterial Inhibition Experiments *in Situ*
 VI. *In Vitro* Assays for Bacterial Adhesion
     A. Bacterial Binding to Immobilized Receptors in Solid Phase
     B. Bacterial Binding to Receptors in Solution
VII. Probing Eukaryotic Cell Glycoconjugates with Purified Bacterial Adhesins
     A. Cloning and Purification of an Adhesin as a Translational Fusion
VIII. Concluding Remarks
     References

I. Introduction

   Attachment is a prerequisite for bacterial colonization of epithelial cell surfaces. This initial host–microbial interaction is mediated by bacterial surface proteins called adhesins (Jones and Isaacson, 1983; Sharon, 1986) that bind to protein (Bliska *et al.*, 1993) or carbohydrate (Karlsson, 1989; Hultgren *et al.*,

1993) epitopes presented on the cell surfaces. Just like antibodies and lectins (Goldstein and Poretz, 1986), bacterial adhesins have very specific structural requirements for recognition and binding of eukaryotic cell surface epitopes (Strömberg *et al.*, 1991). Consequently, bacterial binding is restricted to a set of cell populations carrying the optimal receptor epitopes. This is referred to as tropism and partly determines the niche a bacterium is able to occupy, or the set of clinical symptoms a bacterial pathogen is characterized by. Given the variation in expression of cell surface proteins and carbohydrates known to exist between different species and tissues/organs, as well as among cells originating from the same cell lineage as a function of spatial distribution, developmental and differentiation stage (e.g., Falk *et al.*, 1994) bacteria are valuable probes for specific cell populations, and the distribution of these cell lineages within a selected tissue. This chapter reviews methods where bacteria and bacterial adhesins can be applied as tools for defining specific cell lineages and for charactererizing eukaryotic receptor epitopes, primarily based on our own experiences with *Helicobacter pylori*, *Streptococcus pyogenes*, *Escherichia coli*, and *Actinomyces* spp.

II. *In Situ* Screening of Host Receptor Distribution

In situ screening of histological sections has been used to reveal the cell specific distribution of bacterial receptors among different cell populations in target tissues using histochemical (Wyatt *et al.*, 1990) and immunohistochemical (Andersen *et al.*, 1988) stainings, *in situ* DNA hybridization (van den Berg *et al.*, 1989), or electron microscopy (Hessey *et al.*, 1990). Direct binding of fluorochrome conjugated *E. coli* to frozen sections of human kidney has also been described (Korhonen *et al.*, 1986). A powerful adaptation of this technique developed in our laboratories utilizes the bacteria themselves as probes to study the distribution of receptor epitopes in histological sections of potential target tissues. The advantage with this model is that it allows direct evaluation of the cellular distribution of receptor molecules and that it can be used to further characterize the receptor directly on the tissue by performing double labelings, biochemical modifications, and inhibition assays as outlined below. This approach has been used to characterize the binding patterns of uropathogenic *E. coli* to tissue sections of human kidney (Roberts *et al.*, 1994) and to identify functional tissue receptors for *H. pylori* (Falk *et al.*, 1993; Borén *et al.*, 1993), an organism considered to have an exclusive niche in the human gastric mucosa (Goodwin *et al.*, 1986), and for *S. pyogenes* causing infections in human cutaneous tissue (Okada *et al.*, 1994a).

A. Bacterial Labeling Procedure

Bacteria can be surface-labeled with fluorochromes such as fluorescein isothiocyanate or tetramethyl-rhodamine isothiocyanate (FITC or TRITC, Sigma, St. Louis, MO). In addition, marker molecules such as *N*-hydroxysuccinimide ester activated digoxigenin (Dig-NHS) or biotin (Biotin-x-NHS) (both from Boehringer Mannheim, Mannheim, Germany) can be used for labeling and subsequently detected with a wide range of conjugated antibodies, or, in the case of biotin, streptavidin. This approach has proven very useful, since the same batch of labeled bacteria can be applied in a number of different adhesion assays—for instance, binding *in situ* (e.g., histological sections and tissue culture) detected with fluorescence and binding *in vitro* (e.g., Western blots, and HPTLC immunostainings) detected with enzyme (peroxidase or alkaline phosphatase) or immunogold conjugated antibodies. Furthermore, labeled bacteria can be stored frozen, at $-20°C$ or $-70°C$, and the same batch can be used for extended periods of time (at least up to 6 months), thus circumventing problems with batch to batch variability (see Remarks below).

1. Freshly harvested bacteria are washed twice [5000 rpm ($3000 \times g$) in an Eppendorf table top centrifuge, for 3 min] in 1 ml of 0.2 M carbonate buffer (10% Na_2CO_3, 90% $NaHCO_3$), pH 9.2, and resuspended in 1 ml of the same buffer.

2. 10 μl of 10 mg/ml FITC/TRITC, Dig-NHS, or Biotin-x-NHS freshly prepared in DMSO is added to the bacterial suspension. Samples are uncubated for 1 hr at room temperature in the dark with occasional turning of the tube.

3. Bacteria are washed twice in 1 ml PBS, pH 7.6/0.05% Tween-20 and the bacterial suspension is diluted to a density of 1 OD_{600} and immediately frozen in 100-μl aliquots ($-20°C$), in the same buffer.

No differences in attachment patterns have been observed between strains used fresh or frozen once prior to use. However, this should be established for each bacterial strain that is to be studied.

Remarks

Since labeling of the bacterial surface could possibly modify the adherence characteristics either by interfering with the adhesin proteins themselves or by inducing artifactual binding properties, it may be important to confirm that the binding pattern observed for surface labeled bacteria is representative of the native organism. This could be performed by direct staining of adherent bacteria after binding has been allowed to occur using any one of several fluorochromes that are freely taken up by the bacteria and only become fluorescent following DNA intercalation (e.g., acridine orange or Hoescht dye) or through enzymatic conversion (e.g., 6-carboxyfluorescein diacetate).

For instance, acridine orange staining (Okada *et al.*, 1994a; Color Plate 2) is performed by immersing the slide with a section of tissue or cells containing bound bacteria in a solution of 10 μg/ml of acridine orange in 50 mM sodium acetate buffer, pH 4, for 10 min in the dark. The slide is briefly rinsed once in PBS and analyzed by microscopy as described below.

Detection of bound nonlabeled bacteria could also be performed by antibodies directed bacterial strain-specific surface antigens (e.g., Andersen *et al.*, 1988).

Adherence to red blood cells, as illustrated by hemagglutination, often reflects binding that occurs to glycoconjugates on epithelial cell surfaces. For instance, this trait is found in many bacteria that express sialic acid-specific (Korhonen *et al.*, 1984; Nyberg *et al.*, 1990) or digalactoside-specific (Leffler and Svanborg-Edén, 1980) lectins/adhesins. If hemagglutination correlates to the adhesive properties seen to target tissue, a simple way to asses whether surface labeling has affected binding characteristics is to analyze titers of hemagglutination using appropriate erythrocyte species. Hemagglutination assays also sometimes allow for evaluation of the fine-tuned receptor specificty of a specific bacterial adhesin by assaying the hemagglutination using erythrocytes from different species with well-defined cell surface glycoconjugate compositions (Strömberg *et al.*, 1990).

Growth conditions affect the expression of bacterial adhesins and have to be optimized for the specific binding property and standardized for reproducible binding results. For instance, environmental temperature regulates transcription of the digalactoside binding P-pili (Göransson *et al.*, 1990) and the fibronectin binding curli organelle (Olsén *et al.*, 1989) in *E. coli*. Furthermore, curli is preferentially expressed in stationary phase conditions (Olsén *et al.*, 1993a). The fucosylated blood group antigen binding property of *H. pylori* is in a similar way only expressed in stationary phase cultures, whereas sialic acid binding hemagglutinins are readily detected during log phase growth (Akopyanz *et al.*, 1994). Furthermore, the group A streptococci, *S. pyogenes,* regulates expression of its adhesins (M protein and protein F) in response to environmental concentrations of oxygen and carbon dioxide (Caparon *et al.*, 1992; VanHeyningen *et al.*, 1993), thus demonstrating two completely different receptor specificities and cellular binding patterns depending on the partial pressure of these two gases. (Okada *et al.*, 1994a).

B. Tissue Section Overlay Assay

1. Tissues are fixed in 10% formalin and paraffin embedded, cut in 5-μm-thick sections, and affixed to microscope glass slides, according to standard procedure (e.g., Luna, 1968). Series of consecutive sections should preferably be used for binding analyses to minimize variations in tissue quality that could affect adhesion patterns.

2. Histological sections are deparaffinized in 100% xylene or Hemo-De (Fisher Scientific Corp., Pittsburg, PA) for 10 min, followed by a 3-min separate rinse in the same solvent. After rehydration in isopropanol (3x5 min), sections are washed for 5 min in slowly running distilled water, followed by three washes in PBS, each for 5 min. The reconditioned sections are finally incubated in blocking buffer (PBS/0.5% BSA, 0.05% Tween-20) for 1 hr.

3. Labeled bacterial suspensions are diluted to 0.05 OD_{600} in blocking buffer and 200-μl aliquots applied to the prepared sections. In order to confine the bacterial suspension to the tissue, the glass around the tissue is wiped dry and the tissue is encircled with wax (PAP-Pen, Research Products International Corp., Mt. Prospect, IL). It is crucial that the tissue never dry during the overlay procedures.

For bacteria producing proteases, the following mixture of protease inhibitors have proven useful for optimal binding; 1 mM PMSF, 5 mM EDTA–Na$_2$, and 10 mM benzamidine-HCl (all from Boehringer Mannheim, Mannheim, Germany) (Borén et al., 1993).

4. Bacteria are overlayed onto tissue sections for 1 hr at room temperature, followed by six washes for 5 min each in PBS. Dig/biotin-labeled bacteria are detected with fluorochrome-conjugated (FITC/TRITC) antidigoxigenin or streptocidion/antibiotin (Boehringer Mannheim), diluted 1:100 in blocking buffer and incubated on the section for 1 hr at room temperature, followed by washing 3x5 min in PBS. The tissue is protected by a microscopy cover glass (VWR Scientific Corp., San Fransisco, CA) mounted on the section with PBS/ glycerol (1:1, v/v). The fluorescence is stable for at least a month at 4°C, in the dark.

5. Distribution of bacteria adherent to cells is assessed using fluorescence microscopy.

Remarks

The *in situ* adherence assay was primarily designed as a method for identifying potential bacterial receptors in host target tissues. However, by comparing the cellular distribution patterns of cell lineage-specific carbohydrate and protein markers using lectins and antibodies with the bacterial binding pattern *in situ*, it is possible to correlate bacterial binding to certain cell types as well as to a restricted set of defined epitopes. We have used this approach to establish the cell lineage-specific adhesion patterns of *H. pylori* to human gastric surface mucous cells (Falk et al., 1993), also expressing fucosylated antigens (Sakamoto et al., 1989) (Color Plate 3) and of protein F expressing group A *Streptococci* to Langerhan's cells (Okada et al., 1994a), identified with antibodies to the CD1 antigen (Fithian et al.,1981) in the epidermis of human skin/. Bacteria often have extremely selective receptor specificities and can be used as highly specific probes, like the Le^b-specific adhesin from *H. pylori* (Borén et al., 1993), the MCP protein-specific M protein from *S. pyogenes* (Okada et al., 1994b), and the digalactoside-specific isotypes of P-pilus adhesins from uropathogenic *E. coli* (Haslam et al., 1994). Interpretation of the adherence experiments should be made with caution. Bacterial adherence to cells or tissues *in situ* demonstrates the presence of potential receptor molecules. However, it does not necessarily indicate that the bacteria will have a natural habitat at this location *in vivo*, since a number of environmental factors (only one of which is the presence of proper attachment sites for colonization) taken together will make up the bacterial niche.

III. Bacterial Adherence to Cells in Culture

In addition to their use in probing tissues, bacteria can be also be employed as useful probes for the distribution of specific molecules on cultured cells. The techniques discussed above for the characterization of binding to fixed tissue sections can be directly applied to the analysis of bacterial adherence to cultured cells. In addition, since adherence to cultured cells can be conducted using live and unfixed cells, it can provide a valuable technique for confirming the observations made using fixed tissue in the *in situ* assays. Cells in culture can be used to define receptor presentation in the membrane. As an initial step for defining the presentation of the receptor, protease treatment of the cell surfaces can reveal the topographic localization of the receptor, i.e., whether the receptor structure resides in the membrane-close glycolipid layer or in the peripheral protein/glycoprotein layer (Strömberg and Borén, 1992). The ability to culture large numbers of cells and process multiple samples simultaneously also makes the use of cultured cells attractive. As closely as possible, the cultured cell chosen for analysis should resemble cells naturally colonized, and to which the organism adheres to in tissue as was revealed by the *in situ* assay. It may be possible to prepare and maintain cell populations from explants of normal tissue in primary culture. An example of this is epidermal tissue, which can readily be maintained in primary culture and has as an additional useful feature in the fact that techniques are available that allow the state of differentiation of the cells to be manipulated during culture (Fuchs, 1990). For tissues whose cells cannot be conveniently obtained in primary culture, it is often possible to identify a transformed cell line that is derived from either the tissue of interest or a related tissue. Careful consideration must be made to the type of cultured cell chosen for the assay. Once a suitable cell has been identified, the bacterial binding assays are performed and interpreted in much the same manner as that described for the *in situ* assays. However, when attempting to compare results obtained with cultured cells to host tissue, a number of important limitations should be considered. These include the fact that cultured cells may not represent all the different types of cells present in a particular tissue. Also, the state of differentiation that a given cell type can be maintained at in culture may not reflect the state that is present in the tissue. Cultured cells can frequently lose many of the traits that are characteristic of the cells *in vivo,* especially since most cell lines have been established from malignant tissue. Also, if the cell utilized has been transformed, it may be considerably altered in the repertoire of receptors that it will express in culture. Working with primary cell cultures will circumvent many of these draw-backs, since these cells should reflect the cells found in nondiseased tissues *in vivo* and in histological sections. However, these cells can also be phenotypically altered when they are released from their supportive submucosa and established in monocellular or aggregate form under tissue culture conditions. Thus, assays for distribution of receptors that utilize cultured cells are most powerful when conducted in parallel with an *in situ* assay.

A. Assay for Bacterial Adherence to Cultured Cells

1. Culture conditions will vary and will depend on the particular cell type employed. For adherent cells, plating on glass coverslips in a 24-well plate or

in a slide-based culture chamber from which the slide can be removed (e.g., Nunc Catalog No. 177402) will greatly facilitate subsequent manipulations.

2. Following culture, the cells are washed several times in a suitable maintenance medium (e.g., minimal essential medium, or Hanks' balanced salts) and are left to incubate in the maintenance medium for the remainder of the assay.

3. Bacteria are cultured under the conditions with which optimal adherence was obtained in the *in situ* assay. The bacteria are labeled at this point if a surface labeling procedure is to be used (see above for details of available methods).

4. The bacteria are then washed three times in the same maintenance medium used for incubation of the cells. The bacteria are resuspended in the maintenance medium at several different concentrations that will allow the initial assays to be conducted at multiplicities of infection of 100, 10, 1, and 0.1 bacteria per cell. These determinations are best made by a direct enumeration method, such as the use of a Petroff–Hauser chamber adapted for phase contrast microscopy, rather than by viability determinations. These initial assays should be used to determine the optimal concentration of bacteria for subsequent studies and should be in the range of concentrations where a two- or fourfold increase in concentration is reflected in a concomitant two- or fourfold increase in adherence to the target cells, i.e., in a linear interval.

5. The bacteria–cell mixture is then incubated at a temperature and atmosphere appropriate to maintain viability of the cells for 1 to 4 hr with gentle rotation. Incubation of phagocytic cells at 4°C will inhibit phagocytosis of the bacteria by the cells, which makes subsequent quantitation of bacterial adherence to the cells more accurate.

6. The cells are then washed to remove unbound bacteria, under as gentle conditions as possible. These are most easily determined using a bacterial mutant, strain, or species that is not expected to bind to the cells as a control. If a suitable control is not available, one of the following methods should be employed. For adherent cells, the coverslip or slide is removed from the culture chamber and held at one corner with forceps and gently immersed into a beaker that contains PBS. This procedure is continued for a total of 5 to 10 consecutive immersions. For nonadherent cell types, unbound bacteria are separated from cells by low-speed centrifugation (e.g., $500 \times g$ for 5 min.), which should pellet the cells, but not the unbound bacteria.

7. The cells are now fixed. For adherent cells, the coverslips or slides are immersed in methanol for 30 sec, then allowed to dry in air. The pellet obtained from the assay utilizing nonadherent cells is resuspended in PBS and an aliquot is placed onto the surface of a glass slide where it is allowed to dry in air. The slide is then immersed in methanol and subsequently treated in a manner identical to the adherent cell assay.

Remarks

Assessment of bacterial binding may be done in a number of ways, for instance quantitation of adherent isotope-labeled bacteria by scintillation. However, it is highly recommended that the initial assessment of adherence be conducted by microscopy, to ascertain that the bacteria are actually binding to the cells and are not binding to intercellular spaces, which may indicate that the bacteria are actually adherent to a molecule secreted by the cells or are binding to some component of the culture medium that has coated the cells and slides. In addition, microscopic inspection may reveal that the organisms are binding nonuniformly and are binding to specific area of the cells that may correlate with distribution of the cellular receptor. Since the assay has been conducted with live bacteria and cells, it is also important to determine that the bacteria are not cytotoxic and have not lysed the cells. If this proves to be the case, it may be possible to fix the cells prior to incubation with the bacteria. Visualization of bacterial adherence is conducted as described for the *in situ* assay and will depend on the method used to stain the bacteria. In addition, it is possible to stain the fixed preparations using common staining methods like the Gram stain if the bacterial species used is a Gram-positive organism or the Wright–Giemas stain if the bacterium is a Gram-positive or a Gram-negative organism. For either stain, the cells will appear orange and the bacteria deep blue. The Gram stain is not recommended for Gram-negative bacteria, as it will stain these organisms in a red color that can be difficult to distinguish from the cells. For details of these staining methods, consult any standard clinical microbiological manual (e.g., Ballows, 1991).

IV. Biochemical Characterization of the Molecular Nature of Receptors *in Situ*

As described above, microorganisms may be used to probe intact cells in tissue culture or tissue sections. Once it has been determined that a microorganism is a useful probe for cells in fixed tissues or cultured cells, it may be of interest to further characterize the nature of the cellular receptor structure. As an example, a certain bacteria might adhere only to cells within a tissue at a certain stage of differentiation. This would suggest that the eukaryotic ligand is developmentally regulated and the microbe may thus be a marker for committed but not terminally differentiated cells. The assays for biochemical characterization of receptor molecules are described below for histological tissue sections (Borén *et al.*, 1993; Falk *et al.*, 1993; Okada *et al.*, 1994a), but these methods have also been applied for the analysis of fixed cultured cells (Orlandi *et al.*, 1992; St. Geme, 1994).

A. Periodate Oxidation

The first step in characterization of the receptor is to determine whether the bacteria binds to protein or carbohydrate determinants on the host cell. In order to discriminate between protein- and carbohydrate-mediated binding, selective chemical hydrolysis of carbohydrate epitopes can be achieved by meta-periodate oxidation under mild acidic conditions (Woodward *et al.*, 1985). This treatment will cleave carbon–carbon linkages carrying vicinal hydroxyl groups, as is found in the ring structure of many sugar residues, while leaving peptides intact. Furthermore, by adjusting the oxidation conditions, it is possible to discriminate between sialylated and nonsialylated glycoconjugate receptors (Manzi *et al.*, 1990; Falk *et al.*, 1993).

1. Deparaffinized and rehydrated tissue sections (see above) are washed twice for 5 min in 0.1 M NaAc, pH 4.5, followed by incubation in 10 mM periodate, 0.1 M NaAc, pH 4.5, for 1 hr at room temperature, in the dark.

2. Section are washed once for 5 min in 0.1 M NaAc, pH 4.5, and twice in PBS, pH 7.5. The tissue is then reduced by immersion in 50 mM NaBH$_4$, for 30 min at room temperature, followed by 2x5-min washes in PBS.

3. Sections are then blocked and overlayed with bacteria as described above.

Remarks

To verify the selectivity for carbohydrates, a control experiment was performed using rat gastric mucosa (Falk *et al.*, 1993) (Color Plate 4). An antiserum to intrinsic factor (IF) was used to monitor the integrity of peptide antigens in the tissue, and blood group H-antigen-specific *Ulex europaeus* type 1 (UEAl) lectin was used as carbohydrate marker in the same section. The staining intensity of IF was retained, whereas UEAl staining was completely abolished by the periodate treatment.

By performing the periodate reaction under milder conditions (pH 5.5, 10 min, on ice), a selective cleavage of the unsubstituted side chain (carbons 7–9) of sialic acid can be achieved. This was evident in a control using dog gastric sections by the complete loss of binding of the NeuAcα2,6Gal/GalNAc recognizing *Sambucus nigra* (SNA) lectin. No effect on the ability of Fucα1,2Gal-specific UEAl to bind surface mucous cells was found under the same conditions. Using this approach we could demonstrate the carbohydrate-specific, sialic acid-independent binding of *H. pylori* to human gastric surface mucous cells (Falk *et al.*, 1993) as is illustrated in Color Plate 5, summarizing the effects on fucosylated and epitopes of periodate oxidation under various conditions. In contrast, protein F- and M protein-mediated binding of *S. pyogenes* to human skin Langerhan's cells and keratinocytes, respectively, was unaffected by periodate oxidation (Okada *et al.*, 1994a). The M protein-mediated binding to keratinocytes in the epidermis has been shown to be mediated by the MCP protein (Okada *et al.*, 1994b). It is important to note that a requirement for the mild meta-periodate oxidation to work is that there are carbohydrate residues with vicinal hydroxyl groups and that carbon number 3 of the ring structure is free and not involved in the glycosidic linkage (Woodward *et al.,* 1985). There are also examples of substitutions in the monosaccharide molecule rendering it resistant to periodate oxidation, like O-acetylation of the 7–9 carbon side chain of sialic acids (Manzi *et al.,* 1990). A periodate resistant binding therefore strongly favors a protein receptor but does not entirely exclude the possibility of a unique carbohydrate epitope.

V. Bacterial Inhibition Experiments *in Situ*

Once the cell linage-specific binding patterns and the chemical nature of the receptors in the target tissues have been established, the next step will be to characterize the molecular structure of the binding epitope.

Using data obtained for the distribution of various epitopes as determined by antibody and lectin overlays it is possible to characterize indirectly the receptor active molecules. This is most conveniently done by bacterial inhibition experiments, i.e., preincubating bacteria with defined soluble proteins or glycoconjugates prior to overlay on tissue sections or cells, to prevent attachment in a competitive fashion. To evaluate the activity of different soluble inhibitors in reducing adherence, the degree of bacterial adherence in the presence of inhibitor must be assessed in relation to a noninhibited control.

1. Bacterial suspensions are preincubated with series of 1–100 μg/ml of soluble receptor analogs (i.e., molecules known to be present also on the target cell surfaces) for 1 hr at room temperature.

2. The bacterial suspension is washed once in PBS, pH 7.6, to remove excess receptors and added to sections as described above.

3. Directly counting the number of adherent bacteria at $20\times$ magnification is a convenient approach to evaluate reduction in bacterial binding compared to a noninhibited control (Borén et al., 1993).

Remarks

If there is variability in bacterial adherence, this may be due to heterogeneity in tissue quality, caused by technical (caused by for instance fixation and embedding procedures) and/or phenotypical (e.g., localized inflammatory processes or various stages of neoplastic transformation) factors. This can be circumvented by using consecutively cut and tissues that do not display pathological changes as judged by standard histological stainings, such as hematoxylin, easin, and Alcion blue sections. In addition, all inhibitors including controls should as far as possible be tested at the same time in order to be able to make accurate comparisons of inhibitory activity. The inhibition assay can also be calibrated against a standardized inhibitory substance. These usually consist of commercially available soluble proteins or glycoproteins with well-defined structure and oligosaccharide composition that can be used in the initial screening for structures that inhibit bacterial cell attachment. For instance, common glycoconjugates used for characterization of sialic acid binding bacteria are the highly sialylated glycoproteins fetuin (Spiro and Bhoyroo, 1974) and glycophorin A (e.g., Parkkinen et al., 1986) (both available from Sigma). Mucin fractions, like bovine submaxillary mucins (Sigma), are well characterized and rich sources of complex, both sialylated and fucosylated, carbohydrate eopitopes (Savage *et al.*, 1991), as are protein fractions from human milk (Kitagawa *et al.*, 1990; Amano *et al.*, 1991), and both have proven to be very useful in initial inhibition experiments (Falk *et al.*, 1993; Borén *et al.*, 1993). In addition, immunoglobulins, like human secretory IgA, have well-characterized carbohydrate compositions (Pierce-Crétel *et al.*, 1989) that have been suggested to exert "nonspecific," lectin scavenging mucosal protective function (Davin *et al.*, 1991). They carry fucosylated blood group antigens, making them a valuable source of complex sugar epitopes for inhibition experiments. The final identification of the optimal inhibitory determinant requires highly purified and structurally characterized carbohydrate chains. These sugars can be selected based on the results from the initial inhibition screenings. Conjugation of many (25–50) copies of the same oligosaccharide chain to a nonglycosylated carrier protein like human serum albumin creates a highly defined neoglycoprotein. These synthezised compounds allow for comparison of inhibitory activity among structurally closely related sugar epitopes. They also provide multivalent attachment sites for the bacterial adhesins, which often potentiate bacterial binding to the compound and thereby increase the inhibitory activity (Strömberg and Borén, 1992; Borén *et al.*, 1993).

VI. *In Vitro* Assays for Bacterial Adhesion

In order to obtain detailed structural information on the actual cell surface receptor epitopes it is necessary to purify the receptor active structures from the host target cells or tissue. A number of detailed reviews dealing with the separation and chromatographic purification of lipid/glycolipid (e.g., Karlsson, 1987) and protein/glycoprotein (e.g., Finne and Krusius, 1982; Thomas and McNamee, 1990; Gerard, 1990) fractions from cells and tissues are available and this will not be covered here. *In vitro* assays can provide valuable qualitative and quatitative information of bacterial binding that will be important for studies of the molecular details of the receptors–adhesin interactions (e.g., Hultgren *et al.*, 1989).

A. Bacterial Binding to Immobilized Receptors in Solid Phase

Fractions with potential receptor structures isolated from tissues or cultured cells are initially most conveniently screened for bacterial binding in a solid phase assay system. These include a chromatographic separation of molecules, and consequently allow for identification and partial characterization of single receptor active entities from lipid/glycolipid or protein/glycoprotein mixtures. Solid phase analyses are performed by bacterial overlays either to immobilized proteins/glycoproteins on Western blots or to lipids/glycolipids separated on silica gel high-pressure thin-layer chromatograms (HPTLC).

1. Detection of Bacterial Protein/Glycoprotein Receptors in Western Blots

1. Proteins ($0.1-1\mu g$) are separated by gel electrophoresis, either under native conditions to preserve conformational integrity or alternatively by SDS–PAGE for enhanced presentation of peptide epitopes, and transferred to a nitrocellulose membrane.

2. Membranes are incubated with blocking buffer [Tris-buffered saline, pH 7.5 (TBS), 1% blocking reagents (Boehringer Mannheim), $1mM$ $MnCl_2$, 1 mM $MgCl_2$, 1 mM $CaCl_2$] overnight and washed 2x10 min in TBS.

3. A Dig-labeled bacterial suspension of OD_{600} 0.1 is added and incubated for 8 hr at room temperature, followed by 6x5-min washes in TBS.

4. Alkaline phosphatase conjugated anti-Dig antibody (Boehringer Mannheim) diluted 1:2000 in blocking buffer is added and incubated for 1 hr, washed 5x10 min in TBS, and developed with BCIP/NBT.

Apart from allowing for detection of single-receptor active structures in mixtures of molecules, this assay also makes it possible to compare binding to structurally related purified and well-characterized receptor candidates. This is illustrated in Fig.1 by the binding specificity of *H. pylori* to six different fucosylated blood group active neoglycoproteins.

2. Detection of Bacterial Glycolipid Receptors by HPTLC Overlay

Glycolipids may be extracted from the tissue under study utilizing the technique described by Karlsson (1987) or from small quantities of cells or biopsies as described by Falk *et al.* (1991). In addition, some glycolipid preparations are commercially available, like gangliosides (i.e., sialic acid containing glycolipids) and the globo-series glycolipids found in kidney and gastrointestinal tract from various species (can be purchased, for instance, from Sigma and Accurate Chemical & Scientific Corp., Westbury, NY).

Glycolipids are separated by silica gel HPTLC plates and the presence of specific carbohydrate epitopes can be examined by overlaying the plate with monoclonal antibodies (Brockhaus *et al.*, 1981), polyclonal antisera (Hansson *et al.*, 1985a), lectins (Torres *et al.*, 1988), viruses (Hansson *et al.*, 1984), bacteria (Hansson *et al.*, 1985b), bacterial toxins (Magnani *et al.*, 1980), or bacterial adhesins (Haslam *et al.*, 1994). Detection of bound probe can be done by autoradiography, most commonly used in combination with ^{35}S metabolically

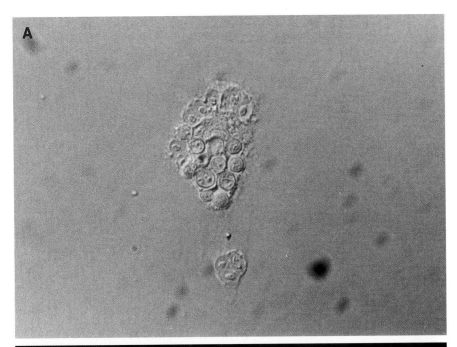

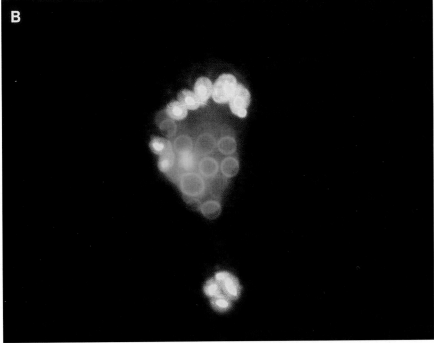

Color Plate 1   (Chapter 7) Phagocytosis of zymosan versus *Cryptococcus neoformans* by murine peritoneal macrophages. (A) Bright field. (B) Double exposure immunofluorescence showing FITC-conjugated zymosan (green) versus TRITC-conjugated *C. neoformans* (red/yellow). (See Chapter 7, Section VI,B for details of experimental procedure.)

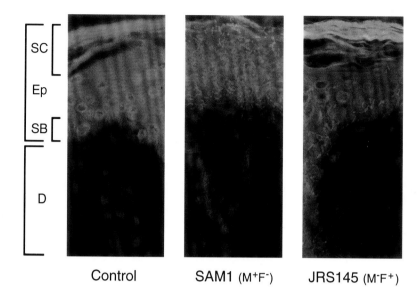

SC
Ep
SB
D

Control   SAM1 (M⁺F⁻)   JRS145 (M⁻F⁺)

Color Plate 2   (Chapter 9) Protein F and M protein direct cell-specfic tropism of *S. pyogenes* in cutaneous tissue: A histological section adherence assay reveals that both M protein and protein F mediate adherence of *S. pyogenes* to cutaneous tissue, but that each protein directs adherence to distinct populations of host cells. At the left of the figure a section of normal human skin has been stained with acridine orange (Control). Note the location of the stratum corneum (SC) and stratum basale (SB) of the epidermis (Ep) and the location of the dermis (D). In the center panel, a section of skin has been overlaid with a streptococcal mutant that expresses M protein, but cannot express protein F (SAM1). The section is washed to remove nonadherent bacteria and stained with acridine orange to reveal both the distribution of the streptococci, which are the small intensely staining orange cocci, and the architecture of the tissue. From the distribution of streptococci in the section and from additional studies using primary cultures of epidermal cells, it can be shown that the M protein mediates attachment to keratinocytes, which are cells that are distributed throughout the epidermis. The right panel illustrates the behavior of a mutant that can express protein F, but can no longer express M protein (JRS145). This mutant adheres to the epidermis, but is exclusively localized to a population of cells that are found only within the basal layer of the epidermis (SB). Additional immunostaining studies using a monoclonal antibody directed against the CD1 antigen have revealed that this population of cells represents epidermal Langerhans cells.

Color Plate 3   (Chapter 9) Immunocytochemical staining showing the coexpression of receptors for *H. Pylori* and fucosylated epitopes in surface mucous cells of the human gastric mucosa. A section of human stomach was overlayed with FITC-conjugated *H. pylori* and TRITC conjugated UEA1 lectin. Inspection of the section in a fluorescence microscope through a blue filter reveals the bound bacteria stained in green (upper panel, *Helicobacter Pylori* 466), whereas the green filter shows the staining of the UEA1 lectin in red (middle panel). The double exposure of the two stainings shows a complete congruence in the two staining patterns (lower panel).

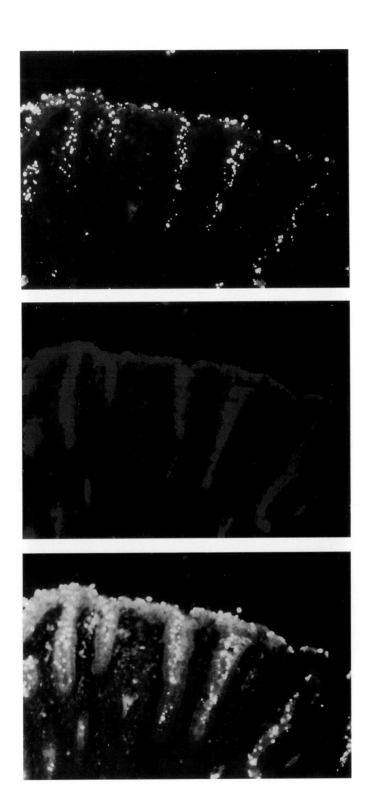

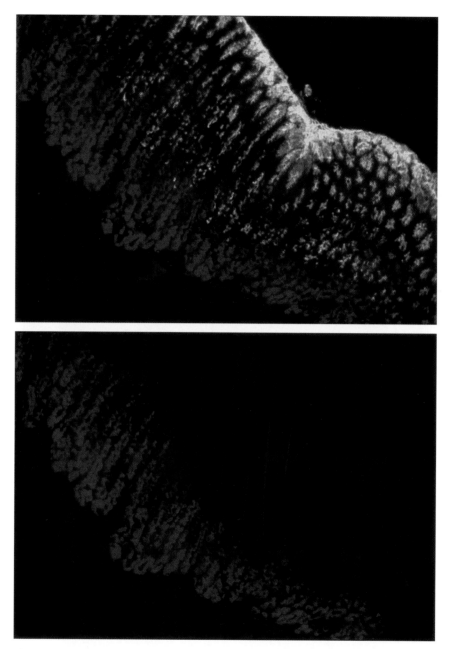

Color Plate 4 (Chapter 9) Ulex europaeus type 1/IF rat stomach. Immunocytochemical staining illustrating the effects of mild periodate oxidation on carbohydrate and peptide epitopes *in situ*. Both sections of rat gastric mucosa have been overlayed with FITC-conjugated UEA1 lectin, acting as a marker for the complex carbohydrates, and an antibody directed to the intrinsic factor protein. The latter is detected by a secondary TRITC-conjugated sheep anti-rabbit antiserum. The section on the bottom was incubated with 10mM NaIO$_4$ in 0.1 M NaAc, pH 4.5, for 1 hr at room temperature, as described in the text. The control slide on the top was incubated under identical conditions in 0.1 M NaAc without NaIO$_4$. The periodate oxidation destroys the receptor epitopes for the lectin in surface mucous cells, but leaves the peptide epitopes, found in chief cells, intact.

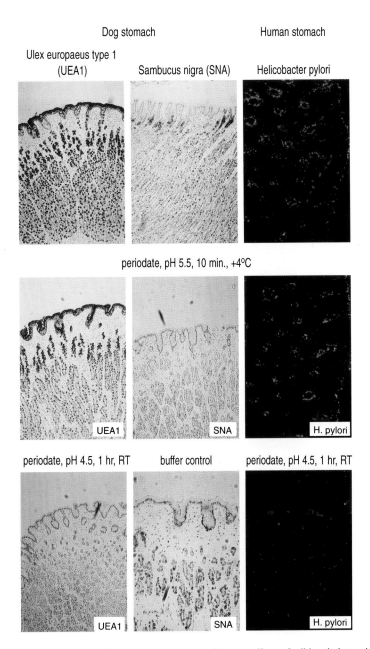

Color Plate 5 (Chapter 9) Immunocytochemical stainings summarizing the effects of mild periodate oxidation under various conditions on sialylated and fucosylated sugar epitopes (as illustrated by lectin staining), and *H. pylori* attachment. α-L-Fucose-specific UEA1 staining to dog gastric mucosa is resistant to $NaIO_4$ oxidation when this is done at pH 5.5, for 10 min on ice (first lane, middle panel), but sensitive to the "harsher" oxidation condition, pH 4.5, 1 hr at room remperature (first lane, lower panel) (see text for details). The sialic acid-specific SNA staining of dog stomach is abolished already by the "milder" oxidation (second lane, middle panel). The control (second lane, lower panel) shows that lectin staining is unaffected by the procedure in the absence of $NaIO_4$. The binding of *H. pylori* to human gastric mucosa (third lane) is resistant to the sialic acid-specific oxidation conditions (middle panel) but is eliminated under the harsher conditions that will also remove UEA1 staining (lower panel).

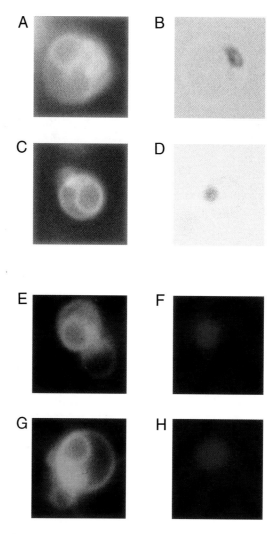

Color Plate 6   (Chapter 12) Perinuclear location of C6-NBD-ceramide in released parasites. Released parasites were labeled with NBD-ceramide and Hoechst as described in Section IV,B,2. Corresponding micrographs of the same cell were obtained using optics appropriate for NBD fluorescence and transmitted light (A and B; C and D) and for NBD and Hoechst fluorescence (E and F; G and H). The digestive food vacuoles are apparent due to the dark hemozoin crystals in images (B) and (D). The nucei are visible in images (F) and (H). Sphingolipid accumulation is clearly seen in a perinuclear region.

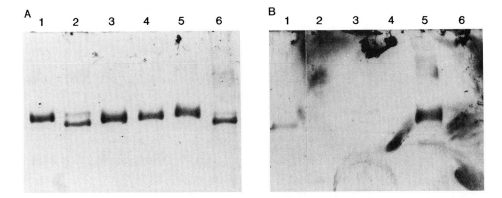

Fig. 1 Western blot immunostaining with whole Dig-conjugated *H. pylori* cells of fucosylated, blood group active antigens. The oligosaccharides chains were purified and structurally characterized by mass spectrometry and NMR, and conjugated to human or bovine serum albumin (see Borén *et al.,* 1993, for details, including structural information). One microgram each of the neoglycoproteins was run in SDS–PAGE, shown with Coomassie B staining in (A). After transfer to nitrocellulose the filters were incubated with Dig-labeled *H. pylori* as described in the text, and adherent bacteria were detected with alkaline phosphatase-conjugated Dig-specific antibodies. (B) Note the strong binding to the Le[b] antigen (lane 1) and the weaker binding to the H-1 antigen (lane 5), whereas none of the other structurally related fucosylated antigens were recognized by the bacteria.

labeled, or ^{125}I surface-labeled microbes (Karlsson and Strömberg, 1987), or by enzyme detection like horseradish peroxidase and alkaline phosphatase (Bethke *et al.,* 1986).

We have adapted HPTLC immunostaining for detection with horseradish peroxidase-conjugated antibodies and enhanced chemiluminescence (ECL) development. ECL detection allows for high sensitivity and rapid development of the immunostainings and, in addition, saves reagents since antibodies usually can be diluted 1–2 orders of magnitude moreas compared to isotope-labeled antibodies.

1. Glycolipids are diluted in a proper solvent (e.g., chloroform : methanol : water, 65 : 35 : 8, v/v/v) and spotted in bands (0.5–1 cm wide) on an alumina-backed HPTLC plate (Si-60, Merck, Darmstadt, Germany), leaving 1 cm on the bottom and on each edge, using a Hamilton syringe. Between 1 ng and 1 μg of pure glycolipid is required for immunological detection (depending on the specificity and affinity of the probe). Approximately 1 μg of glycolipid in a band is required for chemical detection using anisaldehyde, which will stain all glycolipids green, or resorcinol, which stains gangliosides blue to purple (see Stahl, 1967, for a review), and this could be used as a control to verify the purity and mobility of the glycolipids. After the samples have dried for a few minutes, the plate is placed in a developing tank that has been equilibrated with eluent (e.g., chloroform : methanol : water, 65 : 35 : 8; v/v/v; see Karlsson,

1987, and Stahl, 1967, for examples of proper solvents for different glycolipid fractions). The solvent front is allowed to migrate 4–5 cm and the plate is air-dried for 1 to 2 hr.

2. The plate is fixed for 30 sec in 0.25% polyisobutylmetacrylate plastic (P28; Röhm GmBH, Darmstadt, Germany) that has been dissolved in diethyl ether, to preserve the silica gel, and dried at room temperature overnight or for 30 min in a vacuum desiccator. Plastic fixative should be freshly prepared the day before to ensure that it is completely dissolved. If the solution is too concentrated or old, the specificity of the overlay may be altered significantly.

3. The fixed plates are blocked by incubating in 5% BSA in PBS for 1 hr at room temperature. Air bubbles must be removed, since they raise the background staining. In addition, it is essential that the plates not dry from this point onward as this will markedly increase background.

4. Blocking solution is aspirated off the plate, which is then overlaid with a suspension of labeled bacteria (10^7–10^8/ml) in 5% BSA–PBS. Bacteria may either be metabolically radiolabeled with [^{35}S]-methionine, or Dig-labeled for immunologic detection, as described above. Alternatively, purified bacterial adhesin may be utilized at this stage, as will be described below. After 2 hr at room temperature the plates are washed 3 x 5 min with PBS.

5. If radiolabeled bacteria are utilized, the plate is air-dried and exposed to film. If Dig-labeled bacteria or purified bacterial adhesin is used as probe, the plate is now covered with an antibody suspension [horseradish peroxidase-conjugated sheep anti-Dig (Boehringer-Mannheim) or anti-adhesin antibodies diluted 1 : 2500 to 1 : 5000 in 5% BSA-PBS]. The plates are incubated for 1 hr at room temperature, followed by washing 3–5 times in PBS. Secondary antibodies should be HRP conjugated for subsequent ECL detection. They are diluted 1 : 5,000 to 1 : 10,000 in 5% BSA–PBS and incubated on the plates for 1 hr.

6. The plate is washed six times in PBS. We have adapted enhanced chemiluminescence (ECL) for use in HPTLC overlays. The two components of ECL substrate (Detection reagent 1 and 2; Amersham Corp., Arlington Heights, IL) are mixed 1 : 1 and applied on top of the HPTLC plate. The solution is left in place for 1 min. A thin film of substrate solution should always remain on top of the plate. If too little substrate remains, it will be absorbed during film exposure and lead to increased background. The plate is wrapped in cellophane and placed in a film cassette. Autoradiographic film is placed on top of the plate for 1 min and then developed. Preflashed film should be used to increase the linearity in the exposure thus allowing for quantitation of binding intensity (see also Section VI,B,2). Positive bands will appear in black (Fig. 2). Several exposures are usually required to optimize the signal-to-noise ratio. It should be noted that the signal intensity decreases markedly with time.

If a colorimetric substrate is to be used, DAB [e.g., Sigmafast (Sigma) or Immuno-Pure Metal Enhanced DAB Substrate Kit [Pierce, Rockford, IL] appears to provide the greatest sensitivity.

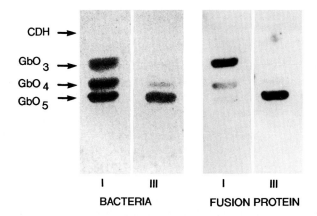

Fig. 2 HPTLC immunostaining of digalactoside containing glycolipids with whole P-fimbriated *E. coli* and the corresponding cloned fusion proteins, as described in the text. (I, III) class I and class III bacteria and fusion proteins, respectively. CDH, ceramide dihexoside; GbO_3, globotriaosylceramide; GbO_4, Globoside; GbO_5, Forssman antigen. Note the similar specificity for the fusion protein and whole bacterial cells.

3. Quantitation of Bacterial Adherence in Solid Phase Using ELISA

Whereas Western blot and HPTLC overlays allow only an approximate comparison of adherence to membrane components, an enzyme-linked immunosorbent assay (ELISA) may prove a simple and reproducible means of quantitating bacterial binding to various cell extracts. The method described below utilizes glycolipids extracted from the tissues under study. The method is easily adapted for adherence to proteins and glycoproteins.

1. Serial dilutions of glycolipids are prepared in methanol and added to a final volume of 50 μl of methanol in each well of a 96-well microtiter plate [U-shaped Maxisorb (Nunc) or Immulon 3 removawell strips (Dynatek) work well for both glycolipids and proteins]. The amount of glycolipid added to each well should range between 1 and 500 ng. Above 200 ng the glycolipids may dry in sheets and subsequently detach during the washing stages, leading to inaccurate readings from these wells (Karlsson and Strömberg, 1987).

2. Plates are place uncovered at 4°C overnight until the methanol dries.

3. Each well is blocked by the addition of 200 μl of 2% BSA in PBS for 2 hr at room temperature. Approximately 1×10^7 bacteria (radiolabeled or Dig-labeled) or purified adhesin diluted to 10 μg/ml, in a total volume of 100 μl, are added to each well and incubated for 1 to 3 hr at room temperature.

4. Wells are washed five times with PBS. Quantitation of radiolabeled bacteria is performed by cutting away each well and placing in liquid scintillation media. If immunologic detection is to be used, antibody dilutions are prepared

as described in the HPTLC overlay section above. Incubations are generally for 1 hr at room temperature.

5. Excess HRP-conjugated antibody is removed by six washes in PBS. Fifty microliters of *ortho*-phenylenediamine dihydrochloride (OPD) (0.4 mg/ml of 0.1 M NaPO$_4$ and 0.1 M Na-citrate, with 0.025% H$_2$O$_2$) is added to each well, and the absorbance at 405 nm is determined in a microwell plate reader (Molecular Devices Corp.).

The ELISA assay can be used to quantitate the amount of bound bacteria as a function of the amount of receptor present in the well by testing bacterial binding to serial dilutions of the receptors. This will allow for relative comparisons of binding affinity between structurally similar receptor candidates. This is illustrated in Fig. 3, showing the binding of purified *E. coli* P-fimbriae class 1 and class 3 lectins and the corresponding cloned fusion proteins to different digalactoside-containing glycolipids. Furthermore, the assay can be standardized against a receptor with known affinity for the studied bacteria.

4. Bacterial Adherence to Glycoproteins Adsorbed onto Hydroxyapatite

Due to the low binding capacity of proteins to plastics, ion-exchange materials like hydroxyapatite offer an efficient alternative for immobilizing a broad range of proteins and glycoconjugates (Gibbons and Hay, 1988). This assay allows for quantitation of binding using radiolabeled bacteria or cells. It has, for instance, been used to identify salivary receptors for *Actinomyces naeslundii* and *A. viscosus* (Strömberg *et al.*, 1992).

1. 40 mg of spheroidal hydroxyapatite (BDH Chemicals Ltd., Poole, UK), equilibrated overnight with adhesion buffer (50 mM KCl, 2 mM KPO$_4$, 1 mM CaCl$_2$, 0.1 mM MgCl$_2$, pH 7.3), are incubated with up to 1 mg of glycoproteins in the same buffer in a total volume of 1 ml. Incubations are done for 1 hr at room temperature, with continuous inversion of the tube.

2. Following two washes in adhesion buffer (add buffer, let beads sediment by gravity, and aspirate the liquid), unreacted binding sites are blocked on treated beads and untreated control beads by incubation with 1 ml of 0.5% bovine serum albumin in adhesion buffer for 30 min.

3. The beads are washed three times with adhesion buffer and then incubated with 1 ml of bacteria metabolically labeled with [^{35}S]-methionine (2 x 10^4 cpm/ml, 4 x 10^8 cells/ml was suitable for the *Actinomyces* binding experiments) in 0.5% bovine serum albumin in adhesion buffer for 60 min.

4. The percentage of bacteria remaining attached after two washes is then determined by scintillation counting.

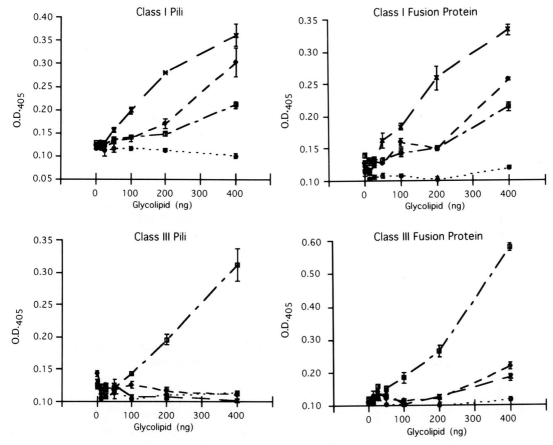

Fig. 3 ELISA showing the affinity of purified P-pili from *E. coli,* and corresponding cloned fusion protein for various digalactoside-containing glycolipids. Glycolipid abbreviations are described in Fig. 6 legend. Regarding the HPTLC data presented in Fig. 6, these results illustrate the retained specificity for the cloned proteins compared to the natural adhesin: ○,CDH; x, Gb03; ◇, Gb04; -□-Gb05.

B. Bacterial Binding to Receptors in Solution

The conventional way to study adhesin–receptor interactions has been to study binding of bacteria to immobilized receptor structures as described above. Due to restrictions in steric flexibility induced by the immobilization of receptor structures in solid phase, these assays might be biased and not representative enought of the *in vivo* situation (Strömberg *et al.,* 1991). It is therefore valuable to compare the solid phase data with adherence experiments using the same receptors in soluble form. This can reveal fine-tuned differences in receptor specificity among structurally related compounds and also allows for quantitation of the adhesin–receptor interaction. Furthermore, many different bacterial species that express protein receptors on their surfaces and have the ability to bind to soluble proteins that are of interest to the cell biologist have been described. These bacteria have proven very useful for the detection and the purification of the protein from both fluids and tissues. Some strains of *Staphylococcus aureus* express protein A, a protein that binds the Fc region of certain immunoglobulin isotypes. A variety of techniques for the detection and purification of immunoglobulins that have utilized *S. aureus* cells or purified protein A provides classic examples of how useful bacteria can be in this regard. In addition, many different bacterial species have been described that can bind to components of the extracellular matrix, including fibronectin, vitronectin, collagen, and laminin (for review, see Doig and Trust, 1993) and there is considerable evidence that bacterial fibronectin binding can contribute to adherence to cells and tissue (Hanski and Caparon, 1992). A partial list of other soluble proteins known to be bound by bacteria includes fibrinogen (Whitnack and Beachey, 1982), immunoglobulin light chain (Kastern *et al.*, 1992), bone sialoprotein (Ryden *et al.*, 1987), plasmin (Lottenberg *et al.*, 1992), β_2-microglobulin (Schönbeck *et al.*, 1981), albumin (Falkenberg *et al.*, 1992), α_2-macroglobulin (Sjöbring *et al.*, 1989), and lactoferrin (Schryvers and Lee, 1993). The following assay has been used extensively in our laboratories for the analysis of fibronectin binding to *S. pyogenes* (a more detailed treatment of the different methods available for bacterial fibronectin binding can be found in Olsén *et al.* (1993b) and can be easily adapted for other soluble proteins.

1. Binding of Soluble Proteins to Bacteria

1. The protein of interest is labeled with ^{125}I. A number of different methods and commercial products are available for radioiodination and the method chosen depends on the protein of interest. For example, fibronectin is labeled to high specific activity (approx. 5×10^6 cpm/μg protein) using the chloramine-T method (Olsén *et al.*, 1993b).

2. Bacteria are cultured under the optimal conditions that promote expression of the receptor for the protein of interest. Specific growth conditions will depend on the specific bacterial species and the binding activity of interest. Keep in mind that expression of bacterial adhesins and surface proteins is frequently tightly regulated in response to both environmental and growth conditions (see discussion above in reference to the *in situ* adherence assay).

3. Bacteria are harvested following growth by centrifugation and resuspended in phosphate-buffered saline supplemented with 1% Tween 20 (PBS/T) to a concentration of at least 1×10^8 bacteria/ml.

4. 100 μl of the cell suspension is added to a microcentrifuge tube along with a 100-μl aliquot of the labeled protein adjusted to contain a total of 1×10^5 cpm

and the total volume adjusted to 1.0 ml with PBS/T. It is usually not necessary to precoat the microcentrifuge tube to prevent binding to tube.

5. The mixture is incubated at room temperature with end-over-end rotation for 1–2 hr.

6. Bacteria in the suspension are then pelleted by centrifugation (13,000 × g for 5 min) and the supernatant fluid is carefully removed by aspiration to remove unbound labeled protein. Further washes of the pellet are generally not required. Quantitative recovery of the bacteria can be facilitated by the addition of a 100-μl aliquot of a concentrated suspension of a carrier bacterium that does not bind the protein of interest. The carrier bacterium is added just prior to centrifugation.

7. The amount of protein bound is then determined by measuring the amount of radioactivity associated with the bacterial pellet using gamma scintillation counting. A control for the amount of fibronectin that binds directly to the tube (background) consists of 1 m of PBS/T to which labeled protein, but no bacteria, is added. If a carrier bacterium is utilized, the amount of labeled protein non-specifically trapped in the pellet is determined by addition of the carrier to the background tubes at the same time it is added to the assay tubes. An additional control for specificity should also be included and consists of a tube to which an excess (10- to 50-fold) of unlabeled protein of interest is added and a tube to which an excess of an unrelated (and unlabeled) protein is added. If binding is specific, it is expected that only the protein of interest and not the unrelated protein will compete for binding with the labeled protein.

2. Determination of Relative Binding Efficiency of Soluble Glycoconjugates

The *in situ* inhibition studies described above usually do not allow for objective quantitation of the degree of reduction in binding. Furthermore, since you can expect to find different binding patterns to immobilized receptors versus receptors in solution (e.g., Borén *et al.*, 1993; Lee *et al.*, 1994), it can be of value to produce quantitative data for binding of soluble receptors to compare with findings on ELISA, Western blots, and HPTLC immunostainings where dilution series of receptors have been applied.

1. Glycoprotein conjugates are labeled with biotin, using biotin-x–NHS as described for bacteria above. Excess biotin is subsequently removed by gel filtration using G-25 columns (Pharmacia).

2. A bacterial suspension (10^8 cfu/200 μl) is incubated with glycoconjugates in titration series from 1–1000 ng/ml of blocking buffer with protease inhibitors (see Section II,2) for 1 hr at room temperature, followed by three washes in blocking buffer.

3. Twofold serial dilutions of bacteria complexed with glycoconjugates are immobilized on nitrocellulose membranes using a slot-blot system (e.g., Schleicher & Schuell Inc., Keene, NH).

4. Membranes are blocked with 1% blocking reagent (Boehringer Mannheim) in TBS for 15 min at room temperature, followed by three washes, 10 min each, in TBS.

5. Immobilized bacterial–glycoconjugate complexes are fixed with 0.4 M perchloric acid for 30 min and subsequently washed five times, 10 min each, in TBS.

6. If endogenous bacterial peroxidase activity is causing background problems, it can be heat inactivated following the fixation by 15-min autoclaving on the dry program, followed by an additional 3-hr blocking step as above and four subsequent washes, 10 min each, in TBS.

7. Glycoconjugates bound to bacteria are then visualized using horseradish peroxidase-conjugated antibiotin antibodies, diluted 1 : 20,000 in blocking buffer, incubated for 1 hr followed by seven washes in TBS, and the enhanced chemoluminiscence (ECL) detection system (Amersham). To ensure linear quantitation, preflashed film should be used.

This method has been used by us to compare binding affinities for various soluble fucosylated neoglycoproteins to *H. pylori* (Borén *et al.*, 1993), as illustrated in Fig. 4.

Remarks

Although all these assays can provide valuable molecular information about the host–bacterial interaction, none of them give a complete picture of the bacterial binding to a host cell surface *in vivo*. Factors like presentation of the receptor epitopes and receptor density will differ between *in vivo* and *in vitro* conditions, and this in turn will influence the binding specificity, avidity, and affinity. It is therefore important to compare results from several different *in vitro* and *in situ* systems, since it might be misleading to extrapolate data obtained in a single *in vitro* assay to the *in vivo* situation. The differences in binding of bacteria to the same receptor candidates is clearly illustrated in Figs. 1 and 4, showing the affinity of *H. pylori* to fucosylated blood group-active antigens in solid phase and in solution, respectively. Furthermore, receptor characterization should also, as far as possible, originate from the natural target cell or tissue, either with material prepared from these cells or with molecules known to be present in the target cell plasma membrane.

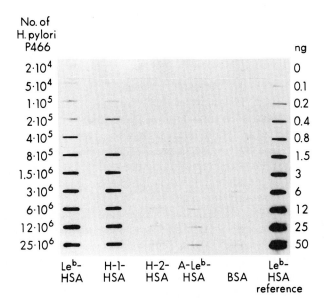

Fig. 4 Analysis of affinity of soluble fucosylated neoglycoproteins to *H. pylori* cells (see text and Borén *et al.*, 1993, for details). The glycoconjugates that bound to a known amount of bacteria in solution are detected with ECL and compared to a standard dilution curve (lane 6). Binding above background is only seen for the Le[b] and H-1 glycoconjugates. Note that the affinities for the soluble Le[b] and H-1 antigens to the bacteria are almost identical, which markedly differs from the clearly higher affinity for the immobilized Le[b] structure showed in the Western blot immunostaining in Fig. 1. This illustrates the different results that can be obtained in different *in vitro* adhesion assays and shows the necessity for comparing data from several *in situ* and *in vitro* assays in the studies of binding events like these.

VII. Probing Eukaryotic Cell Glycoconjugates with Purified Bacterial Adhesins

As described above, the use of whole bacterial cells might have wide applicability to the detection of cell surface glycoconjugates. In some cases, however, it might be desirable to use a more defined probe for the detection of eukaryotic cell surface epitopes. Bacteria could express more than one adhesin at a time, giving rise to a less precise pattern of adherence. Conversely, the desired adhesive capacity displayed by the bacteria may be regulated by environmental factors as described above. Hence, variation in the pattern of bacterial attachment may not be a consequence of host cell expression of ligand, but rather varied bacterial expression of adhesin. In these instances it may be advantageous to use a single bacterial adhesin whose receptor specificity is well characterized. This is illustrated in Figs. 2 and 3, comparing the binding of cloned soluble fusion proteins containing the *E. coli* P-fimbriae, globotriaosylceramide-specific, class 1 lectin or the, Forssman antigen-specific, class 3 lectin to digalactoside containing glycolipids in HPTLC immunostainings and ELISA, respectively.

A large array of bacterial proteins that display adhesive capacity have been identified. The many of these adhesins recognize cell surface glycoconjugates presented as components of glycolipids or glycoproteins. In some cases, the bacterial protein of interest may be overexpressed, purified, and demonstrated to retain ligand specificity. Hence the adhesin may be purified to the same degree as commercially available lectins. Purification of native bacterial adhesin may be tedious and inefficient process, however. This may be greatly simplified by expressing the protein of interest as a translational fusion to a peptide or protein, which enables single-step purification by affinity chromatography. Examples of widely used fusion partners include β-galactosidase, the maltose-binding protein (MBP), glutathione *S*-transferase (GST), and various polyhistidine tags. Each system has distinct advantages and disadvantages, and consideration should be given to the ease of purification, protein yield, size of the fusion partner, ability to cleave it from the protein of interest, availability of antibodies or directed against the fusion partner, and cost of affinity resin. We have used the maltose-binding protein (Kellermann and Ferenci, 1982) as a partner in fusion proteins containing the three different, digalactoside PapG adhesins from uropathogenic *E. coli* (Lund et al., 1988). The MBP has recently been used in fusion proteins with the P1 adhesin from *mutans* (Crowley *et al.,* 1993). The following discussion assumes that the bacterial adhesin has been cloned and the nucleotide sequence of the gene has been defined.

A. Cloning and Purification of an Adhesin as a Translational Fusion

PCR Amplification of the adhesin gene:

1. Primers are designed to amplify the gene of interest. The efficiency of later subcloning steps is improved if the primers incorporate restriction sites that are compatible with the expression vector. The addition of 4 to 6 "G" or "C" nucleotides at the 5′ end of each primer will stabilize the ends of the PCR product, and therefore increase the efficiency of restriction digestion. The primers should be designed such that the gene will be ligated in the correct reading frame to the vector's partner gene.

2. After completion of the PCR reaction the product is gel purified and digested with the appropriate restriction enzyme(s). The product is phenol extracted, passed over a Sephadex G-50 column (Pharmacia), ethanol precipitated, and resuspended in a small volume of water.

3. Preparation of vector DNA, ligation reactions, and screening of transformants follow standard protocols (e.g., Molecular cloning, a laboratory manual "[Sambrook, J., Fritsch, E. F., and Maniatis, T., eds.] Cold Spring Harbor

Lab. Press, 1989). Candidate clones are further screened by growing a 5-ml culture to OD_{600} 0.5, followed by induction for 1–3 hr with 0.1 to 0.5 mM isopropylthio-β-D-galactoside (IPTG). Cells may be pelleted and an aliquot loaded directly into SDS–PAGE buffer. The induced cells are compared by SDS–PAGE to uninduced controls for the production of an appropriately sized protein.

4. Fusion proteins are then purified by affinity chromatograpy. For example, proteins fused to the maltose binding protein are bound to amylose resin (New England Biolabs) and eluted with 10–50 mM maltose in PBS.

5. The concentration and integrity of the purified fusion proteins may then be analyzed by SDS–PAGE.

6. The fusion protein may now be assessed for their ability to act as probes for cell-specific markers, in solid phase assays, in suspension, or by adherence to cultured cells or fixed tissues.

This approach has been applied to the detection of globoseries glycolipids utilizing a portion of the *E. coli* P-pilus, which is illustrated in Figs. 2 and 3. P-pili are composite fibers that mediate adherence to globoseries glycolipids presented on host uroepithelial cells via the PapG adhesin. MBP/PapG fusion proteins were prepared from stains that demonstrate slight differences in glyco-lipid receptor specificity (Lund *et al.*, 1988). The purified PapG fusion proteins were found to retain the variation in receptor specificity in solid phase assays (Haslam *et al.*, 1994).

VIII. Concluding Remarks

The focus of the studies of molecular mechanisms involved in host–microbial cell interaction has traditionally been on the bacteria. This includes indentifying colonization factors like adhesins, adaptation mechanisms that increase bacte-rial chances of survival, and factors that facilitate colonization, as well as components that can give rise to tissue damage and thereby induce symptoms in the host. In chronic infections, the mechanisms by which bacteria invade the host cell and elude its immunological and nonimmunological defence barriers have also been studied. However, without information on how the host cell reacts when challenged by a bacterium, we will still be unable to understand why colonization by certain bacteria gives rise to illness, whereas we can coexist asymptomatically with others.

Over the last few years, we have gained considerable insights into the molecu-lar mechanisms involved in bacterial colonization of epithelial surfaces. The initial contacts between bacteria and cell are mediated by surface proteins on the bacterial surface recognizing and binding to molecules exposed on the plasma membrane of cells. The adhesins recognize their receptors with remark-able specificity, and bacterial binding is therefore restricted to the cell populations that carry the optimal receptor. Thus, bacteria are very useful as tools to target-specific cell surface molecules, and to define specific cell lineages. Due to the strict requirement for optimal stereochemical fit between the adhesin and the receptor, bacteria can sometimes demonstrate unique selectivity. For example, subtle conformational differences in digalactoside-containing glycolipids can be discriminated by three different types of P-fimbriated *E. coli* specifically binding to globotriaosylceramide, globotetraosylceramide (globoside), and globopentaosylceramide (Forssman antigen), respectively (Lund *et al.*, 1988; Strömberg *et al.*, 1990), leading to tissue and species tropism. *H. pylori* is another bacteria expressing highly specific adhesins. This recently described human pathogen (Goodwin *et al.*, 1986) produces adhesins that discriminate between closely related fucosylated epitopes as well as carbohydrate core chains, and binds only to fucosylated H and Lewis antigens expressed on lactoseries type 1, but not type 2, chains (Borén *et al.*, 1993). Furthermore, bacteria can be useful as probes in the studies of cellular response mechanisms, since attachment of bacteria generally leads to activation of transmembrane signaling pathways and cellular defense mechanisms, like tyrosine kinase activation (reviewed by Bliska and Falkow, 1993), interleukin release (reviewed by Blaser, 1993), and cytoskeleton rearrangements (Bliska *et al.*, 1993; Smoot *et al.*, 1993). It is therefore possible to utilize bacteria to induce and study inflammatory responses, internalization process, and, since chronic infection with bacteria like *H. pylori* has been very strongly correlated with the development of carcinoma (Forman *et al.*, 1993), also malignant transformation. Increased knowledge of cellular response triggered by bacterial contact also holds the potential of leading to the development of new generations of antimicrobial drugs that could be more specifically directed toward the molecular events that lead to the development of disease, and thereby circumvent many of the undesirable side effects encountered with today's broad-spectrum antibiotics.

Acknowledgments

This work was supported by grants from Symbicom, Umeå, Sweden, and the American Heart Association. P.F. and T.B. have been supported by the Swedish Institute, the Swedish Medical Research Council, and the Swedish Society for Medical Sciences. P.F. is the recipient of a postdoctoral fellowship from the William Keck Foundation. T.B. is supported by a Public Health Service Fogarty International Research Fellowship and by a fellowship from the Swedish Society of Medicine. D.H. is supported by a grant from the American Academy of Pediatrics, administered through the Pediatric Scientist Development Training Program.

References

Akopyanz, N., Borén, T., Falk, P., and Berg, D. E. (1994). Regulation of transcript abundance in *Helicobacter pylori*. Submitted for publication.

Amano, J., Straehl, P., Berger, E. G., Kochibe, N., and Kobata, A. (1991). Structures of mucin-type sugar chains of the galactosyltransferase purified from human milk. Occurrence of the ABO and Lewis blood group determinants. *J. Biol. Chem.* **266,** 11461–11477.

Andersen, L. P., Holch, S., and Povlsen, C. O. (1988). *Campylobacter pylori* detected by indirect immunohistochemical technique. *Acta Pathol. Microbiol. Pathol. Scand.* **96,** 559–565.

Ballows, A., (1991). "Manual of Clinical Microbiology." Am. Soc. Microbiol., Washington, DC.

Bethke, U., Müthing, J., Schauder, B., Conradt, P., and Mühlradt, P. F. (1986). An improved semi-quantitative enzyme immunostaining procedure for glycosphingolipid antigens on high performance thin layer chromatograms. *J. Immunol. Methods* **89,** 111–116.

Blaser, M. (1993). *Helicobacter pylori:* Microbiology of a "slow" bacterial infection. *Trends Microbiol.* **1,** 255–260.

Bliska, J. B., and Falkow, S. (1993). The role of host tyrosine phosphorylation in bacterial pathogenesis. *Trends Genet.* **9,** 85–89.

Bliska, J. B., Galán, J. E., and Falkow, S. (1993). Signal transduction in the mammalian cell during bacterial attachment and entry. *Cell (Cambridge, Mass.)* **73,** 903–920.

Borén, T., Falk, P., Roth, K. A., Larson, G., and Normark, S. (1993). Attachment of *Helicobacter pylori* mediated by blood group antigens. *Science* **262,** 1892–1895.

Brockhaus, M., Magnani, J., Blaszczyk, M., Steplewski, Z., Koprowski, H., Karlsson K.-A., Larson, G., and Ginsburg, V. (1981). Monoclonal antibodies directed against the human Le^b blood group antigen. *J. Biol. Chem.* **256,** 13223–13225.

Caparon, M. G., Geist, R. T., Perez-Casal, J., and Scott, J. (1992). Environmental regulation of virulence in group A streptococci: Transcription of the gene encoding M protein is stimulated by carbon dioxide. *J. Bacteriol.* **174,** 5963–5701.

Crowley, P. J., Brady, L. J., Piacentini, D. A., and Bleiweis, A. S. (1993). Identification of a salivary agglutinin-binding domain within cell surface adhesin P1 of Streptococcus mutans. *Infect. Immun.* **61,** 1547–1552.

Davin, J.-C., Senterre, J., and Mahieu, P. R. (1991). The high lectin-binding capacity of human secretory IgA protects nonspecifically mucosae against environmental antigens. *Biol. Neonate* **59,** 121–125.

Doig, P., and Trust, T. J. (1993). Methodological approaches of assessing microbial binding to extracellular matrix components. *J. Microbiol. Methods* **18,** 167–180.

Falk, P., Hoskins, L. C., Lindstedt, R., Svanborg, C., and Larson, G. (1991). Deantigenation of human erythrocytes by bacterial glycosidases—evidence for the non-involvement of medium sized glycosphingolipids in the *Dolichos biflorus* lectin hemagglutiantion. *Arch. Biochem. Biophys.* **290,** 312–319.

Falk, P., Roth, K. A., Borén, T., Westblom, T. U., Gordon, J. I., and Normark, S. (1993). An *in vitro* adherence assay reveals that *Helicobacter pylori* exhibits cell lineage-specific tropism in the human gastric epithelium. *Proc. Natl. Acad. Sci. U.S.A.* **90,** 2035–2039.

Falk, P., Roth, K. A., and Gordon, J. I. (1994). Lectins are sensitive tools for defining the differentiation programs of epithelial cell lineages in the developing and adult mouse gastrointestinal tract. *Am. J. Physiol. (Gastrointest Liver Physi. 29)* **266,** G987–G1003.

Falkenberg, C., Björck, L., and Åkerström, B. (1992). Localization of the binding site for streptococcal protein G on human serum albumin. Indentification of a 5.5 kilodalton protein G binding albumin fragment. *Biochemistry* **31,** 1451–1457.

Finne, J., and Krusius, T. (1982). Preparation and fractionation of glycopeptides. *In* "Methods in Enzymology (V. Ginsburg, ed.), Vol. 83, pp. 269–277. Academic Press, New York.

Fithian, E., Kung, P., Goldstein, G., Rubenfeld, M., Fenoglio, C., and Edelson, R. (1981). Reactivity of Langerhan's cells with hybridoma antibody. *Proc. Natl. Acad. Sci. U.S.A.* **81,** 2541–2544.

Forman, D., Webb, P., Newell, D., Coleman, M., Paali, D., Møller, H., Hengels, K., Elder, J., and DeBacker, G. (1993). An international association between *Helicobacter pylori* infection and gastric cancer. *Lancet* **341,** 1359–1362.

Fuchs, E. (1990). Epidermal differentiation: The bare essentials. *J. Cell Biol.* **111,** 2807–2814.

Gerard, C. (1990). Purification of glycoproteins. *In* "Methods in Enzymology" (M. Deutscher, ed.), Vol. 182, pp. 529–539. Academic Press, San Diego.

Gibbons, R. J., and Hay, D. I. (1988). Human salivary acidic proline-rich proteins and statherin promote the attachment of *Actinomyces viscosus* LY7 to apatic surfaces. *Infect. Immun.* **56,** 439–445.

Goldstein, I. J., and Poretz, R. D. (1986). Isolation, physicochemical characterization, and carbohydrate-binding specificity of lectins. *In* "The Lectins: Properties, Functions, and Applications in Biology and Medicine" (I. E., Liener, N., Sharon, and I. J., Goldstein, eds.), pp. 33–247. Academic Press, New York.

Goodwin, C. S., Armstrong, C. S., and Marshall, B. J. (1986). *Campylobacter pyloridis*, gastritis, and peptic ulceration. *J. Clin. Pathol.* **39**, 353–365.

Göransson, M., Sondén, B., Nilsson, P., Dagberg, B., Forsman, K., Emanuelsson, K., and Uhlin, B.-E. (1990). Transcriptional silencing and thermoregulation of gene expression in *Escherichia coli. Nature (London)* **344**, 682–685.

Hanski, E., and Caparon, M. G. (1992). Protein F, a fibronectin-binding protein, is an adhesin of the group A streptococcus. *Proc. Natl. Acad. Sci. U.S.A.* **89**, 6172–6176.

Hansson, G. C., Karlsson, K.-A., Larson, G., Strömberg, N., Thurin, J., Örvell, C., and Norrby, E. (1984). A novel approach to the study of glycolipid receptors for viruses. Binding of Sendai virus to thin-layer chromatograms. *FEBS Lett.* **170**, 15–18.

Hansson, G. C., Karlsson, K.-A., Larson, G., Samuelsson, B. E., and Thurin, J. (1985a). Detection of blood group type glycosphingolipid antigens on thin-layer plates using polyclonal antisera. *J. Immunol. Methods* **83**, 37–42.

Hansson, G. C., Karlsson, K.-A., Larson, G., Strömberg, N., and Thurin, J. (1985b). Carbohydrate-specific adhesion of bacteria to thin-layer chromatograms: A rationalized approach to the study of host cell glycolipid receptors. *Anal. Biochem.* **146**, 158–163.

Haslam, D., Borén, T., Falk, P., Ilver, D., Chou, A., Xu, Z., and Normark, S. (1994). The aminoterminal domain of the P-pilus adhesin determines receptor specificity. *Molec. Micro.*, in press.

Hessey, S. J., Spencer, J., Wyatt, J. I., Sobala, G., Rathbone, B. J., Axon, A. T. R., and Dixon, M. F. (1990). Bacterial adhesion and disease activity in *Helicobacter* associated chronic gastritis. *Gut* **31**, 134–138.

Hultgren, S. J., Lindberg, F., Magnusson, G., Kihlberg, J., Tennent, J. M., and Normark, S. (1989). The PapG adhesin of uropathogenic *Escherichia coli* contains separate regions for receptor binding and for the incorporation into the pilus. *Proc. Natl. Acad. Sci. U.S.A.* **86**, 4357–4361.

Hultgren, S. J., Abraham, S., Caparon, M., Falk, P., St. Geme, J. W., III, and Normark, S. (1993). Pilus and non-pilus bacterial adhesins: Assembly and function in cell recognition. *Cell (Cambridge, Mass.)* **73**, 887–901.

Jones, G. W., and Isaacson, R. E. (1983). Proteinaceous bacterial adhesins and their receptors. *CRC Crit. Rev. Microbiol.* **10**, 229–260.

Karlsson, K.-A. (1987). Preparation of total nonacid glycolipids for overlay analysis of receptors for bacteria and viruses and for other studies. *In* "Methods in Enzymology" (V. Ginsburg, ed.), Vol. 138, pp. 212–220. Academic Press, Orlando, FL.

Karlsson, K.-A. (1989). Animal glycosphingolipids as membrane attachment sites for bacteria. *Annu. Rev. Biochem.* **58**, 309–350.

Karlsson, K.-A., and Strömberg, N. (1987). Overlay and solid phase analysis of glycolipid receptors for bacteria and viruses. *In* "Methods in Enzymology" (V. Ginsburg, ed.), Vol. 138, pp. 220–231. Academic Press, Orlando, FL.

Kastern, W., Sjöbring, U., and Björck, L. (1992). Structure of peptostreptococcal protein L and indentification of a repeated immunoglobulin light chain-binding domain. *J. Biol. Chem.* **267**, 12820–12825.

Kellermann, O. K., and Ferenci, T. (1982). Maltose-binding protein from *Escherichia coli. In* "Methods in Enzymology" (W. Wood, ed.), Vol. 90, 459–463. Academic Press, New York.

Kitagawa, H., Nakada, H., Numata, Y., Kurosaka, A., Fukui, S., Funakoshi, I., Kawaskai, T., Shimada, I., Inagaki, F., and Yamashina, I. (1990). Occurence of tetra- and pentasaccharides with the sialyl-Le(a) structure in human milk. *J. Biol. Chem.* **265**, 4859–4862.

Korhonen, T. K., Väisanen-Rhen, V., Rhen, M., Parkkinen, J., and Finne, J. (1984). *Escherichia coli* fimbriae recognizing sialyl galactosides. *J. Bacteriol.* **159**, 762–766.

Korhonen, T. K., Parkkinen, J., Hacker, J., Finne, J., Pere, A., Rhen, M., and Holthöfer, H.

(1986). Binding of *Escherichia coli* S fimbriae to human kidney epithelium. *Infect. Immun.* **54,** 322–327.

Lee, J., Fogg, G., Gibson, C., Yates, D., and Caparon, M. (1994). An oxygen-induced fibronectin binding activity distinct from protein F, binds soluble, but not immobilized fibronectin. Submitted for publication.

Leffler, H., and Svanborg-Edén, C. (1980). Chemical indentification of a glycosphingolipid receptor for *Escherichia coli* attaching to human urinary tract epithelial cells and agglutinating human erythrocytes. *FEMS Microbiol. Lett.* **8,** 127–134.

Lottenberg, R., Broder, C. C., Boyle, M. D. P., Kain, S. J., Schroeder, B. L., and Curtiss, R., III. (1992). Cloning, sequence analysis and expression in *Escherichia coli* of a streptococcal plasmin receptor. *J. Bacteriol.* **174,** 5204–5210.

Luna, L. G., ed. (1968). "Manual of Histologic Staining Methods of the Armed Forces Institute of Pathology," 3rd ed. McGraw–Hill, New York.

Lund, B., Marklund, B.-I., Strömberg, N., Lindberg, F., Karlsson, K.-A., and Normark, S. (1988). Uropathogenic *Escherichia coli* can express serologically identical pili of different receptor binding specificities. *EMBO J.* **2,** 255–263.

Magnani, J. L., Smith, D. F., and Ginsburg, V. (1980). Detection of gangliosides that bind cholera toxin: Direct binding of ^{125}I-labeled toxin to thin-layer chromatograms. *Anal. Bichem.* **109,** 399–402.

Manzi, A. E., Dell, A., Azadi, P., and Varki, A. (1990). Studies of naturally occuring modifications of sialic acids by fast-atom bombardment-mass spectrometry. *J. Biol. Chem.* **265,** 8094–8107.

Nyberg, G., Strömberg, N., Jonsson, A.-B., Karlsson, K.-A., and Normark, S. (1990). Erythrocyte gangliosides act as receptors for *Neisseria subflava: Identification of the Sia-1 adhesin. Infect. Immun.* **58,** 2555–2563.

Okada, N., Pentland, A. P., Falk, P., and Caparon, M. (1994a). M protein and protein F act as important determinants of cell-specific tropism of *Streptococcus pyogenes* in skin tissue. *J. Clin. Invest.* In press.

Okada, N., Liszewski, M. K., Atkinson, J. P., and Caparon, M. (1994b) Membrane cofactor protein (MCP or CD46) is a keratinocyte receptor for the M protein of the group A streptococcus. Submitted for publication.

Olsén, A., Jonsson, A., and Normark, S. (1989) Fibronectin binding mediated by a novel class of surface organelles on *Escherichia coli. Nature (London)* **338,** 652–655.

Olsén, A., Arnqvist, A., Hammar, M., Sukupolvi, S., and Normark, S. (1993a). The RpoS sigma factor relieves H-NS-mediated transcriptional repression of *csg*A, the subunit gene of fibronectin-binding curli in *Escherichia coli. Mol. Microbiol.* **7,** 523–536.

Olsén, A., Hanski, E., Normark, S., and Caparon, M. G. (1993b). Molecular characterization of fibronectin binding proteins in bacteria. *J. Microbiol. Methods* **18,** 213–226.

Orlandi, P. A., Klotz, F. W., and Haynes, J. D. (1992). A malaria invasion receptor, the 175-kilodalton erythrocyte binding antigen of *Plasmodium falciparum* recognizes the terminal Neu5Ac(α2,3)Gal-sequences of glycophorin A. *J. Cell Biol.* **116,** 901–909.

Parkkinen, J., Rogers, G. N., Korhonen, T. K., Dahr, W., and Finne, J. (1986). Identification of O-linked sialyloligosaccharides of glycophorin A as the erythrocyte receptors for S-fimbriated *Escherichia coli. Infect. Immun.* **54,** 37–42.

Pierce-Crétel, A., Decottignies, J.-P., Wieruszeski, J.-M., Strecker, G., and Montreuil, J. (1989). Primary structure of twenty three neutral and monosialylated oligosaccharides O-glycosidically linked to the human secretory immunoglobulin A hinge region determined by a combination of permethylation analysis and 400-MHz ^{1}H-NMR spectroscopy. *Eur. J. Biochem.* **182,** 457–476.

Roberts, J. A., Marklund, B.-I., Ilver, D., Haslam, D., Kaack, M. B., Baskin, G., Möllby, R., Winberg, J., and Normark, S. (1994). The Galα1,4Galspecific tip adhesin of *Escherichia coli* P-fimbriae is needed for pyelonephritis to occur in the urinary tract. *Proc. Natl. Acad. Sci. U.S.A.,* in press.

Rydén, C., Maxe, I., Franzén, A., Ljungh, Å., Heinegård, D., and Rubin, K. (1987). Selective binding of BSP to *Staphyloccus aureus* in osteomyelitis. *Lancet* **335,** 515–518.

St. Geme, J. W., III (1994). The HMW1 adhsein of nontypable *Haemophilus influenzae* recognizes sialylated glycoprotein receptors on cultured conjunctival cell. *Infect. Immun.*, in press.

Sakamoto, J., Watanabe, T., Tokumaru, T., Tagaki, H., Nakazoto, H., and Lloyd, K. O. (1989). Expression of Lewis[a], Lewis[b], Lewis[x], Lewis[y], sialyl-Lewis[a], and sialyl-Lewis[x] blood group antigens in human gastric carcinoma and in normal gastric tissue. *Cancer Res.* **49**, 745–752.

Savage, A. V., D'Arcy, S. M. T., and Donoghue, C. M. (1991). Structural characterization of neutral oligosaccharides with blood group A and H activity isolated from bovine submaxillary mucin. *Biochem. J.* **279**, 95–103.

Schönbeck, C., Björck, L., and Kronvall, G. (1981). Receptors for fibrinogen and aggregated β_2-microglobulin detected in strains of group B streptococci. *Infect. Immun.* **31**, 856–861.

Schryvers, A. B., and Lee, B. C. (1993). Analysis of bacterial receptors for host iron-binding proteins. *J. Microbiol. Methods* **18**, 255–266.

Sharon, N. (1987). Bacterial lectins, cell-cell recognition and infectious disease. *FEBS Lett.* **217**, 145–157.

Sjöbring, U., Trojnar, J., Grubb, A., Åkerström, B., and Björck, L. (1989). Ig-binding bacterial proteins also bind proteinase inhibitors. *J. Immunol.* **143**, 2948–2954.

Smoot, D. T., Resau, J. H., Naab, T., Desborders, B. C., Gilliam, T., Bullhenry, K., Curry, S. B., Nidiry, J., Sewchand, J., Millsrobertson, K., Fontin, K., Abebe, E., Dillon, M., Chippendale, G. R., Phelps, P. C., Scott, V. F., and Mobley, H. L. T. (1993). Adherence of *Helicobacter pylori* to cultured human gastric epithelial cells. *Infect. Immun.* **61**, 350–355.

Spiro, R. G., and Bhoyroo, V. D. (1974). Structure of the O-glycosidically linked carbohydrate units of fetuin. *J. Biol. Chem.* **249**, 5704–5717.

Stahl, E. (1967). "Dünnschichtschromatographie." Springer-Verlag, Berlin.

Strömberg, N., and Borén, T. (1992). *Actinomyces* tissue specificity may depend on differences in receptor specificity for GalNAcβ-containing glycoconjugates. *Infect. Immun.* **60**, 3268–3277.

Strömberg, N., Marklund, B.-I., Lund, B., Ilver, D., Hamers, A., Gaastra, W., Karlsson, K.-A., and Normark, S. (1990). Host-specificity of uropathogenic *Escherichia coli* depends on differences in binding specificity to Galα1,4Gal-containing isopeceptors. *EMBO J.* **9**, 2001–2010.

Strömberg, N., Nyholm, P.-G., Pascher, I., and Normark, S. (1991). Saccharide orientation at the cell surface affects glycolipid receptor function. *Proc. Natl. Acad. Sci. U.S.A.* **88**, 9340–9344.

Strömberg, N., Borén, T., Carlén, A., and Olsson, J. (1992). Salivary receptors for GalNAcβ-sensitive adherence of *Actinomyces spp.*: Evidence for heterogenous GalNAcβ and proline-rich protein receptor properties. *Infect. Immun.* **60**, 3278–3286.

Thomas, T. C., and McNamee, M. G. (1990). Purification of membrane proteins. *In* "Methods in Enzymology" (M. Deutscher, ed.), Vol. 182, pp. 499–520. Academic Press, San Diego.

Torres, B. V., McCrumb, D. K., and Smith, D. F. (1988). Glycolipid-lectin interactions: Reactivity of lectins from *Helix pomatia*, *Wisteria floribunda*, and *Dolichos biflorus* with glycolipids containing N-acetylgalactosamine. *Arch. Biochem. Biophys.* **262**, 1–11.

van den Berg, F. M., Ziljmans, H., Langenberg, W., Rauws, E., and Schipper, M. (1989). Detection of *Campylobacter pylori* in stomach tissue by DNA *in situ* hybridisation. *J. Clin. Pathol.* **42**, 995–1000.

VanHeyningen, T., Fogg, G., Yates, D., Hanski, E., and Caparon, M. (1993). Adherence and fibronectin binding are environmentally regulated in the group A streptococci. *Mol. Microbiol.* **9**, 1213–1222.

Whitnack, E., and Beachey, E. H., (1982). Antiopsonic activity of fibrinogen bound to M protein on the surface of group A streptococci. *Infect. Immun.* **69**, 1042–1048.

Woodward, M. P., Young, W. W., Jr., and Bloodgood, R. A. (1985). Detection of monoclonal antibodies specific for carbohydrate epitopes using periodate oxidation. *J. Immunol. Methods* **78**, 143–153.

Wyatt, J. I., Rathbone, B. J., Sobala, G. M. Shallcross, T., Heatly, R. V., Axon, A. T. R., and Dixon, M. F. (1990). Gastric epithelium in the duodenum: Its association with *Helicobacter pylori* and inflammation. *J. Clin. Pathol.* **43**, 981–989.

CHAPTER 10

Cytoadherence and the *Plasmodium falciparum*-Infected Erythrocyte

Ian Crandall and Irwin W. Sherman

Department of Biology
University of California at Riverside
Riverside, California 92521

I. Introduction
II. Ligands for Adherence
    A. ICAM-1
    B. ELAM-1 and VCAM-1
    C. Thrombospondin (TSP)
    D. CD36
III. *P. falciparum*-Infected Red Cell Adhesins
    A. PfEMP1
    B. Pfalhesin
IV. Cytoadherence, an *in Vitro* Model of Sequestration: Practical Considerations of Cytoadherence Assays
    A. Parasite Lines
    B. Target Cells
    C. Assay Conditions
    D. An Example of a Cytoadherence Assay
References

I. Introduction

The hemoglobin-containing erythrocyte is a free-floating, non-adherent cell. Were erythrocytes to become adhesive, clumping and microvessel occlusion could result in diminished blood flow and the transport function of the circulatory system would be impaired. With certain diseases such as thalassemia (Butthep *et al.,* 1992), diabetes (Wautier *et al.,* 1981), and sickle cell anemia (Hoover *et al.,* 1979) red cells do become adhesive (Berendt, 1991), and, when erythrocytes are infected with the human malaria parasite, *Plasmodium falciparum,* these cells adhere to the endothelial cells lining the blood vessels and block capillaries and venules (MacPherson *et al.,* 1985; Warrell, 1987; Pongponratn *et al.,* 1991). Adhesion of the *P. falciparum*-infected red cell is a developmentally regulated process. Only red cells bearing mature pigmented forms of the parasite, called trophozoites and schizonts, attach to the capillary and postcapillary venules. Red cells containing young, unpigmented parasites, called ring stages, are nonadhesive, and circulate freely; consequently they are found in the peripheral blood along with uninfected cells (Bignami and Bastianelli, 1890). Sequestration is the term applied to the absence of mature forms in the peripheral blood; the biological significance of sequestration is not completely understood; however, sequestered red cells can by mechanical obstruction lead to organ dysfunction, coma, and death (Howard and Gilladoga, 1989; MacPherson *et al.,* 1985).

The asexual developmental cycle of *P. falciparum* within the red cell is 48 hr; mature forms appear at ~24 hr and persist through the next day. Protuberances (visible by scanning and transmission electron microscopy, and called knobs) appear on the outer surface of red cells bearing the pigmented stages (Gruenberg *et al.,* 1983; Langreth *et al.,* 1978); thus, knobs are correlated with the adhesiveness of the malaria-infected erythrocyte (Aikawa *et al.,* 1990). Some parasite lines lose their capacity to form knobs; knobless (K−) lines are a consequence of a spontaneous loss of a subtelomeric portion of chromosome 2, have a partial or complete deletion of the gene encoding the knob-associated histidine rich protein (KAHRP), and do not sequester (Biggs *et al.,* 1989a). A subtelomeric deletion of chromosome 9 results in a reduction in cytoadherence as well as lack of expression of a putative adhesin, PfEMP 1 (Day *et al.,* 1993). However, it is not clear whether this deletion encodes a structural or a regulatory gene that affects binding. Knoblessness is irreversible and K− lines are rare in natural infections (Biggs *et al.,* 1989b).

II. Ligands for Adherence

The surface of endothelial cells or other target cells contains adhesion molecules that play a critical role in sequestration and cytoadherence. Four families of adhesion molecules have been identified as ligands for *P. falciparum*-infected red cells: (1) the immunoglobulin gene superfamily (ICAM-1, VCAM), (2) the selectin family (ELAM-1), (3) the thrombospondins, and (4) the CD36/LIMP II family.

A. ICAM-1

Berendt *et al.* (1989) have shown that the binding of the ITO4 line (selected from the Ituxi isolate) was significantly inhibited by anti-ICAM antibodies different from those that recognized the sites for LFA-1 and Mac-1 (and which are involved in the binding of rhinoviruses) (Ockenhouse *et al.*, 1992c; Staunton *et al.*, 1990). The binding region for malaria-infected cells has been localized to domains 1 and 2 of the five tandemly arranged immunoglobulin-like domains (Berendt *et al.*, 1992). Stimulation of HUVEC by the cytokines IL-1 and TNF, as well as LPS, enhanced the adhesiveness of these target cells for the ITO4 line, but did not affect the binding of FCR-3 (Berendt *et al.*, 1989). The ITO4 line, but not FCR-3, bound to ICAM-1-transfected cells; since FCR-3 adheres only to CD36-transfected COS cells, and ITO4 binds to both ICAM-1 and CD36 transfectants, it is clear that the ability of a plasmodial line to bind to ICAM-1 is not a universal feature of *P. falciparum* isolates. Indeed, the binding of erythrocytes obtained from Thai patients with uncomplicated and severe malaria to immobilized purified ICAM-1 or CD36 showed the number of bound infected red cells to be in the ratio 1:4 (Singh *et al.*, 1988).

B. ELAM-1 and VCAM-1

Ockenhouse *et al.* (1992b) have shown that erythrocytes containing a cloned line of parasites with enhanced binding compared to the parental wild-type isolate could be obtained by sequential panning over purified ELAM-1 and VCAM-1. The binding was specific to ELAM and VCAM and independent of other CAMs. Since these ligands can be upregulated, and their expression is more frequently associated with brain tissue from patients who died from cerebral malaria than from other causes, it may be that coexpression of these ligands contributes, in part, to the pathologic consequences of *P. falciparum* infections.

The binding of infected red cells to these ligands, in addition to TSP, ICAM-1, and CD36 (see below), suggests that the surface of the infected erythrocyte possesses multiple adhesins.

C. Thrombospondin (TSP)

Parasite lines bind to TSP immobilized onto plastic as well as target cells that express this large (420-kDa) glycoprotein (Roberts *et al.*, 1985; Lawler and Hynes, 1986). The region on TSP that mediates adhesion is the C-terminal calcium binding globular domain, although the heparin-binding domain is also involved in adhesion (Sherwood *et al.*, 1990). Cytoadherence is specifically inhibited by soluble TSP and anti-TSP monoclonal antibodies under both static and shear flow conditions (Barnwell *et al.*, 1989). Although TSP may contribute to cytoadherence, it cannot be a principal ligand since there are parasite lines that do not adhere to melanoma cells that secrete TSP, and some anti-TSP antibodies neither bind to nor inhibit the attachment of infected cells to amelanotic melanoma cells (Panton *et al.*, 1987; Sherwood *et al.*, 1989). Further, it was found that CD36-transfected COS cells supported the binding of infected red cells, yet there was no binding of TSP to the transfectants (Oquendo *et al.*, 1989).

D. CD36

The sialylated 88-kDa glycoprotein CD36 (also termed GP88, GPIIIb and PASIV) is found on platelets, monocytes, amelanotic melanoma cells, fetal erythrocytes, U937 myelmonocytic cells, human erythroleukemia cells, and certain endothelial cells (Tandon *et al.*, 1989). There is tissue-specific expression of CD36 on endothelial cells; for example, it is not present on HUVECs, but is present on HDMEC and HBEC (Smith *et al.*, 1992). Nothing is currently known about the development/regulation of CD36 biosynthesis. Although CD36 has been claimed to be a ligand for TSP (Silverstein *et al.*, 1989), this has recently been questioned (Tandon *et al.*, 1991).

In vitro, the expression of CD36 (and ICAM-1) on HDMEC can be upregulated using γ-interferon and downregulated by the protein kinase C agonist phorbol 12,13-dibutyrate (Wick *et al.*, 1992; Johnson *et al.*, 1993).

The FC27 and FCR-3 lines do not bind to ICAM-1; instead their binding is exclusively via CD36 (Berendt *et al.*, 1989). The HB3 line binds to both ICAM-1 and CD36; however, since upregulation of ICAM-1 by TNF did not lead to an increase in the binding of infected red cells to HDMEC, it was suggested that the major ligand for adherence of *P. falciparum*-infected erythrocytes to the microvascular endothelium is CD36 (Wick *et al.*, 1992).

The CD36-specific monoclonal antibodies OKM5 and OKM8 inhibit the adhesion of infected erythrocytes to amelanotic melanoma cells, and the presence of the OKM5 antigen is directly correlated with the adherent capacity of such cells (Panton *et al.*, 1987); suppression of CD36 synthesis by exposure to tunicamycin or castanospermine-6-butyrate disrupts cytoadherence (Wright *et al.*, 1991). Malaria-infected red cells bind specifically to purified human [but not bovine (Greenwalt *et al.*, 1990)] CD36 immobilized onto plastic or to COS cells transfected with CD36 cDNA (Oquendo *et al.*, 1989). (Binding to bovine CD36 required selection by "panning" parasitized cells on bovine CD36.) Because short peptides based on the CD36 amino acid sequence fail to block cytoadherence, it has been suggested that the CD36 binding site may be composed of discontinuous stretches of amino acids that are exposed only in its tertiary structure (Greenwalt *et al.*, 1990). Although purified human (platelet) CD36 binds to all culture-adapted parasite lines and to most field isolates, the degree of binding is highly variable and no correlation was found between severity of disease and the extent of *in vitro* binding to this ligand (Singh *et al.*, 1988).

III. *P. falciparum*-Infected Red Cell Adhesins

Although there are several well-characterized candidates for the ligand(s) on the target cell, the identification and characterization of the adhesin(s) on the *P. falciparum*-infected red cell have not advanced to that stage. There are two broad categories of adhesins on the *P. falciparum*-infected red cell: parasite-encoded proteins and host-related proteins.

A. PfEMP1

The best characterized parasite encoded adhesin is *P. falciparum* erythrocyte membrane protein 1 (PfEMP-1). This strain-specific high-molecular-weight (>240,000) protein can be metabolically labeled, radioiodinated, and immunoprecipitated from SDS extracts of adherent infected red cells by hyperimmune sera (Howard *et al.*, 1988). PfEMP-1 is protease sensitive (a property that results in abrogation of adhesiveness (Leech *et al.*, 1984) and its expression is correlated with the cytoadherent behavior of the red cell (Howard *et al.*, 1990).

Using the anti-CD36 antibody OKM 8, Ockenhouse *et al.* (1991) prepared an anti-idiotype rabbit antibody. This antibody was neither strain- nor knob-specific, but did react (by immunofluorescence) only with cytoadherent lines and inhibited binding of infected cells to amelanotic melanoma cells. The anti-idiotype antibody immunoprecipitated ~270-kDa radioiodinated protein from Triton X-100-soluble lysates of a knobless cytoadherent line and a protein of similar size was labeled by radioactive isoleucine. Such proteins were not in evidence if intact infected red cells were treated with trypsin. This protein, called sequestrin, may be identical to PfEMP-1; however, there are some differences: PfEMP-1 is Triton X-100 insoluble, is antigenically diverse amongst strains, and is present in knobby but not knobless lines. There is evidence for considerable and rapid antigenic diversity within a single parasite isolate (Roberts *et al.*, 1993), a factor that complicates the study of parasite-encoded surface adhesins. The gene(s) encoding PfEMP-1 and/or sequestrin has not been cloned or sequenced.

B. Pfalhesin

A protein called pfalhesin is the only host-related adhesin that has been identified and characterized (Crandall *et al.*, 1993). Pfalhesin consists of amino acids 546–553 and 824–829 of the human anion transport protein, band 3. Although pfalhesin is present in uninfected red cells, it is normally cryptic (Crandall and Sherman, 1994). Antibodies directed against pfalhesin and synthetic peptides based on the amino acid sequence of pfalhesin inhibit CD36-mediated adhesion (Crandall *et al.*, 1991, 1994). Therefore it has been concluded that pfalhesin and CD36 form an adhesin/ligand pair. The CD36-mediated (and pfalhesin-mediated) adhesion is found in both knobby and knobless parasite lines and is present in all isolates (I. Crandall, unpublished data). The expression of parasite-encoded adhesins (such as PfEMP-1 or sequestrin) can be selected for or against—this appears not to be the case for pfalhesin/CD36-mediated adhesion.

≡≡≡≡ IV. Cytoadherence, an *in Vitro* Model of Sequestration: Practical Considerations of Cytoadherence Assays

A. Parasite Lines

P. falciparum can be maintained in *in vitro* culture by using a simple tissue culture medium (RPMI 1640), human red blood cells, and human serum (Trager and Jensen, 1976). N.B. Care must be taken to avoid spillage since *P. falciparum* and human blood are biohazards. Parasites can be cultured in any one of several blood types (i.e., O^+ or A^+), but care must be taken to avoid the use of a serum that will agglutinate red cells. Routinely we employ O^+ red cells; this prevents agglutination and the potential for rosetting (David *et al.*, 1988), phenomena that can nonspecifically affect cytoadherence. Since the adherent stages of the parasite are the pigmented forms of *P. falciparum*, infected red cells are often manipulated in order to obtain synchronous cultures with parasites at the same developmental stage, such that adhesin expression is consistent; in addition, with the use of a defined parasitemia and a fixed hematocrit, variation in binding is minimized.

Stage synchronization of a culture may be achieved by collecting the mature knob-bearing forms by gelatin flotation (Pasvol *et al.*, 1978), or by synchronization of the cultures by sorbitol lysis of erythrocytes containing mature forms of the parasite on the previous day (Lambros and Vanderberg, 1980). Knobless lines of parasites cannot be concentrated by gelatin flotation and therefore sorbitol lysis is the only practical method of synchronization. Cultured parasite lines tend toward asynchronous growth; therefore the periodic use of gelatin flotation maintains the synchrony of the culture and provides the added benefit of removing knobless erythrocytes and older uninfected red cells. It is possible to select lines for binding to defined ligands by "panning" parasite cultures over target cells and collecting adherent parasites (Ockenhouse *et al.*, 1992b). Such a method may lead to parasite lines with an increased expression of a particular adhesin; since the panning process must be performed with target cells under sterile conditions, this method can be time-consuming and technically demanding.

The parasitemia of the blood sample used for cytoadherence studies can be varied to suit the experiment; however, in our experience a parasitemia of ~5% provides acceptable results and has the benefit of allowing samples to be taken directly from *in vitro* cultures. If a defined parasitemia is required, the mature parasite population can be isolated by gelatin flotation, the parasitemia of the enriched fraction can be determined, and the enriched fraction can then be suitably diluted with uninfected blood to produce parasitemias between 0 and 90%. Producing a parasitemia higher than that found in *in vitro* cultures (i.e., 5–10%) increases the preparation time by about 1 hr.

As a general rule healthy cultures produce infected erythrocytes that display higher levels of adherence. Supplementation of culture media with 5 m*M* hypoxanthine appears to reduce variability in parasite growth. Some reports indicate that long term *in vitro* culture leads to a loss of adhesiveness despite the presence of knobs (Day *et al.*, 1993). We have not found this to be the case for CD36-mediated adhesion and have used the same parasite line (FCR-3), continuously passaged for several years in the laboratory, for adhesion experiments.

B. Target Cells

The binding of an infected erythrocyte to a target cell may involve more than a single ligand. For example, the parasite line ITO4 may adhere through both CD36 and ICAM-1 ligands, whereas the FCR-3 line binds exclusively to CD36. As a consequence, the degree to which a target cell line is adherent or nonadherent depends on the parasite line being studied. Some target cell lines express many different kinds of adhesion molecules (i.e., C32 amelanotic melanoma cells), whereas others may be more limited (i.e., HUVEC express ICAM-1 but not CD36). If a specific adhesin/ligand pair is to be studied, it is necessary to determine the kind of ligand on the target cell by either confirming the presence or absence of a particular molecule with a ligand-specific antibody or resorting to a cell type that normally lacks the ligand but has been transfected with the ligand of interest [e.g., CHO cells have been successfully transfected with cDNA for CD36 or ICAM-1 and used for cytoadherence assays (Hasler *et al.*, 1993)]. Cell lines that naturally express ligand molecules on their surface have the advantage of remaining stable through many cell divisions, but may be difficult and/or expensive to maintain *in vitro* (i.e., HUVEC and HBEC); transfected cell lines are easier to propagate, but should they lose the transfected element a population of nonadherent target cells may arise. Transfected cell lines can be stored frozen in liquid nitrogen and samples retrieved easily; this helps to overcome the problem of the loss of ligand expression and also permits testing of lines of defined age.

Most target cell lines adhere to glass (either uncoated or coated with a protein such as gelatin or fibronectin) and therefore it is convenient to seed these cells onto glass slides. Glass slides with a Teflon coating that creates depression "spots" are particularly useful. A 50-µl aliquot of a suspension of cells (10^4 cells/ml) is allowed to settle and attach for 24–48 hr, the tissue culture medium is flicked off, and then the target cells are fixed to prevent detachment. Slides containing fixed target cells can be stored in an isotonic medium (i.e., RPMI 1640) at 4°C for a few weeks. The method of fixation can affect the activity of the target cell's ligand; one method of fixing target cells that appears not to impair the function of some of the more frequently studied ligands (i.e., CD36 and ICAM-1) is a 30-min exposure at room temperature to 1% (w/v) paraformaldehyde /0.1 *M* sodium cacodylate/4% (w/v) sucrose. Fixation with 0.5–1% (v/v) glutaraldehyde frequently destroys the ligand, and in our experience ICAM-1 is more sensitive to methanol fixation than is CD36.

One means of eliminating target cell variability is by the use of isolated ligand molecules bound to an inert substrate such as plastic. This method frequently produces unambiguous results, but requires a source of purified ligand and proper configuration of the immobilized ligand.

C. Assay Conditions

1. Temperature and Incubation Buffer Composition

The cytoadherence of infected erythrocytes is a temperature-dependent process; infected erythrocytes do not adhere at 4°C, give diminished adherence at 15–20°C, and show optimal binding at 25–37°C. Since binding studies are frequently performed in replicates with negative controls (i.e., no antibody or peptide added), the effect of temperature on individual experiments may be discounted; however, uneven temperature control of the environment in which the experiments are performed can result in significant day-to-day variations between control values; the use of a heat lamp, incubator, or warming table may be useful to reduce such temperature-related variablility.

Cytoadherence of a particular parasite line to a target cell may vary with environmental conditions other than temperature. Since pH can affect the degree of adherence (Fig. 1), it may be necessary to alter the pH to optimize binding. For example, the pH optimum for CD36-mediated adherence is lower (pH 6.6–6.8) than that of serum (Crandall et al., 1991). The medium in which binding is carried out may also be a factor; we find Bis Tris to be a good choice of a buffer salt since its pK_a (6.5) provides maximal protection from pH drift during the assay, and its composition does not appear to inhibit cytoadherence. (MES, PIPES, HEPES, ACES, MOPS, and MOPSO inhibit cytoadherence to varying degrees.) The inclusion of Ca^{2+} in the incubation medium results in increased adherence, but the level used (25–50 mM) is nonphysiological. However, Ca^{2+} is not critical for binding and may be eliminated. Serum does not appear to be a necessary component of the binding medium and its omission helps to reduce the effect of batch-to-batch variability (Sherman and Valdez, 1989).

ICAM-1-mediated adherence is relatively pH insensitive (Fig. 1), and although less is known about the optimum buffer conditions for ICAM-1 mediated adherence, it would appear that the buffer conditions are less critical. We have adopted the practice of using a medium consisting of 50 mM Bis Tris/100 mM NaCl/25 mM calcium lactate (pH 6.6) for most adhesion assays, since it promotes CD36-mediated adherence without inhibiting ICAM-1-mediated adherence. It should be emphasized that as new adhesin/ligand pairs are discovered the composition of the binding medium may require optimization.

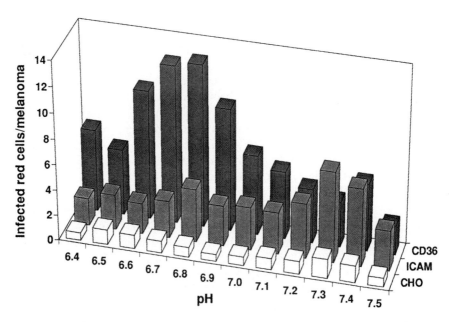

Fig. 1 Adherence versus pH. CHO cells expressing no ligand (front row), ICAM-1 (middle row), or CD36 (back row) were assayed for cytoadherence at various pH values.

2. Physical Agitation

Cytoadherence assays are frequently performed using a suspension of infected and uninfected erythrocytes that will settle out within 10–15 min. To overcome this, the binding medium is gently agitated. One method is to use a mechanical rocking motion (side to side) of sufficient strength to keep the blood cells suspended and to provide good contact between the infected erythrocytes and the target cells attached to a glass or plastic surface. It is critical to ensure that the slowest speed necessary for red cell suspension be used, that shear forces are minimized, and that the volume of medium is compatible with the shape and size of the container. We have found that a $10 \times 3 \times 2$-cm vessel with 5 ml of blood made up to a 5% (v/v) suspension should be rocked at $\pm 10°$ at 15 sec/cycle. Petri dishes can be rocked in a circular motion at 3–5 sec/cycle and the fluid volume adjusted so that agitation creates a gentle "wave" that promotes suspension of the red cells.

Some researchers feel that binding studies should be carried out under flow conditions that mimic the *in vivo* condition, and therefore an *ex vivo* system that uses artificially perfused rat mesocecum microvasculature (Raventos-Suarez *et al.*, 1985; Rock *et al.*, 1988), and *in vitro* flow systems have been developed (Wick *et al.*, 1992; Rowland *et al.*, 1993; Cooke *et al.*, 1993).

3. Quantitation of Cytoadherence

Methods for the quantitation of cytoadherence can be divided into three categories: (a) quantitation of a radioactive label introduced into the infected erythrocyte, (b) colorimetric assay of an infected erythrocyte associated dye, or (c) microscopic observation.

A radioactive label can be introduced into infected erythrocytes prior to their use in a cytoadherence assay. A common method for labeling the parasitized cells is to introduce a metabolic label [such as tritiated or ^{14}C-labeled isoleucine or hypoxanthine (Wright et al., 1990); N. B. malaria parasites do not incorporate thymidine!)] or to use a radioisotope such as ^{51}Cr that will label red cells (Sherman and Valdez, 1989).

The introduction of a nonspecific (i.e., one that does not label parasites) radiolabel prior to use requires that infected red cells be isolated prior to labeling so that adherence of uninfected red cells is not quantitiated. This step can be accomplished by gelatin flotation and the extra time required for the isolation, labeling, and dilution should be taken into account when planning experiments.

Incorporation of a metabolic label (for nucleic acids) is straightforward in P. falciparum cultures since the parasite is grown in suspensions of human red cells that lack a nucleus. Unincorporated label can be washed out of the culture and the cells may be used directly in a cytoadherence assay. Care should be taken to avoid the possibility of color quenching (by hemoglobin) when the samples are prepared for liquid scintillation counting, and since variation in isotope incorporation is dependent on the stage of development this too must be taken into account.

If cells are to be labeled immediately before use, then ^{51}Cr is a useful isotope because of its high energy emissions (which allows quantitation in a gamma counter) and its ease of introduction. ^{125}I can be a useful protein labeling reagent; however, the introduction of iodine onto the surface of an infected red cell results in complete loss of CD36-mediated adherence (Sherman and Valdez, 1989) because the adhesin pfalhesin contains a tyrosine residue [the major site of iodine incorporation (Crandall et al., 1993)]. Since ^{51}Cr and ^{125}I labeling are not specific for infected red cells, measuring radioactivity bound to target cells may be misleading if both uninfected and infected red cells bind to target cells.

The use of colorimetric assays to quantitate the number of infected erythrocytes is potentially limited by the relatively small number of adherent cells that may be present in an assay. Two approaches to colorimetric assays are either to introduce a dye with a high specific absorbance into the infected erythrocytes prior to binding or to incubate the bound infected cells with a metabolic dye (such as MTT) that forms a chromogen when living cells are present (Mosmann, 1983). Adaptation of a colorimetric assay to a microplate format would be particularly desirable since there are a wide variety of automated tools (e.g., pipettors and plate readers) available. Unfortunately, the use of a chromogen usually requires that target cells be fixed onto wells. In our experience the fixation of target cells to wells is relatively simple, but it may be difficult to achieve effective cell suspension and rinsing off of unbound cells. Considerable care must be taken to remove unbound cells at the very edge of the well (an area with reduced shear forces) without removing the adherent cells from the center of the well. By the use of 24-well plates or 12-well plates instead of 96-well plates a larger surface area is provided in each well and this partially overcomes the "rim effect," however, it requires that material be transferred to a 96-well plate before it can be read by a standard plate reader.

Care should be taken that equal numbers of target cells be seeded into each well since colorimetric assays are based on the assumption that each well contains the same amount of ligand. Multiple determinations using several wells reduces the potential variability introduced by this assumption.

The most sensitive and simple quantitation method for measuring adhesion is microscopic observation (Udeinya *et al.*, 1985). In order to determine the degree of binding, the target cells are grown on a transparent substrate (plastic or glass depending on the target cell type), the adherent red cells are fixed to prevent them from detaching, and then both kinds of cells are stained to aid visualization. The sample container (i.e., Petri dish) or slide is placed on the stage of a microscope and the ratio of infected red cells to target cells determined by counting at $1000 \times$ or $400 \times$ magnification. This method has the advantage of being relatively inexpensive and simple, and also allows the experimenter to detect any conditions (rosetting, agglutination, cell lysis, adherence of uninfected material, bacterial contamination, nonspecific adherence of infected material, etc.) that will give results that are not an accurate reflection of a specific adhesin/ligand interaction. The primary disadvantage of direct microscopic examination of target and infected red cells is that it must be done manually and counting can be tedious.

Radiolabeling and colorimetric assays provide values that represent the sum of many individual infected cell/target cell interactions and therefore are less prone to variations introduced by single target cells. Microscopic observation provides the sum of results from observed target cells and therefore its statistical accuracy depends on the number of target cells counted. Counting a large sample size improves the accuracy of the assay, but this increases the amount of time and effort required for quantitation. It is frequently necessary to determine the level of accuracy desired in the assay and plan the counting process accordingly. Experience indicates that more consistent results are obtained when (1) multiple determinations (i.e., replicate slides or Petri dishes) are performed for each point in the assay, (2) target cell/infected cell determinations are made at multiple points on a single slide with results averaged, (3) the target cell area is constant, and (4) a reasonable number of infected cells/target cells are counted. Some target cells (i.e., C32 amelanotic melanoma and transfected CHO cells) lend themselves to the determination of an average number of infected cells/target cell because they are of a convenient size (i.e., they have a surface area that is equivalent to the surface area of <100 infected red cells) and a large proportion of the target cell population is adherent. Other target cells are significantly larger in size (i.e., HBEC); however, only a subpopulation of these cells may express ligand. This may result in highly variable averages that do not accurately reflect the cytoadherent potential of either the infected cell or the target cell. The results of such assays can be more accurately expressed using a "bin method" where the number of cells expressing a given range of cytoadherent ratios (i.e., 0–50, 51–100, 101–150) is determined (Smith *et al.*, 1992).

D. An Example of a Cytoadherence Assay

The basis of any cytoadherence assay is to put two cell types, one that expresses adhesins and a second that expresses ligand, together (Udeinya *et al.*, 1981). In order to maximize binding we routinely employ the following method.

Malaria parasites are cultured in human O^+ blood and are kept as a synchronous culture. The experiment is planned for a day when the cultures contain mature (trophozoite or schizont) stages of parasite development. If additions are to be made to the parasite cultures to determine the effect of a substance on cytoadherence then cultures can be treated 24–48 hr before use. It is critical to check that the added material does not alter the development or morphology of the parasites; this can be monitored by Giemsa staining thin blood films from the cultures and determining the developmental stage and morphology of the parasites in the culture. In general we find 12 target slides per experiment to be a convenient number (12 = 6 duplicates = 4 triplicates = 3 quadruplicates), therefore we plan to have six culture plates (0.5-ml packed cell volume/10-cm Petri plate) at ~5% parasitemia to be available for use. Twelve target slides are removed from storage and placed in individual compartments of a transparent "tackle box" with 4.5 ml of BTC buffer [50 mM Bis Tris/100 mM NaCl/25 mM calcium lactate (pH6.6)]. The slides have a frosted portion that can be used to write an identifying mark on each slide with a pencil. If additions are to be made to an individual compartment (antibody, peptide, etc.,), the added material is dissolved in BTC at a high concentration (i.e., 1 mg/ml for peptides) and additions are made in the minimum volume possible. (Peptide solutions are frequently added in the 0- to 100-μl range.) Care should be taken to avoid additions that will drastically alter the tonicity, pH, or viscosity of the incubation medium. Additions of large amounts (1–4 ml) of tissue culture media may result in a pH or composition shift; therefore it is advisable to ensure that these shifts are monitored and/or controlled for by altering the composition of the control buffer.

Blood suspensions at ~5% parasitemia are pelleted at 700 g for 3 min; all but 6 ml (i.e. 12 × 0.5 ml) of the supernatant is removed by aspiration. (If the blood suspension is being pooled in two tubes the contents of the tubes *must* be mixed together.) One half milliliter of the cell suspension is added to each compartment of the "tackle box." The tackle boxes are then placed on a rocking table and allowed to rock for 90 min (Fig. 2). The room temperature is noted and if below 22°C heat should be supplied. After 90 min the blood suspension

Fig. 2   A cytoadherence assay in progress. Multispot slides bearing target cells are exposed to a blood suspension that contains *P. falciparum*-infected erythrocytes. The separate sections of a "tackle box" allow the experimental conditions to be varied between slides. The rocker platform keeps the red cells in suspension.

is removed by *gentle* aspiration and 5 ml of BTC is added to each compartment. The boxes are then agitated for a further 5 min. This prerinse step removes nonadherent erythrocytes and allows for fewer subsequent rinses. The slides are removed and "dipped" once or twice into a beaker filled with 150 ml of BTC. Care must be taken to avoid excessive agitation of the slide in the beaker, and in particular side-to-side motions should be avoided. When the slide is removed from the beaker a blood film will run down the slide surface; the removal of this film can be hastened by gently touching the bottom of the slide to the lip of the beaker. The blood film that is visible on the slide results from

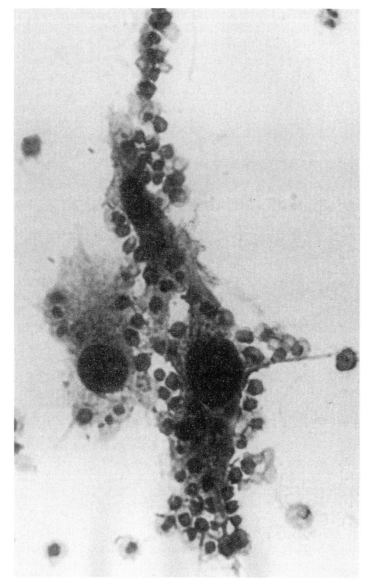

Fig. 3 *P. falciparum*-infected erythrocytes adhered to C32 amelanotic melanoma cells.

the slight attraction between uninfected red cells and the glass surface. In most instances the bound infected cells will not be apparent to the unaided eye. Since repeated dipping results in slides with greatly reduced numbers of infected erythrocytes bound to the target cells the minimum number of dippings sufficient to remove all nonadherent cells should be used. (We find two dippings to be effective.) The rinsed slide can be placed in a rack and should be kept moist in preparation for the next step. After slides have been rinsed they are placed in a fixing solution (2% glutaraldehyde, 0.1 M sodium cacodylate, 4% sucrose) for 30 min at room temperature. After fixing, the slides are placed in Geimsa stain (5%) for at least 1 hr (or overnight, which is more convenient). The slides are then removed from the stain and rinsed with distilled water and allowed to air dry. The slides are observed at $400\times$ or $1000\times$ magnification (the samples are stable under oil immersion) and the ratio of target cells to infected erythrocytes determined for multiple fields across the slide (Fig. 3). A hand-operated counting device is used to record the number of bound erythrocytes and counting is continued until a predetermined number of cells is reached. We frequently count 200 target cells or 300 infected red cells on each slide. If uninfected red cells constitute a significant proportion of the population (i.e., >5%), then the entire experiment is rejected.

Acknowledgments

We thank Jacques Prudhomme for maintaining the parasite cultures and preparing the target cells, and providing excellent advice and input. This research was supported in part by grants from the UNDP/World Bank/WHO Special Programme for Research and Training in Tropical Diseases and from the National Institute of Allergy and Infectious Diseases (National Institutes of Health).

References

Aikawa, M., Iseki, M., Barnwell, J., Taylor, D., Oo, M. M., and Howard, R. (1990). The pathology of human cerebral malaria. *Am. J. Trop. Med. Hyg.* **43**, 30–37.

Barnwell, J. W., Asch, A. S., Nachman, R. L., Yamada, M., Aikawa, M., and Ingravallo, P. (1989). A human 88kDa membrane glycoprotein (CD36) functions in vitro as a receptor for a cytoaderence ligand on *Plasmodium falciparum*-infected erythrocytes. *J. Clin. Invest.* **84**, 765–772.

Berendt, A. R. (1991). Erythrocyte-endothelial interactions in *Plasmodium falciparum* malaria, sickle cell anaemia and diabetes. *In* "Vascular Endothelium: Interactions with Circulating Cells" (J. L. Gordon, ed.), pp. 253–275. Elsevier, Amsterdam.

Berendt, A. R., Simmons, D. L., Tansley, J., Newbold, C. I., and Marsh, K. (1989). Intercellular adhesion molecule 1 is an endothelial cell adhesion receptor for *Plasmodium falciparum*. *Nature (London)* **341**, 57–59.

Berendt, A. R., McDowall, A., Craig, A. G., Bates, P. A., Sternberg, M. J., Marsh, K., Newbold, C. I., and Hogg, N. (1992). The binding site on ICAM-1 for *Plasmodium falciparum*-infected erythrocytes overlaps, but is distinct from, the LFA-1-binding site. *Cell (Cambridge, Mass.)* **68**, 71–81.

Biggs, B. A., Culvenor, J. G., Ng, J. S., Kemp, D. J., and Brown, G. V. (1989a). *Plasmodium falciparum:* Cytoadherence of a knobless clone. *Exp. Parasitol.* **69**, 189–197.

Biggs, B. A., Kemp, D. J., and Brown, G. V. (1989b). Subtelomeric chromosome deletions in field isolates of *Plasmodium falciparum* and their relationship to loss of cytoadherence in vitro. *Proc. Natl. Acad. Sci. U.S.A.* **86**, 2428–2432.

Bignami, A., and Bastianelli, G. (1890). Osservazioni sulle febbri malariche estive autunnali. *Riforma Med.* **223**, 1334–1335.

Butthep, P., Bunyaratvej, A., Kitaguchi, H., Funahara, Y., and Fucharoen, S. (1992). Interaction between endothelial cells and thalassemic red cells in vitro. *Southeast Asian J. Trop. Med. Public Health* **23**, Suppl. 2, 101–104.

Cooke, B. M., Morris-Jones, S., Greenwood, B. M., and Nash, G. B. (1993). Adhesion of parasitized red blood cells to cultured endothelial cells: A flow-based study of isolates from Gambian children with falciparum malaria. *Parasitology* **107**, 359–368.

Crandall, I., and Sherman, I. W. (1991). *Plasmodium falciparum* (human malaria)-induced modifications in human erythrocyte band 3 protein. *Parasitology* **102**, 335–340.

Crandall, I., and Sherman, I. W. (1994). Cytoadherence-related neoantigens on *Plasmodium falciparum* (human malaria)-infected human erythrocytes result from the exposure of normally cryptic regions of the band 3 protein. *Parasitology* **108**, 257–267.

Crandall, I., Smith, H., and Sherman, I. W. (1991). Plasmodium falciparum: The effect of pH and Ca^{2+} concentration on the in vitro cytoadherence of infected erythrocytes to amelanotic melananoma cells. *Exp. Parasitol.* **73**, 362–368.

Crandall, I, Collins, W. E., Gysin, J., and Sherman, I. W. (1993). Synthetic peptides based on motifs present in human band 3 protein inhibit cytoadherence/ sequestration of *Plasmodium falciparum* (human malaria). *Proc. Natl. Acad. Sci. U.S.A.* **90**, 4703–4707.

Crandall, I., Land, K. M., and Sherman, I. W. (1994). *Plasmodium falciparum:* Pfalhesin and CD36 form an adhesin/receptor pair that is responsible for the pH dependent portion of cytoadherence/ sequestration. *Exp. Parasitol.* **78**, 203–209.

David, P. H., Handunnetti, S. M., Leech, J. H., Gamage, P., and Mendis, K. (1988). Rosetting: A new cytoadherence property of malaria-infected erythrocytes. *Am. J. Trop. Med. Hyg.* **38**, 289–297.

Day, K. P., Karamalis, F., Thompson, J., Barnes, D. A., Peterson, C., Brown, H., and Kemp, D. J. (1993). Genes necessary for expression of a virulence determinant and for transmission of *Plasmodium falciparum* are located on a 0.3 megabase region of chromosome 9. *Proc. Natl. Acad. Sci. U.S.A.* **90**, 8292–8296.

Greenwalt, D., Watt, K., Hasler, T., Howard, R., and Patel, S. (1990). Structural, functional, and antigenic differences between bovine heart endothelial CD36 and human platelet CD36 and human platelet CD36. *J. Biol. Chem.* **265**(27), 16296–16299.

Gruenberg, J., Allred, D. R., and Sherman, I. W. (1983). Scanning electron microscope-analysis of the protrusions (knobs) present on the surface of *Plasmodium falciparum*-infected erythrocytes. *J. Cell. Biol.* **97**, 795–802.

Hasler, T., Albrecht, G. R., Van Schravendijk, M. R., Aguiar, J. C., Morehead, K. E., Pasloske, B. L., Ma, C., Barnwell, J. W., Greenwood, B., and Howard, R. J. (1993). An improved microassay for *P. falciparum* cytoadherence using stable transformants of Chinese hamster ovary cells expressing CD36 or intercellular adhesion molecule-1. *Am. J. Trop. Med. Hyg.* **48**, 332–347.

Hoover, R., Rubin, R., Wise, G., and Warren, R. (1979). Adhesion of normal and sickle erythrocytes to endothelial monolayer cultures. *Blood* **54**, 872–876.

Howard, R. J., and Gilladoga, A. D. (1989). Molecular studies related to the pathogenesis of cerebral malaria. *Blood* **74**, 2603–2618.

Howard, R. J., Barnwell, J. W., Rock, P. E., Neequaye, J., Oforiadjei, D., Maloy, W. L., Lyon, J. A., and Saul, A. (1988). Two approximately 300 kilodalton *Plasmodium falciparum* proteins at the surface membrane of infected erythrocytes. *Mol. Biochem. Parasitol.* **27**, 207–224.

Howard, R. J., Handunnetti, S., Hasler, T., Gilladoga, A. D., de Aguiar, J., Pasloske, B., Morehead,

D., Albrect, G., and van Schravendijk, M. (1990). Surface molecules on *Plasmodium falciparum*-infected erythrocytes involved in adherence. *Am. J. Trop. Med. Hyg.* **43**(2), 15–29.

Johnson, J. K., Swerlick, R. A., Grady, K. K., Millet, P., and Wick, T. M. (1993). Cytoadherence of Plasmodium falciparum-infected erythrocytes to microvascular endothelium is regulatable by cytokines and phorbal ester. *J. Infect. Dis.* **167**(3), 698–703.

Lambros, C., and Vanderberg, J. P. (1980). Synchronization of *Plasmodium falciparum* erythrocytic stages in culture. *J. Parasitol.* **65**, 418–420.

Langreth, S. G., Jensen, J. B., Reese, R. T., and Trager, W. (1978). Fine structure of human malaria *in vitro*. *J. Protozool.* **25**, 443–452.

Lawler, J., and Hynes, R. O. (1986). The structure of human thrombospondin, an adhesive glycoprotein with multiple calcium-binding sites and homologies with several different proteins. *J. Cell. Biol.* **103**, 1635–1648.

Leech, J. H., Barnwell, J. W., Miller, L. H., and Howard, R. J. (1984). Identification of a strain specific malarial antigen exposed on the surface of *Plasmodium falciparum*-infected erythroytes. *J. Exp. Med.* **159**, 1567–1575.

MacPherson, G. G., Warrell, M. J., White, N. J., Looareesuwan, S., and Warrell, D. (1985). Human cerebral malaria: A quantitative ultrastructural analysis of parasitized erythrocyte sequestration. *Am. J. Pathol.* **119**, 385–401.

Mosmann, T. (1983). Rapid colorimetric assay for cellular growth and survival: Application to proliferation and cytotoxicity assays. *J. Immunol. Methods* **65**, 55–63.

Ockenhouse, C. F., Ho, M., Tandon, N. N., Van Seventer, G. A., Shaw, S., White, N. J., Jamieson, G. A., Chulay, J. D., and Webster, H. K. (1991). Molecular basis of sequestration Sequestrinin severe and uncomplicated *Plasmodium falciaprum* malaria: Differential adhesion of infected erythrocytes to CD36 and ICAM-1. *J. Infect. Dis.* **164**, 163–169.

Ockenhouse, C. F., Klotz, F. W., Tandon, N. N., and Jamison, G. A. (1992a). Sequestrin, a CD36 recoginition protein on *Plasmodium falciparum* malaria-infected erythrocytes identified by anti-idiotype antibodies. *Proc. Natl. Acad. Sci. U.S.A.* **88**, 3175–3179.

Ockenhouse, C. F., Tegoshi, T., Maeno, Y., Benjamin, C., Ho, M., Kan, K., Thway, Y., Win, K., Aikawa, M., and Lobb, R. R. (1992b). Human vascular endothelial cell adhesion receptors for plasmodium falciparum-infected erythrocytes: Roles for endothelial leukocyte adhesion molecule 1 and vascular cell adhesion molecule 1. *J. Exp. Med.* **176**, 1183–1189.

Ockenhouse, C. F., Betageri, R., Springer, T. A., and Staunton, D. E. (1992c). *Plasmodium falciparum*-infected erythrocytes bind ICAM-1 at a site distinct from LFA-1, Mac-1 and human rhinovirus. *Cell (Cambridge, Mass.)* **68**, 63–69.

Oquendo P., Hundt, E., Lawler, J., and Seed, B. (1989). CD36 directly mediates cytoadherence of *Plasmodium falciparum* parasitized erythrocytes. *Cell (Cambridge, Mass.)* **58**, 95–101.

Panton, L. J., Leech, J. H., Miller, L. H., and Howard, R. J. (1987). Ctyoadherence of *Plasmodium falciparum*-infected erythrocytes to human melanoma cell lines correlates with surface OKM5 antigen. *Infect. Immun.* **55**, 2754–2758.

Pasvol, G., Wilson, J. M., Smalley, M. E., and Brown, J. (1978). Separation of viable schizont-infected red cells of *Plasmodium falciparum* from human blood. *Ann. Trop. Med. Parasitol.* **72**, 87–88.

Pongponratn, E., Riganti, M., Punpoowong, B., and Aikawa, M. (1991). Microvascular sequestration of parasitized erythrocytes in human falciparum malaria: A pathological study. *Am. J. Trop. Med. Hyg.* **44**(2), 168–175.

Raventos-Suarez, C., Kaul, D. K., Macaluso, F., and Nagel, R. L. (1985). Membrane knobs are required for the microcirculatory obstruction induced by *Plasmodium falciparum*-infected erythrocytes. *Proc. Natl. Acad. Sci. U.S.A.* **82**, 3829–3833.

Roberts, D. D., Sherwood, J. A., Spitalnik, S. L., Panton, L. J., Howard, R., Dixit, V. M., Frazier, W. A., Miller, L. H., and Ginsburg, V. (1985). Thrombospondin binds *falciparum* malaria parasitized erythrocytes and may mediate cytoadherence. *Nature (London)* **318**, 64–66.

Roberts, D. J., Biggs, B. A., Brown, G., and Newbold, C. I. (1993). Protection, pathogenesis and phenotypic plasticity in *Plasmodium falciparum* malaria. *Parasitol. Today* **9**(8), 281–286.

Rock, E. P., Roth, E. F., Rojas-Corona, R. R., Sherwood, J. A., Nagel, R. L., Howard, R. N., and Kaul, D. K. (1988). Thrombospondin mediates the cytoadherence of *Plasmodium falciparum*-infected red cells to vascular endothelium in shear flow conditions. *Blood* **71**, 71–75.

Rowland, P. G., Nash, G. B., Cooke, B. M., and Stuart, J. (1993). Comparative study of the adhesion of sickle cells and malaria-parasitized red cells to cultured endothelium. *J. Lab. Clin. Med.* **121**(5), 706–713.

Sherman, I. W., and Valdez, E. (1989). In vitro cytoadherence of *Plasmodium falciparum*-infected erythrocytes to melanoma cells: Factors affecting adhesion. *Parasitology* **98**, 359–369.

Sherwood, J. A., Roberts, D. D., Spitalnik, S. L., Marsh, K., Harvey, E. B., Miller, L. H., and Howard, R. J. (1989). Studies of the receptors on melanoma cells for *Plasmodium falciparum* infected erythrocytes. *Am. J. Trop. Med. Hyg.* **40**, 119–127.

Sherwood, J. A., Roberts, D. D., Spitalnik, S. L., Lawler, J. W., Miller, L. H., and Howard, R. J. (1990). *Falciparum* malaria parasitized erythrocytes bind to a carboxy-terminal thrombospondin fragment and not the amino-terminal heparin binding region. *Mol. Biochem. Parasitol.* **40**, 173–182.

Silverstein, R. L., Asch, A. S., and Nachman, R. L. (1989). Glycoprotein IV mediates thrombospondin-dependent platelet-monocyte and platelet-U937 cell adhesion. *J. Clin. Invest.* **84**, 546–552.

Singh, B., Ho, M., Looareesuwan, S., Mathai, E., Warrell, D., and Hommel, M. (1988). *Plasmodium falciparum*: Inhibition/reversal of cytoadherence of Thai isolates to melanoma cells by local immune sera. *Clin. Exp. Immunol.* **72**, 145–150.

Smith, H., Nelson, J. A., Gahmberg, C. G., Crandall, I., and Sherman, I. W. (1992). *Plasmodium falciparum*: Cytoadherence of malaria-infected erythrocytes to human brain capillary and umbilical vein endothelial cells— a comparative study of adhesive ligands. *Exp. Parasitol.* **75**, 269–280.

Staunton, D. E., Dustin, M. L., Erickson, H. P., and Springer, T. A. (1990). The arrangment of the immunoglobulin-like domains of ICAM-1 and the binding sites for LFA-1 and rhinovirus. *Cell (Cambridge, Mass.)* **61**, 243–254.

Tandon, N. N., Lipsky, R. H., Burgess, W. H., and Jamieson, G. A. (1989). Isolation and characterization of platelet glycoprotein IV (CD36). *J. Biol. Chem.* **264**, 7570–7575.

Tandon, N. N., Ockenhouse, C. F., Greco, N. J., and Jamieson, G. A. (1991). Adhesive functions of platelets lacking glycoprotein IV (CD36). *Blood* **78**(11), 2809–2831.

Trager, W., and Jensen, J. B. (1976). Human malaria parasites in continuous culture. *Science* **193**, 673–675.

Udeinya, I. J., Schmidt, J. A., Aikawa, M., Miller, L. H., and Green, I. (1981). *Falciparum* malaria-infected erythrocytes specifically bind to cultured human endothelial cells. *Science* **213**, 555–557.

Udeinya, I. J., Leech, J., Aikawa, M., and Miller, L. H. (1985). An *in vitro* assay for sequestration: Binding of *Plasmodium falciparum*-infected erythrocytes to formalin-fixed endothelial cells and amelanotic malanoma cells. *J. Protozool.* **32**, 88–90.

Warrell, D. A. (1987). Pathophysiology of severe *falciparum* malaria in man. *Parasitology* **94**, 853–876.

Wautier, J.-L., Paton, R. C., Wautier, M.-P., Pintigny, D., Abadie, E., Passa, P., and Caen, J. P. (1981). Increased adhesion of erythrocytes to endothelial cells in diabetes mellitus and its relation to vascular complications. *N. Engl. J. Med.* **305**, 237–242.

Wick, T., Johnson, J., Swerlick, R., Grady, K., and Millet, P. (1992). Cytokine and pharmacologic regulation of *Plasmodium falciparum*-infected red cell adhesion to microvascular endotheial cells under shear conditions. Suppl., to *Am. J. Trop. Med. Hyg.* **47**(4), 149.

Wright, P. S., Cross-Doersen, D. E., McCann, P. P. and Bitonti, A. J. (1990). *Plasmodium falciparum*: A rapid assay for cytoadherence of [³H] hypoxanthine-labeled infected erythrocytes to human melanoma cells. *Exp. Parasitol.* **71**, 346–349.

Wright, P. S., Cross-Doersen, D. E., Schroeder, K. K., Bowlin, T. L., McCann, P. P., and Bitonti, A. J. (1991). Disruption of *Plasmodium falciparum*-infected erythrocyte cytoadherence to human melanoma cells with inhibitors of glycoprotein processing. *Biochem. Pharacol.* **41**, 1855–1861.

PART III

The Study of
Intracellular Pathogenesis

CHAPTER 11

Purification of *Plasmodium falciparum* Merozoites for Analysis of the Processing of Merozoite Surface Protein-1

Michael J. Blackman

Division of Parasitology
National Institute for Medical Research
London NW7 1AA
England

I. Introduction
II. Parasite Culture and Synchronization
    A. *In Vitro* Culture of *P. falciparum*
    B. Synchronization of *P. falciparum* Cultures
III. Merozoite Isolation
IV. Assay for Secondary Processing of the Merozoite Surface Protein-1 (MSP-1)
    References

I. Introduction

This chapter describes the preparation of highly purified merozoites of *Plasmodium falciparum* and the use of the isolated cells to study the proteolytic processing of a particularly well characterized merozoite surface protein. The merozoite isolation method is based, with slight modification, on that of Mrema *et al.* (1982). Acrylic membrane sieves are used to deplete culture supernatants, containing naturally released merozoites, of parasitised and uninfected erythrocytes. Merozoites are then recovered from the filtered medium by centrifugation. The method is simple, rapid, and efficient and yields merozoite preparations completely free of erythrocyte contamination, although some contamination with hemozoin pigment is often evident. Merozoites prepared in this way have been used for surface radioiodination studies (e.g., Heidrich *et al.*, 1983; McBride and Heidrich, 1987) exploring the composition of the merozoite surface, and the interactions between proteins on it. They have also been used to demonstrate the presence of sphingomyelin synthase in the plasma membrane of extracellular merozoites (Elmendorf and Haldar, 1994), and to investigate and characterize proteolytic processing of the merozoite surface protein-1 (MSP-1) (Blackman *et al.*, 1993). All these studies have indicated that merozoites produced using this method are morphologically intact at the level of light and electron microscopy (see Fig. 1), and structurally intact at the molecular level. However, in our hands the cells are completely noninvasive to normal erythrocytes. This is in contrast to *P. knowlesi* merozoites prepared using a similar technique (e.g., Dennis *et al.*, 1975), and may reflect fundamental differences between the two species.

II. Parasite Culture and Synchronization

A. *In Vitro* Culture of *P. falciparum*

The FCB-1 isolate of P. falciparum is maintained in human A^+ erythrocytes at a 1–2% hematocrit in RPMI 1640 medium supplemented with 25 mM HEPES, 24 mM NaNCO$_3$, 0.2% w/v glucose, 25 μg/ml gentamicin, 20 μg/ml hypoxanthine, and 10% v/v decomplemented human A^+ serum. This is referred to as complete medium. The serum can be replaced by 0.5% w/v Albumax (GIBCO)

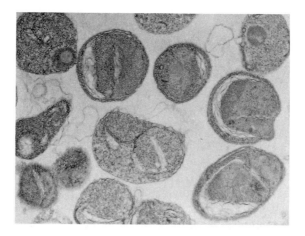

Fig. 1 Electron micrograph of *Plasmodium falciparum* merozoites isolated using the method described. (kindly provided by Dr. K. Haldar, Department of Microbiology and Immunology, Stanford University, CA).

(H. Matile, personal communication) for routine cultivation—however, this must *not* be used for merozoite production since released merozoites adhere strongly to the membrane sieves in the presence of Albumax. Cultures are gassed with a mixture containing 7% CO_2, 5% O_2, and 88% N_2, and incubated at 37°C. The medium is usually changed daily. Cultures are resuspended, transferred to 250-ml conical-bottomed polypropylene centrifuge tubes (Corning 25350-250), and pelleted in a Beckman J-6B centrifuge at 2200 rpm for 5 min. The supernatant is aspirated and the pelleted cells are simply resuspended and returned to culture. We routinely use 175-cm^2 (800 ml) Nunclon flasks (Nunc), 100 ml of culture per flask. With daily medium changes, parasitemias of 10–15% can be easily achieved under these conditions. The T9/94 and T9/96 *P. falciparum* clones are also cultivated in our laboratory using the above conditions, and are also used for merozoite production. However, these clones do not have as high a replication rate as FCB-1, and their schizonts cannot be concentrated on Plasmagel. The following methodology therefore refers specifically to FCB-1.

B. Synchronization of *P. falciparum* Cultures

Parasite cultures must be closely synchronized prior to merozoite production. We use a combination of flotation on gelatin (Pasvol *et al.*, 1978) and disruption with sorbitol (Lambros and Vanderberg, 1979). All media and solutions are prewarmed to 37°C before use.

1. Cultures containing mature, segmented schizonts are pelleted in a Beckman J-6B centrifuge at 2200 rpm for 5 min at room temperature, and the supernatant is discarded. Up to 5 ml of packed cells are resuspended in 30–40 ml of warm complete medium, transferred to a 50-ml polypropylene screw-cap tube, and pelleted again at 560 g for 5 min on a bench centrifuge.

2. The supernatant is aspirated to just above the cell pellet; caution must be observed at this stage as the pellet is rather loose. The volume of the pellet is estimated as accurately as possible. The pellet volume is then multiplied by 2.4, and warm medium is added to the tube to bring the volume (i.e., pellet plus liquid) up to the calculated figure. Now a volume of warm Plasmagel (obtained from Laboratoire Roger Bellon, 159, Avenue A.-Peretti, 92200 Neuilly sur Seine, France) equal to the total volume of cells plus medium is added to the tube. The contents of the tube are carefully but completely resuspended in the medium–Plasmagel mix and transferred to a fresh tube. The tube is sealed (without gassing) and placed in a 37°C incubator for 30 min without disturbing.

3. The tube is carefully removed from the incubator. At this stage two phases should be visible; the top, brownish layer contains mature forms of the parasite, whereas the lower, red layer contains uninfected red cells and immature forms of the parasite. The upper layer is removed and transferred to a fresh tube. An equal volume of warm complete medium is added, and the cells are pelleted

at 450 g for 5 min on a bench centrifuge (Labofuge M, Heraeus). The pellet is carefully resuspended in fresh medium and added to a culture flask containing fresh medium and fresh, washed red cells such that that parasitemia is 2–10% and the final hematocrit, is 20–25%. The flask is now gassed and incubated without disturbing at 37°C for 3–4 hr. During this period schizont rupture and erythrocyte reinvasion occur.

4. Following the incubation step, the culture is treated with sorbitol to remove residual schizonts. The cells are pelleted at 560 g for 5 min, the supernatant is aspirated, and the pellet is resuspended into at least 10 vol of 5% w/v sorbitol in water. The suspension is left for 10 min at room temperature and then pelleted at 560 g for 5 min, washed once in serum-free medium, and recultured. The culture now contains only ring forms of the parasite, the most mature of which are only 3–4 hr old.

5. The above synchronization step is repeated as necessary. We routinely treat our stock cultures once weekly in order to maintain synchrony. When bulking up cultures for merozoite preparation, the parasites are synchronized at 48-hr intervals (i.e., once per cycle) until 4 days before merozoite isolation, from which point they are simply fed daily. This is to prevent overstressing the parasites, and to allow maximal growth and reinvasion rates.

III. Merozoite Isolation

1. Acrylic-supported membrane filters, 3- and 1.2-μm-pore size and 47-mm diameter, are obtained from Gelman Sciences (Ann Arbor, MI 48106, Versapor membranes, Product Nos. 66394 and 66387). The filters can vary somewhat in hydophilicity, so it is recommended that different batches be tested before making a large order. The membranes should wet immediately when placed in complete medium. Before use, the membranes are soaked in two changes of complete medium for at least 10 min each in order to remove any residual detergents and to wet the filters fully. Filter units used are Sterifil Aseptic System (47 mm), obtained from Millipore.

2. Mature schizonts from synchronous cultures are purified on Plasmagel as described above. The schizonts should be 40–44 hr old at this point, containing about 8 nuclei. The parasitemia of such enriched preparations should be at least 80%. The schizonts are recultured in warm complete medium at a 0.25% hematocrit, without the addition of any fresh erythrocytes, changing the medium at 4-hr intervals until merozoite release begins; this is assessed by making Giemsa-stained smears of the culture at 1- to 2 hr intervals.

3. Cultures are transferred to 50-ml tubes and centrifuged at 440 g for 6 min at room temperature to pellet the schizonts. Meanwhile, two filter units are assembled, one with a 3-μm membrane and the other with a 1.2-μm membrane.

4. The supernatant from the pelleted culture, which contains free merozoites,

is poured into the filter unit containing the 3-μm filter. The schizont pellet is immediately resuspended in warm complete medium and recultured for further merozoite production (this can be continued for the next 6–10 hr, depending on the synchronicity of the starting culture).

5. By applying gentle suction to the receiver flask of the filter unit, the supernatant is passed under negative pressure through the 3-μm filter at a rate of approximately 50 ml min^{-1}, then the flow-through is transferred to the next filter unit and similarly passed through the 1.2-μm filter. A maximum of 100 ml of supernatant is passed through each membrane. The filter membranes often appear dark-brown to black at this stage, due to trapped pigment that adheres preferentially to the filters.

6. The filtered supernatant, which should now appear clear, is transferred to 50-ml flanged polycarbonate tubes (Beckman, tube Catalog No. 03146) and centrifuged in a Beckman HB-4 swing-out rotor at 3000 g for 10 min at 4°C. Note that use of a fixed-angle rotor results in reduced merozoite yields, as the cells tend to adhere to the sides of the tubes. The resulting pellet should appear grayish-white, often with a black center due to contaminating particles of hemozoin pigment.

7. The supernatant is aspirated from the tubes, and the merozoite pellet resuspended in ice-cold phosphate-buffered saline pH 7.2 (PBS). We routinely use PBS that is calcium and magnesium-free. The buffer may be supplemented with protease inhibitors as required [we add leupeptin, antipain, and aprotinin, each at 10mg ml^{-1}, and tosyl-L-lysyl chloromethyl ketone (TLCK) at 100 μM]. The preparation can be checked (by Giemsa staining) at this stage to ensure the complete absence of contaminating erythrocytes. Merozoites are usually washed three times before use. For surface labeling procedures the preparations must be used immediately. For use in MSP-1 processing assays the merozoites may be used at once or stored in aliquots at −70°C for up to 6 months.

IV. Assay for Secondary Processing of the Merozoite Surface Protein-1 (MSP-1)

Merozoites prepared using the above protocol posess little or no intact MSP-1 on their surface. In what has been termed primary processing, the protein is cleaved before merozoite release into four major fragments. These are present on the merozoite surface, held together in a protein complex by noncovalent interactions. The complex also contains other, as yet poorly characterized proteins not derived from MSP-1, but it does not contain another major merozoite surface protein, MSP-2 (McBride and Heidrich, 1987; Blackman and Holder, 1992; Stafford *et al.*, 1994). Current evidence suggests that only one component of the complex, a 42-kDa protein (MSP-1$_{42}$) corresponding to the C-terminal region of the MSP-1 precursor, is membrane-bound. Secondary processing of the MSP-1, which occurs at some point following merozoite release from the schizont, is defined as a single proteolytic cleavage within MSP-1$_{42}$ to produce two fragments of 30–33 kDa (MSP-1$_{33}$) and 19 kDa (MSP-1$_{19}$). MSP-1$_{33}$, which derives from the N-terminal region of MSP-1$_{42}$, is released from the merozoite surface, taking with it the remainder of the complex in a soluble form, whereas MSP-1$_{19}$ remains membrane-bound. Work in this laboratory has shown that secondary processing of the MSP-1 goes to completion when a merozoite successfully invades an erythrocyte, suggesting that the proteolysis is intimately connected to the invasion event. However, activation of secondary processing does not appear to require interaction between merozoite and host cell. Thus, if purified isolated merozoites are simply incubated at 37°C in the presence of calcium, the processing continues, and can be detected by assaying for *de novo* production of the MSP-1$_{33}$ fragment. The following assay has been used to characterize partially the protease responsible for secondary processing of the MSP-1 as a membrane-associated, calcium-dependent parasite serine protease, probably located on the merozoite surface (Blackman and Holder, 1992; Blackman *et al.*, 1993). The activity is exquisitely sensitive to inhibition by phenylmethylsulfonyl fluoride (PMSF), so this is usually included in processing assays as a control inhibitor.

1. Washed merozoites are resuspended in ice-cold PBS containing 1 mM CaCl$_2$, 1 mM MgCl$_2$, and protease inhibitors antipain, leupeptin, aprotinin, and TLCK at the concentrations described above (PBS Ca/Mg). Note that none of these protease inhibitors affects secondary processing of MSP-1. The merozoites are dispensed into aliquots of about 2×10^9 merozoites each in 1.4 ml Eppendorf tubes (on ice) and pelleted in a microfuge at about 12,000 g for 2 min at 4°C.

2. The buffer is aspirated from the pellets, which are kept on ice. It is important that the merozoite pellets are not allowed to warm up above 4°C, since at this point it is only the low temperature that is preventing processing from taking place and producing significant "background" levels of MSP-1$_{33}$.

3. The merozoite pellets are then resuspended in 30 μl cold PBS Ca/Mg further supplemented with the appropriate protease inhibitors, antibodies, etc. At least three controls are always included, as follows: A positive control sample, resuspended in PBS Ca/Mg only; a negative control sample, resuspended in PBS Ca/Mg containing 1 mM PMSF; and a "zero time" control, containing PBS Ca/Mg only, in which processing is immediately stopped by solubilization in SDS (see below). All other samples are placed in a 37°C water bath for 1 hr.

4. Processing is stopped by the addition of 30 μl of 2× SDS–PAGE sample buffer, with or without the addition of a reducing agent (dithiothrietol at a final concentration of 100 mM). The samples are boiled for 4 min and clarified by centrifugation for 5 min at 12,000 g in a microfuge and 5–20 μl of each sample is subjected to electrophoresis on 12.5% minigels (SE250, Hoefer Scientific

Instruments, San Francisco). The inclusion on each gel of a track containing prestained molecular weight markers (GIBCO BRL) aids in assessing the efficiency of the subsequent blotting step.

5. Following electrophoresis, separated proteins are transferred to nitrocellulose. Blotting is carried out in a Hoefer Scientific Transphor tank, using a Tris–glycine buffer, pH 8.3, containing 20% methanol, for 4 hr at 75 V.

6. Blots are blocked for 1 hr at 37°C in 5% nonfat milk powder in PBS 0.05% Tween 20 (PBS/T). Blots are washed, then incubated with a rabbit antiserum raised against a recombinant protein corresponding to a 93 amino acid region within the N-terminal half of MSP-1$_{42}$ (Blackman *et al.*, 1993). The antiserum therefore reacts with both MSP-1$_{42}$ and MSP-1$_{33}$. The serum is used at a 1/100 dilution in PBS/T, and incubation is usually for 2 hr at room temperature.

7. The blots are washed and incubated for 1 hr with a horseradish peroxidase-conjugated goat anti-rabbit IgG (Bio-Rad, Catalog No. 170-6515) at a 1/1000 dilution in PBS/T. Blots are then washed three times in PBS/T and once in 20 mM Tis–HCl, pH 7.4, before developing in a substrate buffer containing 4-chloro-1-naphthol (dissolve 30 mg 4-chloro-1-naphthol into 10 ml methanol and

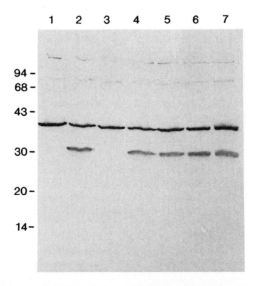

Fig. 2 Western blot assay of secondary processing of the *Plasmodium falciparum* (FCB-1) MSP-1. Tracks 1–3 correspond to the "zero time" control, positive control, and negative (1 mM PMSF-containing) control respectively. Tracks 4–7 indicate processing in the presence of, respectively, 10, 5, 2, and 1 mM diisopropyl fluorophosphate (DFP), which is a relatively poor inhibitor of the processing. The gel was run under nonreducing conditions; MSP-1$_{42}$ therefore migrates with an apparent M$_r$ of 36,000. The band at 30 kDa in tracks 2, 4, 5, 6, and 7 corresponds to MSP-1$_{33}$. Sizes in kDa of molecular weight markers are indicated on the left.

add to 50 ml 20 mM Tris–HCl, pH 7.4; add 30 μl of 30% w/v H_2O_2, mix, and use at once).

8. Secondary processing of the MSP-1 is seen as a conversion of MSP-1$_{42}$ to MSP-1$_{33}$ (see Fig. 2). Due to the nonquantitative nature of Western blots, an increase in levels of MSP-1$_{33}$ is not always concomitant with an obvious decrease in the level of the MSP-1$_{42}$ signal. However, quantitative assays based on immunoprecipitation from metabolically radiolabeled (Blackman *et al.*, 1993) or radioiodinated (M. Blackman, unpublished data) merozoites have clearly shown stoichiometric conversion of MSP-1$_{42}$ to MSP-1$_{33}$.

Acknowledgments

I am grateful to Hans Heidrich for invaluable help and advice with merozoite preparations and to Tony Holder for critical reading of the manuscript. This work was supported by the Medical Research Council, UK.

References

Blackman, M. J., and Holder, A. A. (1992). Secondary processing of the *Plasmodium falciparum* merozoite surface protein-1 (MSP1) by a calcium-dependent membrane-bound serine protease: Shedding of MSP1$_{33}$ as a noncovalently associated complex with other fragments of the MSP1. *Mol. Biochem. Parasitol.* **50,** 307–316.

Blackman, M. J., Chappel, J. A., Shai, S., and Holder, A. A. (1993). A conserved parasite serine protease processes the *Plasmodium falciparum* merozoite surface protein-1. *Mol. Biochem. Parasitol.* **62,** 103–114.

Dennis, E. D., Mitchell, G. H., Butcher, G. A., and Cohen, S. (1975). *In vitro* isolation of *Plasmodium knowlesi* merozoites using polycarbonate sieves. Parasitology **71,** 475–481.

Elmendorf, H. G., and Haldar, K. (1994). *Plasmodium falciparum* exports the Golgi marker sphingomyelin synthase into a tubovesicular network in the cytoplasm of mature erythrocytes. *J. Cell Biol.* **124,** 449–462.

Heidrich, H.-G., Strych, W., and Mrema, J. E. K. (1983). Identification of surface and internal antigens from spontaneously released *Plasmodium falciparum* merozoites by radio-iodination and metabolic labeling. *Z. Parasitenkd.* **69,** 715–725.

Lambros, C., and Vanderberg, J. P. (1979). Synchronisation of *Plasmodium falciparum* erythrocytic stages in culture. *J. Parasitol.* **65,** 418–420.

McBride, J. S., and Heidrich, H.-G. (1987). Fragments of the polymorphic Mr 185 000 glycoprotein from the surface of isolated *Plasmodium falciparum* merozoites form an antigenic complex. *Mol. Biochem. Parasitol.* **23,** 71–84.

Mrema, J. E. K., Langreth, S. G., Jost, R. C., Rieckmann, K. H., and Heidrich, H.-G. (1982). *Plasmodium falciparum:* Isolation and purification of spontaneously released merozoites by nylon membrane sieves. *Exp. Parasitol.* **54,** 285–295.

Pasvol, G., Wilson, R. J., Smalley, M. E., and Brown, J. (1978). Separation of viable schizont-infected red blood cells of *Plasmodium falciparum* from human blood. *Ann. Trop. Med. Parasitol.* **72,** 87–88.

Stafford, W. H. L., Blackman, M. J., Harris, A., Shai, S., Grainger, M., and Holder, A. A. (1994). N-terminal amino acid sequence of the *Plasmodium falciparum* merozoite surface protein-1 polypeptides. *Mol. Biochem. Parasitol.* **66,** 157–160.

CHAPTER 12

In Vitro Secretory Assays with Erythrocyte-Free Malaria Parasites

Kasturi Haldar, Heidi G. Elmendorf, Arpita Das, and Wen lu Li

Department of Microbiology and Immunology
Stanford University School of Medicine
Stanford, California 94305

David J. P. Ferguson[*] and Barry C. Elford[†]

[*] Nuffield Department of Pathology
and
[†] Molecular Parasitology Group
Institute of Molecular Medicine
John Radcliffe Hospital
Oxford University
Oxford OX3 9DU
United Kingdom

I. Introduction
II. Release and Separation of Late Ring and Trophozoite Stage Parasites from the Erythrocyte Membrane (EM) and Tubovesicular Membrane (TVM) Network
    A. Preparation of Released Rings/Trophozoites
    B. Isolation of the EM/TVM Fraction
III. Synthesis and Secretion of Proteins by Intact, Ring/Trophozoite Parasites
    A. Transmission Electron Microscopy
    B. Protein Synthesis in Intact Released Parasites
    C. Biosynthetic Export of Proteins from Released Parasites
    D. Concluding Remarks
IV. Organization of Secretory Activities at Different Stages of the Asexual Life Cycle
    A. Distribution of Secretory Markers in Released Rings/Trophozoites and Isolated EM/TVM Fractions Using Western Blots
    B. Distribution of Sphingomyelin Synthase Activity in Released Parasites and the EM/TVM Fraction
    C. Distribution of Sphingomyelin Synthase in Merozoites

D. Comparison of Sphingomyelin Synthase Activity in Merozoites or Ring-, Trophozoite-, and Schizont-Infected Erythrocytes
E. Concluding Remarks
V. Release of Pigmented Trophozoites and Schizonts from Infected Erythrocytes by Osmotic Shock in Isoosmolar Dipeptide-Based Media
A. Mature Asexual Stages Freed by Exposure to Glycyl-L-serine
B. Differential Hemolytic Effects of Diverse Dipeptides
References

I. Introduction

In this chapter we describe methods to isolate asexual stages of the human malaria parasite *Plasmodium falciparum* and characterize *in vitro* their secretory activities. During its asexual life cycle *P. falciparum* develops in a red cell. The intracellular parasite exports proteins and lipids to induce the formation of tubovesicular membranes (TVM) in the erythocyte cytoplasm (reviewed by Barnwell, 1990; Elmendorf and Haldar, 1993a). By transmission electron microscopy, these structures appear as "clefts" and "loops" (see Fig. 1A). However, recent studies by laser confocal and scanning electron microscopy suggest that they exist in a novel membrane network that is attached to the parasite at one end (Elmendorf and Haldar, 1993a, 1994; Elford and Ferguson, 1993; see Figs. 1B and 2). Further, the network contains the Golgi enzyme, sphingomyelin synthase (Elmendorf and Haldar, 1994). Numerous parasite encoded proteins are also detected in the erythrocyte membrane (reviewed by Howard *et al.*, 1987). Current models of secretion in eukaryotes (Rothman and Orci, 1992) do not account for membrane development and protein transport beyond the plasma membrane of a cell. *P. falciparum*-infected erythrocytes therefore provide a model system to study (i) novel mechanisms which regulate secretory export beyond the parasite plasma membrane and (ii) *de novo* membrane formation in a relatively simple eukaryotic cell, the mature erythrocyte.

Secretory processes appear to play key roles at all stages of plasmodial entry and development in the red cell. The extracellular merozoite stage invades the red cell and extrudes the contents of its apical secretory complex (Bannister and Mitchell, 1989; Perkins, 1992). This, along with the recruitment of lipids from the parasite and the red cell bilayer (Dluzewski *et al.*, 1992; Haldar and Uyetake, 1992; Ward *et al.*, 1993), appears to be closely linked to the formation

Fig. 1 (A) Transmission electron microscopy. A *P. falicparum* trophozoite-infected red cell. P, parasite; E, periphery of the erythrocyte; l, loop; c, cleft. (B) Three-dimensional reconstruction of serial, optical sections of a trophozoite-stage parasite labeled with C5-DMB-ceramide. Twenty consecutive images taken at 400-nm intervals through the depth of the infected erythrocyte were used to recalculate the three-dimensional morphology of the parasite. The images were first processed through a Gaussian data filter to eliminate nonspecific signal and noise, and then the image intensity threshold was set to exclude signal from uninfected erythrocytes in the same sample. Sequential images (A–F) are shown as the model rotates (into the page) around the *x*-axis at 60° intervals.

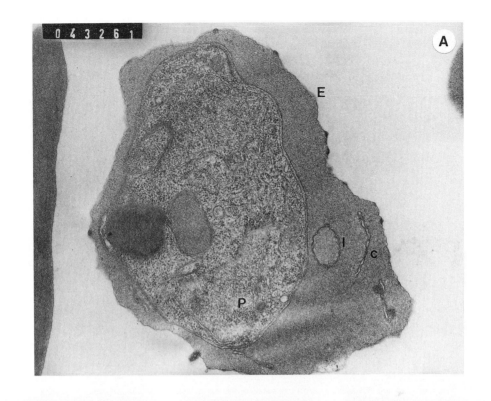

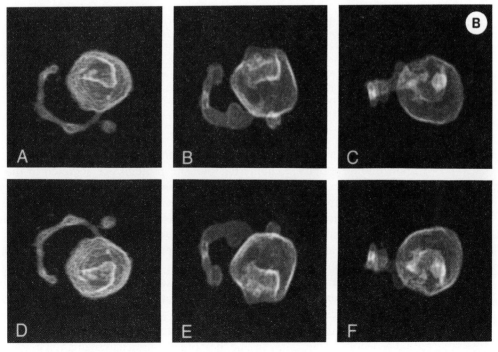

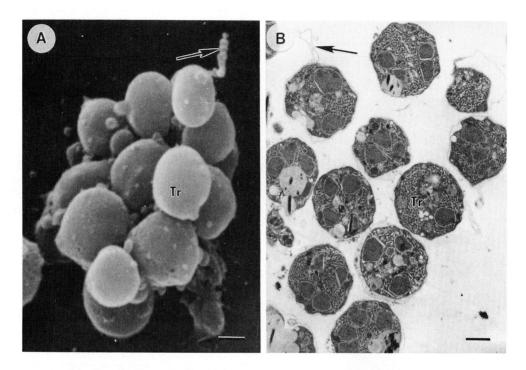

Fig. 2 Scanning (A) and transmission (B) electron micrographs showing homotypic aggregates of naked trophozoites of *Plasmodium falciparum* released from parasitised red blood cells (PRBCs) exposed for about 20 min at 37°C in 5% (w/v) glycyl-L-serine (A) or glycyl-glycine (B). PRBCs were then packed, the supernatant was aspirated, and the cells were resuspended in growth medium (RPMI 1640). (A) Freed parasites (Tr) were allowed to settle for 30 min onto polylysine-coated coverslips before fixation in 2.5% glutaraldehyde-cacolylate buffer; (B) parasites (Tr) were fixed within 5 min of release. Note the presence of a vacuolated pilus-like structure attached to certain trophozoites (arrows). Bars are 1 μm.

of the parasitophorous vacuole. In subsequent intracellular stages of ring and trophozoite development, numerous parasite proteins are secreted to the red cell, and are presumably required to effect essential modifications in the cytoplasm and membrane of the host red cell (reviewed by Haldar and Holder, 1993). Schizogony is a period of mitosis: secretory organelles such as the Golgi multiply (Elmendorf and Haldar, 1993b, 1994) and the parasite plasma and vacuolar membranes enlarge rapidly. Each merozoite is formed when the parasite plasma membrane pinches off around a daughter nucleus and a set of organelles. Approximately 48 hr after the onset of invasion, mature schizonts lyse to release daughter merozoites, which in minutes reinvade red cells, to reinitiate the cycle.

Although the parasite is replete with secretory activity, very little is understood of secretory mechanisms and their organization and regulation in *Plasmo-*

dium. A major limitation is that the organism is intracellular for all of its cycle (except for a few minutes at invasion). Studies on protein transport in rings, trophozoites, and schizonts have therefore historically been limited to immunolocalization of exported proteins with little information on the mechanisms of transport. Further, the examination of secretory processes in merozoites has been limited by the relative inaccessibility of this particular stage to experimental manipulation. Hence, methods for the quantitative isolation of (i) intraerythrocytic stages freed from red cells and (ii) merozoites are central to biochemical studies of plasmodial secretory functions and parasite and membrane development in the erythrocyte.

Elmendorf *et al.* (1992) therefore devised a method of controlled mechanical homogenization to release intact, transport active, late ring, and trophozoite stage parasites from the red cell. The free parasites are intact by rigorous biochemical criteria, capable of active protein synthesis (Elmendorf *et al.*, 1992) and have been used to define specific steps of secretory export of a "cleft" membrane protein out of the parasite (Das *et al.*, 1994; in press). They are also relatively denuded of the TVM. The TVM can be recovered with the erythrocyte membrane (EM) in a seperate membrane fraction (shown schematically in Fig. 3). Analysis of secretory markers in the released parasites and EM/TVM fraction led to the surprising discovery that the Golgi enzyme sphingomyelin synthase is exported to the TVM (Elmendorf and Haldar, 1994).

Unlike the intraerythrocytic stages, merozoites are naturally released from the infected red cell at the end of schizogony. It is therefore possible to synchronize cultures of *P. falciparum* to 4 hr in the cycle, purify mature schizonts, and prepare merozoites released by natural lysis and completely free of contaminating schizonts (Blackman *et al.*, 1993; M. Blackman, this volume). This procedure was successfully used by Elmendorf and Haldar (1994) to demonstrate that sphingomyelin synthase in merozoites was entirely intracellular. When the merozoite enters a red cell and becomes a ring, a fraction of the enzyme is exported to the TVM, thereby enabling sphingomyelin synthesis and membrane growth there. Thus in addition to the release of its apical secretory components, the parasite also exports components of its Golgi in a cell cycle-dependent fasion.

The procedure for the isolation of ring/trophozoites freed from red cells and an EM/TVM fraction are provided in Section II. Biochemical assays used to demonstrate the intactness of the released parasites and their ability to synthesize and export protein are in Section III. The determination of the relative distribution of protein and lipid secretory activities between the intact, released parasites, and the TVM is covered in Section IV. Also included in Section IV are assays for the latency of sphingomyelin synthase in purified merozoites and the comparison of enzyme levels in rings, trophozoites and schizonts.

The rigidity, morphology, and permeaselectivity to a number of ions and solutes of the erythrocyte membrane of trophozoite/schizont-infected red cells are markedly altered compared to their uninfected counterparts. Elford and

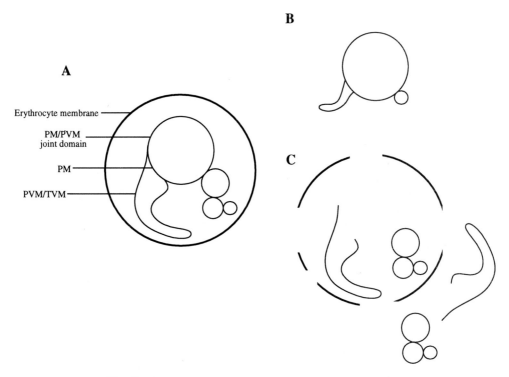

Fig. 3 Schematic of fractionation of *P. falciparum*-infected erythrocytes. Infected erythrocyte (A); released parasite with parasitophorous vacuolar membrane (B); erythrocyte membrane (EM), and tubovesicular membrane fraction (TVM) (C). The TVM is found associated with the EM and also free in this fraction.

Ferguson (1993) were able to exploit this selective permeability and load mature trophozoites and schizonts (under isotonic conditions) with the dipeptide Gly–Ser. Resuspension of the loaded cells in warm growth medium leads to a rapid efflux of the dipeptide and the release of the parasite from the red cell. This release procedure produces "naked trophozoites," which develop through to merogony (and it is possible then that these parasites may be useful in transformation studies). As shown by scanning electron microscopy, naked trophozoites contain long tubules and small membrane spheres projecting from their vacuolar surface (Elford and Ferguson, 1993) and are strikingly similar in their morphology and organization to the tubovesicular network detected by laser confocal microscopy in live infected erythrocytes (Elmendorf and Haldar, 1994).

Does the TVM network catalyze protein transport from the parasite to the red cell membrane? In attempting to answer this question Gormley *et al.* (1992) fragmented the TVM using formaldehyde fixation in a low osmotic strength buffer, to simultaneously fix and permeablize infected erythrocytes. In contrast, naked trophozoites have intact TVM structures accessible on their surface, and may provide a suitable system to investigate protein export to the red cell. The isolation of naked trophozoites is covered in Section V.

II. Release and Separation of Late Ring and Trophozoite Stage Parasites from the Erythrocyte Membrane (EM) and Tubovesicular Membrane (TVM) Network

A. Preparation of Released Rings/Trophozoites

1. Growth of *P. falciparum*

We have made released parasites from the clonal line of *P. falciparum* FCR-3/A₂ (Trager *et al.*, 1981). This clone is cultured *in vitro* according to a modification of the method of Trager and Jensen (1976; Haldar *et al.*, 1985). The culture media contains RPMI 1640 (GIBCO/BRL) supplemented with 25 mM HEPES, pH 7.4, 11 mM glucose, 92 μM hypoxanthine, 0.18% NaHCO₃, 25 μg/ml gentamicin, and 10% AB$^+$ human serum (Gemini Bio-Products). The parasites are grown in A$^+$ red blood cells (0–4 weeks old) at 2.5–5% hematocrit to parasitemias of 15–25%. Parasite cultures are synchronized to within 10 hr by separation of the early and late stages over Percoll gradients and the subsequent separate reincubation of these stages in culture. It should be possible to use any strain or clone of *P. falciparum*, which can grow to parasitemeas >15%.

2. Homogenizer

The use of a stainless-steel ball homogenizer has proven very effective for the isolation in intact and transport competent intracellular organelles in higher eukaryotes (Balch *et al.*, 1984). We therefore used this method to release late ring and trophozoite stages of *P. falciparum* from erythrocyte host cells (Elmendorf *et al.*, 1992). A stainless-steel ball homogenizer (internal bore diameter, 0.2500 inches; ball diameter, 0.2494 inches) was custom built at the Department of Biochemistry, Stanford University. These homogenizers are no longer made at Stanford but can be obtained on order from the EMBL, Heidelberg. The homogenization procedure selectively released late ring and early trophozoite stage parasites. Lysis of uninfected erythrocytes varies, depending on the age of the red cells and the length of their time in culture. The success of our release procedure is probably facilitated by destabilization of the red cell membrane during intraerythrocytic parasite maturation. The red cell membrane of uninfected erythrocytes and of infected erythrocytes at the earlier ring stage of development is more deformable and therefore resistant to the mechanical stresses of homogenization. Interestingly, schizont-infected red cell membranes are lysed during homogenization, but the parasites are not released; perhaps their much larger size does not permit escape from the perforated erythrocyte membranes.

3. Protocol for Preparation of Released Rings/Trophozoites

Culture should be between 15 and 25% parasitemia—mostly late ring to trophozoites (20–36 hr). Start with growing up 600 μl of packed infected and uninfected erythrocytes at 20% parasitemia in 50 ml of RPMI containing 10% human serum. All solutions to be used, the homogenizer, and the table-top centrifuge should be chilled to 0°C.

1. Feed culture and return to incubator for ~1 hr.

2. Save amount of culture required for control samples of internalized parasites. Keep on ice.

3. Spin down rest of culture in a 50-ml conical tube for 5 min at 1000 × g. Pull off and discard supernatant. Estimate cell pellet size.

4. Resuspend pellet to 6% hematocrit in 1X homogenization buffer. Spin down for 5 min at 1000 × g. Pull off and discard supernatant.

5. Resuspend pellet to 6% hematocrit in 1X homogenization buffer. Transfer to a 12-ml syringe with leur tip—not leur lock (use ≤8 ml/syringe).

6. Fill one 6-ml syringe with 3 ml 1X homogenization buffer.

7. Perform homogenization in cold room. Attach one empty and one full 6-ml syringe (containing buffer alone) to the homogenizer and depress each syringe twice to flush out the homogenizer.

8. Attach one empty and one full 12-ml syringe (containing cell suspension to be homogenized) to the homogenizer and depress each syringe five times. **USE CAUTION during this step—excessive, incorrect pressure can cause the syringes to break. To avoid accidents: insert both syringes firmly into the homogenizer and hold them straight and steady while *slowly* depressing the plunger. If the homogenizer gets "stuck" remove syringes, disassemble homogenizer, and wash out with ddH$_2$O.**

9. Make a wet mount with ~3 μl of the homogenized solution. Count the number of uninfected RBC, the number of still-internalized parasites, and the number of free parasites. If >80% of the parasites have been freed, then proceed. If not, repeat the homogenization step and recount.

10. Dilute the homogenate 1 in 10 into buffer A (i.e., 5 ml 10X buffer A + 5 ml homogenate + 40 ml ddH$_2$O per 50-ml tube) and spin down 10 min at 2000 x g.

11. Pull off and respin supernatant for 10 min at 2000 x g. Remove supernatant, which should contain the EM/TVM fraction and store for further concentration (see protocol for isolation of EM/TVM fraction). The pellet should be combined with that from step 11—both contain released late ring and trophozoites. Resuspend these pellets in 1X buffer A including vitamin/amino acid mix (vit/aa).

12. Make up Percoll gradient. Add 5 ml 40% Percoll to 15-ml conical tube. Layer 5 ml 10% Percoll over 40% Percoll. Layer homogenate over 10% Percoll. Centrifuge for 10 min at 2000 x g.

13. Carefully pull off and discard the 0–10% interface. Pull off the 10–40% interface, distribute among three to four microfuge tubes and bring to 1.5 ml in 1X buffer A including vit/aa. Centrifuge for 10 min at 2000 x g. Pull off and discard supernatant.

14. Resuspend and pool pellets in 1 ml 1X buffer A (or B—depending on experiment planned) including vit/aa. Centrifuge for 10 min at 2000 x g. Pull off and discard supernatant. Repeat wash twice more.

15. Count total cells in a hemacytometer—use 20X objective and look hard—the parasites are small.

16. Parasite viability is determined by mixing parasites at ~1 × 10^8 cells ml^{-1} 1:1 with Evan's blue dye; count at least 100 cells at 100x magnification.

17. Estimate yield relative to number of parasites in original culture. Maximum expected yield is 30%; minimum is 10%.

Buffers for released ring/trophozoite production. Store all buffers at 4°C, preferably in sterile aliquots to discourage contamination.

10X homogenization buffer, pH to 7.4: 650 mM sucrose, 850 mM HEPES, 150 mM K$^+$EDTA, 10 mg/ml BSA, 3 mM DTT.

10X buffer A, pH 7.4 (high EDTA buffer—reduces parasite "stickiness"): 950 mM K$^+$ acetate, 150 mM Na$^+$EDTA, 80 mg/ml glucose, 65 mM sucrose, 200 mM HEPES, 10 mg/ml BSA, 3 mM DTT.

10X buffer B, pH 7.4 (low EDTA, with Mg^{2+}; allows enzyme assays that require Mg^{2+}): 950 mM K$^+$ acetate, 300 mM HEPES, 100 mM sucrose, 50 mM MgCl$_2$, 10 mM Na$^+$EDTA, 80 mg/ml glucose, 10 mg/ml BSA, 3 mM DTT: It may be convenient to make up 10X buffer A and 10X buffer B stocks without BSA and/or DTT if you will be running samples on SDS–PAGE or if you need to quantitate protein. The homogenization buffer may also be used without BSA and DTT, but the yields never seem as high.

5X vitamins and amino acid (vit/aa) cocktail, pH -7.4, from RPMI 1640 Select Amine Kit (GIBCO/BRL): Use equal volumes of all amino acids + vitamins + phenol red: this will automatically give you a 5X stock. You may choose not to include an amino acid(s) of your choosing for the purpose of metabolic labeling.

Percoll: Make up 40% and 10% Percoll in buffer A with vit/aa.

Note

We also add an ATP reconstitution system (ARS) to the released parasite buffer for certain experiments. This increases protein synthesis and protein secretion. ARS recipe: 2 mM ATP, 20 mM creatine phosphate, 40 U/ml creatine phosphokinase.

B. Isolation of the EM/TVM Fraction

1. Make up a 2 M sucrose solution in buffer A. Add 10 ml to Beckman Ultra-Clear Tubes, 1 x 3 $\frac{1}{2}$ inches.

2. Layer 27 ml of the cell-free homogenate obtained from step 11 of the ring/ trophozoite preparation on a 2 *M* sucrose cushion. Spin in ultracentrifuge at 100,000 x *g* for 60 min at 2°C in a Beckman SW 28 rotor.

3. Remove membranes at the sucrose interface in a minimum volume of liquid. Pool and dilute into buffer A as needed, or freeze.

4. The upper loading zone should also be stored to analyze cytosolic components.

III. Synthesis and Secretion of Proteins by Intact, Ring/Trophozoite Parasites

A. Transmission Electron Microscopy

The released parasites detected by transmission electron microscopy are surrounded by two membranes (Fig. 4). As indicated in the literature the released parasites are difficult to preserve for TEM. Optimal preparations are obtained when parasites are fixed for 30 min in 100 m*M* K$^+$cacodylate, 1% sucrose, 5 m*M* CaCl$_2$, 2% glutaraldehyde, and 1.5% acrolein. Fixations are started at 4°C and allowed to reach room temperature during the course of the incubation. Cells are washed, postfixed for 60 min in 1% osmium tetroxide in buffer, washed, and stained with 0.25% uranyl acetate in water for 12 hr. Cells are dehydrated through a series of ethanol steps, transferred in propylene oxide, and embedded in Poly/Bed 812. Sections are cut with a diamond knife, transferred onto copper grids, and stained with 0.5% uranyl acetate/lead citrate or with bismuth. Grids are viewed on a Phillips EM 300 electron microscope.

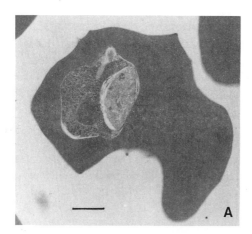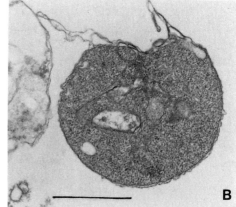

Fig. 4 Transmission electron microscopy. (A) Infected erythrocyte of trophozoite stage. (B) Released parasite.

B. Protein Synthesis in Intact Released Parasites

We have investigated the integrity of the released parasite population by a variety of biochemical criteria. First, it should be noted that parasite protein synthesis in the presence of 15 mM EDTA in buffer A is a strong indication that parasite integrity is not compromised. If the parasite membrane is not intact, the high concentrations of EDTA used would deplete the free magnesium pool essential for protein synthesis. The integrity of the released parasites may also be tested by determining the effects of exogenously added enzymatic activities on protein synthesis. In one assay, exogenous hexokinase may be used: it has no effect on protein synthesis (Elmendorf *et al.*, 1992): the actual activity of hexokinase in labeling reactions is 20 U/ml, a 1000-fold in excess over that required to deplete, if accessible, all malarial ATP pools in a standard incubation period. In an additional assay, RNaseA may be added at concentrations from 1 to 100 μg/ml with no effect on parasite metabolic activity (Elmendorf *et al.*, 1992: as a control, exogenously added RNA should be completely degraded). Hence the parasite surface is impermeant to charged molecules ranging in size from proteins like hexokinase to metabolites as small as from ATP or magnesium, and the observed protein synthesis occurs within intact organisms.

1. Metabolic Labeling of Parasites

1. Incubate 1–5 x 10^7 released parasites in 1 ml of buffer A with 50 μCi/ml [^{35}S]methionine for 60 min at 37°C.

2. Centrifuge at 2000 x g for 10 min and separate supernatant from pellet.

3. Save the supernatant on ice at 4°C

4. Wash the pellet three times in buffer A without BSA and containing 5 mg/ml L-methionine.

5. Centrifuge supernatant from step 3 for 30 min, 100,000xg, TLA100.2 rotor, 2°C, to separate high-speed pellet and supernatant fractions. Precipitate proteins in the supernatant by addition of TCA (10% final concentration) for 2–16 hr at 4°C.

6. Take all samples from steps 4 and 5 and boil 50–100 μl 1X SDS sample buffer.

7. Analyze on 10% acrylamide gels by SDS–PAGE and fluorography.

8. Incorporation of radiolabel into proteins can also be determined by hot TCA (acid-insoluble) precipitation.

9. Incorporation of radiolabel into infected erythrocytes may be performed as described above for released parasites; the only differences are the use of RPMI without methionine as the incubation buffer, and RPMI + 5 mg/ml L-methionine as the wash buffer.

2. Hexokinase Treatments

1. Resuspend 1–5 x 10⁷ released parasites in 1 ml Buffer B (95 mM KOAc, 10 mM sucrose, 1 mM Na⁺EDTA, 5 mM MgCl₂, 30 mM HEPES, pH 7.4, 0.3 mM DTT, 8 mg/ml glucose, and 1X vitamin/amino acid mix from a methionine-minus RPMI 1640 Select Amine Kit)

2. Add 4–40 U/ml hexokinase (Sigma) to the incubation. Incubate 15 min at 37°C.

3. Add [³⁵S]methionine (final concentration 50 μCi/ml) for an additional 60 min in buffer B. Collect fractions and analyze by SDS–PAGE/fluorography and by acid insoluble incorporation as described in Section III,B,1.

4. Independently measure hexokinase activity using a modification of an enzymatic assay for creatine phosphokinase by the reduction of NADP at $A_{340 \text{ nm}}$, according to the manufacturer's (Sigma) specifications.

3. RNaseA Protection Assays

1. Incubate parasites in buffer A with 1–100 μg/ml RNase A for 15 min.

2. Add [35S]methionine, and proceed as described in step 2, above.

3. The activity of the RNase A under these conditions should also be independently tested. We use control incubations containing parasites plus exogenous poly(A) RNA from *Trypanosoma brucei*. The complete degradation of exogenous RNA may be confirmed by analyzing the samples on 5% polyacrylamide/ 7 M urea gels.

C. Biosynthetic Export of Protein from Released Parasites

Biosynthetic export of newly synthesized proteins is carried out in the presence of an ATP reconstitution system (ARS), which increases cell-associated incorporation of radiolabel. The increase varies slightly between experiments, usually reflecting a two- to threefold stimulation. Newly synthesized proteins are detected in both cell pellet and exported supernatant fractions. The relative intensity of radiolabeled profiles of newly synthesized and released proteins varies between experiments due to the stage-specific protein synthesis. Due to the limitations of culturing *Plasmodium in vitro,* there is heterogeneity in the synchrony and stage of the population and therefore of protein synthesis between different populations of released parasites.

The drug Brefeldin A (BFA), and incubation at 15 and 20°C can result in specific blocks on the export of newly synthesized proteins. Inhibition of a majority of the secreted signal in the presence of brefeldin (Fig. 5) and at 15 and 20°C (not shown) points to the existence of a classical secretory system in

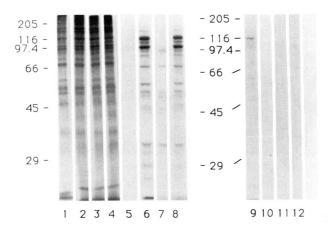

Fig. 5 Biosynthetic protein secretion from released parasites: stimulation by an ARS. Released parasites were radiolabeled in buffer B with [^{35}S]methionine for 60 min at 37°C. Cell pellets (lanes 1–4) were collected by centrifugation at 2000 × g for 10 min. Supernatants were further fractionated into high-speed supernatant (lanes 5–8) and pellet (lanes 9–12) fractions by centrifugation at 100,000 × g for 30 min. Samples were analyzed by SDS–PAGE and fluorography. Lanes 1, 5, and 9, incubations in buffer B; lanes 2, 6, and 10, incubations with ARS; lanes 3, 7, and 11, incubations containing ARS and BFA; lanes 4, 8, and 12, incubations containing ARS and mock-treated with methanol.

the released parasites that is similar to that found in higher eukaryotes. Of the 15 bands seen in a complete secreted profile shown in Fig. 5, five are brefeldin-insensitive. Four of these bands appear consistently resistant in multiple experiments, independent of the stage variation of the preparations. The appearance of these bands is not merely the result of an incomplete brefeldin block—the profile is not a lighter version of the complete secretory profile. Since the brefeldin-insensitive exported profile is limited to so few bands and since protein synthesis occurs within an intact parasite population, it is unlikely that this brefeldin-insensitive signal is the result of cell disintegration or leakage. It may represent secretion through an alternate pathway.

1. ATP Regenerating System

The ATP regenerating system consists of 2 m*M* ATP, 20 m*M* creatine phosphate, and 40 U/ml creatine phosphokinase. All experiments using the ATP regenerating system use buffer B. Fractions may be collected by differential centrifugation and analyzed by electrophoresis and fluorography and by acid insoluble incorporation as described in Section III,B,1.

2. Protocol for Secretory Protein Export: The Effects of Brefeldin, 15 and 20°C Secretory Blocks

a. Brefeldin

1. Prepare BFA stocks at a concentration of 10 mg/ml in methanol and store at −20°C. Prepare a working dilution of 1 mg/ml BFA in methanol.

2. Resuspend 1–5 x 10^7 released parasites in 1 ml of buffer B.

3. Add 5 μl, 1 mg/ml BFA to the incubation (final concentration 5 μg/ml), incubate for 15 min at 37°C (Control incubations should contain identical concentrations of methanol: 0.05%.)

4. Add [^{35}S]methionine and continue as described from step 1 in Section III,B,1. **If you want to test that a particular protein is synthesized and/or exported, cell pellet and supernatant fractions may be subjected to immunoprecipitation with the relevant antibodies.**

b. 15 and 20°C Temperature Blocks

Incubate released parasites with [^{35}S]methionine at these temperatures for 60 min. Separate supernatant and pellet fractions and proceed from step 2 in Section III,B,1.

D. Concluding Remarks

We have used the released parasite system to investigate the biosynthetic export of a 45-kDa protein localized to "clefts" of the TVM (Das *et al.*, 1994, in press). Our results indicate that within 60 min the newly synthesized, membrane bound protein is exported to the surface of the released parasites. This export is blocked by the drug brefeldin A and at 15 and 20°C (summarized in Fig. 6). These results indicate that membrane transport blocks seen in the Golgi of mammalian cells are conserved in the protozoan *P. falciparum*. Further, the newly synthesized protein traffics through parasite Golgi compartments prior to its export to clefts in the red cell. Hence the parasite modulates a classical secretory pathway to support membrane export beyond its plasma membrane (see Fig. 6). Further experimentation is required to define the molecular determinants underlying the formation and targeting of Golgi-derived transport vesicles to and beyond the parasite plasma membrane.

IV. Organization of Secretory Activities at Different Stages of the Asexual Life Cycle

A. Distribution of Secretory Markers in Released Rings/Trophozoites and Isolated EM/TVM Fractions Using Western Blots

A number of parasite-encoded, secretory protein markers have now been identified. The distribution of these proteins between the released ring/trophozoites and EM/TVM fraction may be examined in Western blots. Standard gel

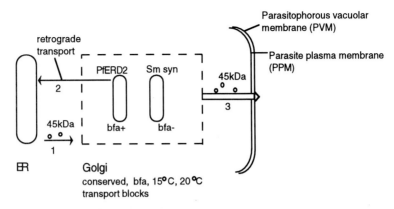

Fig. 6 Secretory export of the 45-kDa protein in rings and trophozoites of *Plasmodium falciparum*. 1 and 3 are likely steps of vesicular transport, although this remains unstudied. 2 is expected to be retrograde transport from the Golgi to the ER. Sm syn, sphingomyelin synthase; bfa+, reorganized to ER by brefeldin A; bfa−, not reorganized by brefeldin A; open circles, vesicles, arrows, directions of transport.

electrophoresis and immunoblotting procedures may be used. Shown in Table I, is the distribution of the major merozoite surface protein (MSP1; parasite plasma membrane marker), PfERD2 (Golgi protein), and a 45/47-kDa protein of the Maurer's clefts, in these fractions. The 45- and 47-kDa proteins are detected in the erythrocyte cytosol as well as the EM/TVM. Biosynthetic

Table I

Distribution, Percentage Recovery of Parasite Markers, and Sphingomyelin Synthase between Released Parasites and the EM/TVM

	SM synthase activity	PfERD2 (Golgi)	MSP1 (plasma mb)	45 kDa (intraerythrocytic cisternae)	47 kDa (intraerythrocytic cisternae)
	% of infected erythrocytes				
A. Infected erythrocyte	100	100	100	100	100
B. Released parasites	51	90	75	2	—
C. EM/TVM	39	5	17	49*	51
D. Erythrocyte cyto.	—	—	—	36*	41
E. % recovery B–D	90	95	92	87	92

Note. For the sphingomyelin synthase assays, cell equivalents of fractions were incubated in buffer A supplemented with 2 mg/ml defatted BSA and 10 μM C6-NBD-ceramide for 60 min at 37°C. Lipids were extracted and analyzed by thin-layer chromatography. Spots migrating at the appropriate Rf for sphingomyelin were scraped and quantitated. For PfERD2, MSP1, and the 45/47-kDa proteins, cell equivalents of fractions were probed in Western blots with the appropriate antibody and subjected to densitometry analysis. *The mature 45/47-kDa proteins are detected in a membrane form associated with the intraerythrocytic cisternae, as well as a soluble form in the erythrocyte cytosol (Li *et al.*, 1991; Das *et al.*, 1994).

studies confirm that both soluble and membrane forms of the 45- and 47-kDa proteins are detected in infected erythrocytes (Das *et al.*, 1994, in press). The relative distribution of the 45- to 47-kDa proteins and PfERD2 indicate a good separation of the EM/TVM membranes from internal parasite Golgi membranes. Seventy-five percent of MSP1 is detected in the released parasites, with 17% in the EM/TVM fraction. This again confirms an effective separation between the EM/TVM fraction and the released parasite membranes (including the parasite plasma membrane). Hence the isolated EM/TVM membranes are relatively depleted of contaminating, internal parasite membranes and may be analyzed for their constituent activities.

B. Distribution of Sphingomyelin Synthase Activity in Released Parasites and the EM/TVM Fraction

The fluorescent lipid probes C_5-DMB-ceramide {N-[5-(5,7-dimethylBODIPY)-1-pentanoyl]-D-erythro-sphingosine; Pagano *et al.*, 1991} and C_6-NBD-ceramide {N-[7-(4-nitrobenzo-2-oxa-1,3-diazole)]-6-aminocaproyl-D-erythro-sphingosine; Pagano, 1989} are metabolized to their sphingomyelin analogues in infected erythrocytes, indicating the presence of sphingomyelin synthase activity in these cells (Haldar *et al.*, 1991; Elmendorf and Haldar, 1994). Uninfected erythrocytes have no detectable activity, indicating that enzyme is parasite encoded. Because no antibodies or even gene sequence are available, identification of the enzyme must be performed by direct analysis of the enzymatic activity. Our microscopy studies suggested an accumulation of fluorescent sphingomyelin and the export of sphingomyelin synthase in the TVM (Haldar *et al.*, 1991; Elmendorf and Haldar, 1994). We therefore assayed released parasites and their corresponding isolated EM/TVM fractions for sphingomyelin biosynthetic activity. Sphingomyelin synthase is a membrane enzyme: both its substrates are lipids, and it is therefore difficult to determine accurately their concentrations in the bilayer. In order to assay correctly for the enzyme activity, the reactions must be carried out in an excess of C_6-NBD-ceramide and over a period of time in which the kinetics of product formation are linear. An increase in parasite material should result in a corresponding increase in the levels of sphingomyelin product formed and 80–90% of the orginal activity must be recovered in the isolated subcellular fractions.

1. Procedure for the Detection of Sphingomyelin Synthase Activity in Released Parasites and the EM/TVM Fraction

1. Use 10^8 parasite equivalents of infected erythrocytes, released rings/trophozoites, and the EM/TVM membrane fraction from a given preparation.

2. Make up a stock of 10 mM C_6-NBD-ceramide in ethanol. Store at $-20°C$. The stock should be stable for 3 to 6 months.

3. For released parasite or EM/TVM membrane fractions: (a) Resuspend cell or EM/TVM fraction in 1 ml of buffer A; (b) Inject 2 μl of 10 mM C_6-NBD-ceramide (ethanol stock) into a separate aliquot of 1 ml of buffer A; (c) Mix (a) and (b) (final concentration of NBD-ceramide 10 μM) and follow instructions from step 4. For infected erythrocytes: (a) Resuspend cells in 1 ml of RPMI 1640 containing 1 mg/ml BSA; (b) Inject 2 μl of 10 mM C_6-NBD-ceramide (ethanol stock) in 1 ml RPMI 1640 containing 1 mg/ml BSA; (c) Mix (a) and (b) (final concentration of NBD-ceramide 10 μM) and proceed to step 5.

4. Incubate samples from (3i) and (3ii), for 60 min at 37°C.

5. Immediately add 8 ml of chloroform:methanol (2:1) to the aqueous solution. Vortex 30 sec. Centrifuge 1 min at 1000 x g (=2000 rpm in GPKR centrifuge).

6. Collect lower organic phase with a Pasteur pipette and transfer to a clean glass borosilicate tube.

7. Reextract upper aqueous phase in 4 ml water-saturated chloroform:methanol (2:1). Vortex 30 sec and centrifuge 1 min at 1000 x g.

8. Collect lower organic phase with a Pasteur pipette and pool with first organic phase.

9. Dry COMPLETELY under nitrogen gas and chromatograph on TLC plates.

10. Bake HPTLC Si60 (without fluorescent indicator) plate for 1 hr at 80°C. Let cool 15 min at room temperature. Using a pencil, draw a line 1.5 cm up from the bottom of the plate and mark 1 to 1.5-cm spaces across this line.

11. Prepare TLC tank. Add 65 ml $CHCl_3$: 25 ml CH_3OH : 4 ml 28% NH_4OH to tank. Insert an 8 x 8-inch sheet of Whatman 3MM paper into tank and cover tightly. Whatman paper should be completely soaked before running TLC plates.

12. Resuspend dried samples from step 9 in ~50–100 μl $CHCl_3$ (centrifuge briefly if necessary to collect liquid) and spot onto TLC plates at the marks drawn in step 9. Use a Pasteur pipette without a bulb to collect ~10 μl in the tip; expel by pressing against plate. Blow gently on the plate to dry spot between applications. Alternatively use a Hamilton syringe

13. Spot known lipid standard (NBD-ceramide and NBD-sphingomyelin) on outermost lanes of plate as markers. Use a separate lane for each lipid standard.

14. Let plate dry 5 min and place into tank. Chromatograph for ~2 hr or until front has reached top of plate.

15. Mark solvent front with a pencil while plate is still in tank. Remove; let dry 5 min.

16. View on UV light box. Photograph plate. Circle fluorescent spots with pencil. Measure relative R_f of spots. R_f of NBD-sphingomyelin is 5.1.

17. To elute, GENTLY scrape spots from plates. Use the end of a small

spatula and scrape in a draft-free environment. Collect resin in labeled 10 x 75-mm borosilicate glass tubes.

18. Add 1 ml 100% gold-shield ethanol to tubes. Vortex 30 sec and centrifuge 1 min at 1000 x *g*. Remove supernatant to fresh glass tube.

19. Repeat step 18 x 3.

20. Pool supernatants: dry under nitrogen gas.

21. Resuspend in 1.5 ml ethanol. Read on fluorometer. We use a Perkin–Elmer 650-10S fluorescence spectrophotometer. To convert relative fluorescence units to moles of sphingomyelin, a standard curve may be generated using commercially available NBD-sphingomyelin.

Note

Do not use plastic tubes, tips, or syringes when working with lipid probes. From step 2 on use only glass.

Our results (shown in Table I; Elmendorf and Haldar, 1994; Das *et al.*, 1994. in press) indicate that an average of 51% of the original sphingomyelin synthase activity is in the released ring/trophozoite parasites, with 39% in the EM/TVM fraction. If we assume that all of the MSP1 in the EM/TVM fraction reflects contamination, and correct for it, an average of the 25% of the sphingomyelin synthase is exported to the EM/TVM fraction. Although it is clear that the enzyme is exported, there is variability in the relative distribution of sphingomyelin synthase activity between the two fractions (released parasite and EM/TVM). This likely reflects the range of parasite stages in the cell populations used for the different experiments. It is well established that synthesis and distribution of proteins change within the 24- to 36-hr parasite stages (Howard *et al.*, 1987; Haldar and Holder, 1993) used in these studies. The morphology of the parasite also changes dramatically as the parasite enlarges and the TVM alters from a vesicular to a more tubular conformation (Elmendorf and Haldar, 1994). These differences suggest either that the enzyme distribution may vary during ring and trophozoite parasite development or that the homogenization procedure might differentially disrupt the TVM–parasite connection at different developmental stages.

2. Labeling of Released Ring and Trophozoite Parasites with C$_6$–NBD–Ceramide for Microscopy

Resuspend released parasites at 5 x 10^7 cells/ml in buffer B containing 1 mg/ml defatted BSA and 10 μM C$_6$-NBD-ceramide. Incubate 30 min at 37°C and wash three times in buffer B containing 1 mg/ml defatted BSA. Bisbenzamide dye (Hoechst 33258) should be included in the final wash at a concentration of 5 μg/ml. Samples may be viewed in a conventional fluorescence microscope with fluorescein filter settings (see Color Plate 6). Bisbenzamide requires UV filters. The sphingolipid stain concentrates in a perinuclear region, consistent with its localization to a Golgi compartment within the parasite.

C. Distribution of Sphingomyelin Synthase in Merozoites

1. Merozoite Preparation

Merozoites were prepared as described by M. Blackman (Chapter 11). *No schizonts* were detected in the preparations despite exhaustive examination of 10–50,000 cells in thick smears. Assuming that each schizont produced 16 daughter merozoites, the net yield varied between 25 and 35% of the starting material. The protein content of the samples was measured by the BioRad method. Samples were resuspended at 1 mg merozoite protein/ml of PBS. Any standard formulation of PBS will do.

2. Protease Protection Assay

Make up 1% CHAPS (Sigma) w/v in PBS, 10% saponin (Sigma) in PBS, 1 mg/ml proteinase K in PBS (store on ice); 1 mM NBD-ceramide in ethanol (keep on ice). Set up the following incubations, **adding reagents in the order in which they appear in the table.** All sample sizes indicated are in microliters.

	A	B	C	D
Merozoite preparation	10	10	10	10
PBS	89	79	69	59
1% CHAPS			10	10
10% saponin			10	10
1 mg/ml proteinase K		10		10
1 mM NBD-ceramide	1	1	1	1

Incubate for 30 min at 37°C. Stop the reaction by adding 1 mM PMSF (final concentration) and chilling the samples on ice. Extract lipid and analyze by thin-layer chromatography.

D. Comparison of Sphingomyelin Synthase Activity in Merozoites or Ring-, Trophozoite-, and Schizont-Infected Erythrocytes

To compare levels of sphingomyelin synthesized in equal numbers of merozoites or infected erythrocytes at the ring, trophozoite, or schizont stage, incubate 1 or 2 x 10^8 parasites of the appropriate stage with 10 μM C_6-NBD-ceramide or (C_5-DMB-ceramide) in either RPMI 1640 or PBS at 37°C for 15, 30, or 60 min. At the indicated times, extract samples for lipid analysis. Since the amount of product detected is directly proportional to the number of parasites or time of incubation in an assay, it may used as a measure of enzyme activity.

To determine the effects of protein synthesis inhibitors on sphingomyelin synthase activity, incubate ring, trophozoite, and schizont-infected red cells with 100 μg/ml cycloheximide in RPMI 1640 for 3 hr. Collect cells by centrifugation and label with C_6-NBD-ceramide for 30 min at 37°C in the continued presence of cycloheximide. For "mock" incubations both the pretreatment and the labeling should be carried out in RPMI 1640 alone. The effects of cycloheximide on protein synthesis and parasite growth may be determined as described by Elmendorf *et al.* (1992).

E. Concluding Remarks

The results of the protease protection studies in merozoites indicated that enzyme sphingomyelin synthase is entirely within the parasites in these extracellular stages (Elmendorf and Haldar, 1994). Rings and trophozoites do not synthesize new sphingomyelin synthase, but export a fraction of the merozoite activity to the TVM. Thus a classic Golgi enzyme sphingomyelin synthase is exported to the TVM and the *de novo* synthesis of sphingomyelin may provide a mechanism of membrane growth in the intraerythrocytic structures (Elmendorf and Haldar, 1994; see Fig. 7). This export is particularly interesting when compared to the quantitative retention of PfERD2 within the parasite. It suggests that the TVM is not an extruded form of the Golgi within the parasite, but a unique organelle, endowed with a subset (at least one) but not all of the characteristics of the Golgi.

V. Release of Pigmented Trophozoites and Schizonts from Infected Erythrocytes by Osmotic Shock in Isoosmolar Dipeptide-Based Media

A. Mature Asexual Stages Freed by Exposure to Glycyl-L-serine

1. Introduction

The intraerythrocytic location of the asexual stages of *P. falciparum* contained within the membrane of the parasitophorous vacuole presents difficulties in the isolation of viable malaria parasites entirely free of host erythrocyte membrane. A straightforward method for generating large numbers ($>10^9$ or 100 μl packed cells) of viable erythrocyte-free parasites is a prerequisite for the study of many aspects secretory mechanisms in the developmental regulation of this important human pathogen. Since the introduction of continuous culture methods for *P. falciparum*, many techniques have been described for the release and isolation of human malaria parasites from the host red cell (Kanaani and Ginsburg, 1989; Choi and Mikkelson, 1990; Trager *et al.*, 1990; Elmendorf *et al.*, 1992; Elford and Ferguson, 1993). In this section, we describe a method that relies on a difference in the permeability to dipeptides between the membrane of the parasitophorous vacuole, which contains the trophozoite, and the plasma membrane of the parasitised red blood cell, which undergoes permeabilization about mid-stage during parasite maturation. This difference in permeability allows the cytoplasm of the infected red cell to be loaded with certain small dipeptides (and certain linear tripeptides), resulting in the osmotic shock and the release of free parasites when the cells are returned to normal culture media (Elford and Ferguson, 1993).

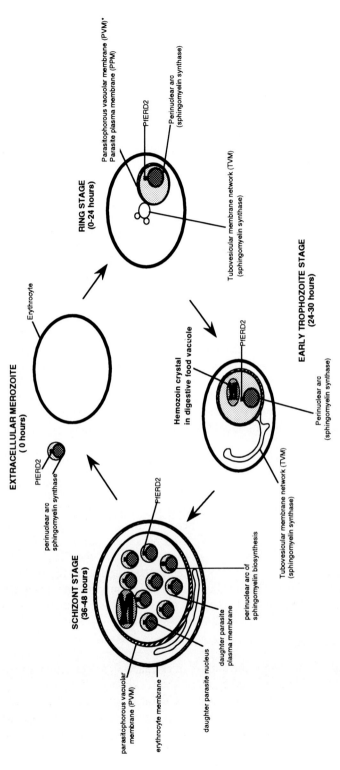

Fig. 7 Model for the stage-specific regulation of the Golgi enzyme sphingomyelin synthase. The life cycle of *P. falciparum* is shown starting at the left with the mature schizont stage. Following the cycle in a clockwise direction, merozoites are released and reinvade new erythrocytes. During the mature schizont stage both sphingomyelin synthase (perinuclear arc) and PfERD2 (perinuclear closed square) are localized within the parasite. After invasion, the parasite matures into ring and then trophozoites stages. Here the parasite induces the development of a tubovesicular network beyond its plasma membrane and exports the sphingomyelin synthase enzyme, originally present solely within the merozoite, partially outward into these membranes. PfERD2 is not exported. The asexual life cycle is completed with the maturation of the parasite into the mitotically dividing schizont stage. As expected, the parasite replicates sphingomyelin synthase and PfERD2 as it divides.

2. Stage-Specific Release of *P. falciparum* Parasites

1. Parasitized red blood cells (PRBCs) infected with ring-form parasites are not susceptible to release in dipeptide-based media. For optimal yields of free parasites, use cultures of synchronized parasites at least at the pigmented trophozoite stage of maturation.

2. Enrich for parasitized red cells either by plasmagel flotation or by centrifugation on a Percoll cushion (60–65% in RPMI) for knobby and knobless parasite phenotypes, respectively.

3. Pellet the concentrated washed PRBCs, aspirate the supernatant, and resuspend the cells at 37°C in an isoosmolar (5% w/v) solution of dipeptide buffered with HEPES (40 mM) and containing D-glucose (about 10 mM). Forty microliters of stock 1 M HEPES (pH 7.2–7.4) raises the pH of 1 ml 5% glycyl-L-serine from about 6.6 to 7. Fifty microliters of packed PRBCs can be conveniently loaded with dipeptide in 1-ml solution in an Eppendorf centrifuge tube.

4. After 15–45 min of exposure at 37°C, centrifuge the cell suspension, aspirate the supernatant, and resuspend the pellet in growth medium. Release of trophozoites occurs mainly on resuspension and is markedly dependent on temperature. Work only with prewarmed solutions close to 37°C. Resuspension of the cell pellet in ice-cold RPMI (after loading with glycyl-L-serine) not only lyses a higher proportion of the parasites per se but leaves more ghosted red cell membrane attached to partially freed intact parasites. A better yield of parasites completely free of membrane ghosts from infected red cells is produced by working close to physiological temperatures at all times.

3. Suspension of Released Trophozoites from Uninfected and the Surviving Intact PRBCs

Layer the partially released parasites over Ficoll 400 (20% in RPMI 1640). This can be done conveniently in Eppendorf tubes by centrifugation at 10000 x g for about 10–20 sec. Recover released parasites, parasites within ghosted erythrocytes, and empty ghosts from the layer above the Ficoll. Intact uninfected red cells and intact hemoglobinized PRBCs collect in the pellet. Further separation of free parasites from ghosted red cell membranes of previously intact parasitized erythrocytes can be carried out on sucrose gradients as required.

4. Viability of Naked Trophozoites

1. Optimal conditions for the culture of released *P. falciparum* parasites have not yet been fully defined. Naked trophozoites released by various dipeptides (Figs. 2 and 8) are able to mature to schizonts with good morphology after culture in standard RPMI-based culture medium (i.e., a Na-rich medium). However, to date, reinvasion rates of merozoites that have developed from naked

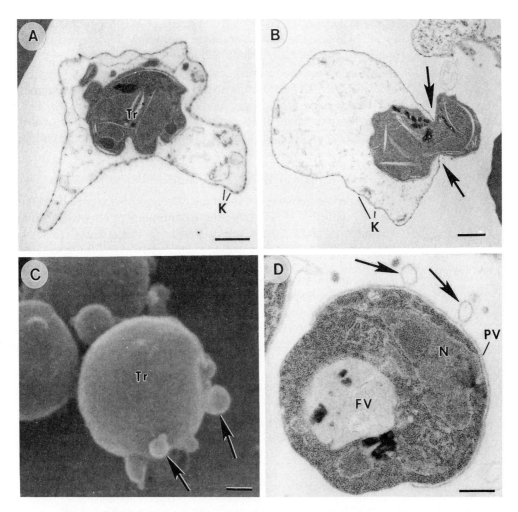

Fig. 8 Examples of the various forms of naked trophozoites seen after PRBCs are exposed to glycyl-L-serine as described in the text. Exposure of PRBCs to glycyl-D-serine produced similar morphologies (not illustrated). (A) Transmission electron micrograph showing a trophozoite within a ghosted red blood cell. About half of the hemolyzed infected cells retained their red cell membrane in this fashion. Bar: 1μm. (B) Parasite showing a partially shed red cell ghost with a constriction around the exit region (arrows). Characteristic knobs on the red cell membrane are marked (K). Bar: 1μm. (C) Scanning electron micrograph of a naked trophozoite (Tr) with several budding vacuole-like structures associated with the surface (arrows). Bar: 0.5 μm. (D) Transmission electron micrograph of a naked trophozoite (Tr) with its surrounding parasitophorous vacuole (PV) and associated vacuoles (arrows). N, nucleus; FV, food vacuole. Bar: 0.5 μm.

schizonts are far lower that those derived from parent intact PRBCs. It should be stressed that even after purification on Ficoll, a few intact PRBCs always contaminate the released trophozoite preparation; this contamination makes it difficult to assess the reinvasion index for merozoites derived solely from free trophozites generated by osmotic shock in dipeptides.

2. Naked pigmented parasites freed after exposure to glycyl-L-serine are viable, as indicated by the comparison of SDS–PAGE profiles of metabolically labeled parasite proteins from the original and the released parasites (not shown). Morphological changes also show that apparently normal maturation from mid-trophozoite stage to late schizonts occurs in standard culture medium.

B. Differential Hemolytic Effects of Diverse Dipeptides

The marked difference in the degree of hemolysis induced by glycyl-L-serine compared with that induced by glycyl-L-threonine, both of which have similar pK_a values (3 and 3.25), seems to reside in the H for CH_3 substitution in the hydroxylated side chain, which significantly slows the uptake of glycyl-L-threonine into *P. falciparum*-infected erythrocytes and defines, to some extent, the molecular exclusion limit for parasite inducible influx pathways through the host erythrocyte. This differential contrasts with the relative influx rates for serine and threonine individually in *P. falciparum*-infected red cells (Kirk *et al.*, 1994). Table II gives the categories which some of the commercially available dipeptides fit into in terms of their ability to induce hemolysis of PRBCs and release free *P. falciparum* parasites.

Table II
Relative Effects of Different Water–Soluble Dipeptides on Erythrocytes Parasitized with Mature *P. falciparum* Trophozoites

Permeable	Relatively impermeable
Glycyl-glycine	Glycyl-L-glutamine
Glycyl-L-serine	Glycyl-L-threonine
Glycyl-L-alanine	Glycyl-L-asparagine
Glycyl-β-alanine	Glycyl-L-tyrosine

Note. These categories are based on hemolysis of PRBC and partial release of mature parasites on resuspension in RPMI—10% human serum after a 20-min incubation in HEPES-buffered media. No difference was observed between the effects of glycyl-D-serine and glycyl-L-serine. Many other dipeptides are taken up more slowly by parasitized red cells and may be used less effectively, e.g., glycyl-valine, glycyl-proline, glycyl-L-leucine, glycyl-D-leucine, D-leucyl-glycine, and L-alanyl-L-serine.

Compared with many techniques for the release of parasites from PRBCs, e.g., the controlled homogenisation process developed by Elmendorf *et al.* (1992), the main advantage offered by osmotic shock in dipeptide-based medium is that it readily generates large numbers (10^9 or more) of viable, free, parasites. These may potentially be used in the study of developmental regulation and secretory mechanisms associated with the export of parasite proteins in pigmented trophozoite and schizont stages, and also as targets in the transfection of foreign DNA into asexual blood stages of the parasite.

Acknowledgments

This work was supported in part by grants from the NIH (RO1 AI26670) and Smith Kline and Beecham Co., Biomedical Research Support Grant RR05353 awarded by the Biomedical Research Support Program, Division of Research Resources, National Institutes of Health, and the MacArthur Foundation, to K.H., and an NSF and a Cell and Molecular Biology Predoctoral Training Fellowship to H.G.E. D.F. is supported by the Wellcome Trust.

References

Balch, W. E., Dunphy, W. G., Braell, W. A., and Rothman, J. E. (1984). Reconstitution of the transport of protein between successive compartments of the Golgi measured by the copupled incorporation of N-acetylglucosamine. *Cell (Cambridge, Mass.)* **39**, 405–416.

Bannister, L. H., and Mitchell, G. H. (1989). The fine structure of secretion by *Plasmodium knowlesi* merozoites during red cell invasion. *J. Protozool.* **36**, 362–367.

Barnwell, J. W. (1990). Vesicle mediated transport of membrane proteins in malaria infected red blood cells. *Blood Cells* **16**, 379–395.

Blackman, M., Chappel, J. A., Shai, S., and Holder, A. A. (1993). A conserved parasite serine protease processes the *Plasmodium falciparum* merozoite surface protein-1. *Mol. Biochem. Parasitol.* **62**, 103–114.

Choi, I., and Mikkelson, R. B. (1990). *Plasmodium falciparum:* ATP/ADP transport across the parasitophorous vacuolar and plasma membranes. *Exp. Parasitol.* **71**, 452–462.

Das, A., Elmendorf, H. G., Li, W-l., and Haldar, K. (1994). Biosynthesis, export and processing of a 45 kDa protein detected in membrane clefts of erythrocytes infected with *Plasmodium falciparum*. *Biochem. J.* **302**, in press.

Dluzewski, A. R., Mitchell, G. H., Fryer, P. R., Griffiths, S., Wilson, R. J. M., and Gratzer, W. B. (1992). Origins of the parasitophorous vacuolar membrane of the malaria parasite *Plasmodium falciparum,* in human red blood cells. *J. Cell Sci.* **102**, 527–532.

Elford, B. C., and Ferguson, D. J. P. (1993). Secretory processes in *Plasmodium falciparum*. *Parasitol. Today* **9**, 80–81.

Elmendorf, H. G., and Haldar, K (1993a). Secretory transport in *Plasmodium. Parasitol. Today* **9**, 98–102.

Elmendorf, H. G., and Haldar, K. (1993b). Identification and localization of ERD2 in the malaria parasite *Plasmodium falciparum:* Separation from sites of sphingomyelin synthesis and implications for organization of the Golgi. *EMBO J.* **12**, 4763–4773.

Elmendorf, H. G., and Haldar, K. (1994). *Plasmodium falciparum* exports the Golgi marker sphingomyelin synthase into a tubovesicular network in the cytoplasm of mature erythrocytes. *J. Cell Biol.* **124**, 449–462.

Elmendorf, H. G., Bangs, J. D., and Haldar, K. (1992). Synthesis and secretion of proteins by released malarial parasites. *Mol. Biochem. Parasitol.* **52**, 215–230.

Gormley, J. A., Howard, R. J., and Tarashi, T. F. (1992). Trafficking of malarial proteins to the host cell cytoplasm and erythrocyte surface membrane involves multiple pathways. *J. Cell Biol.* **119,** 1481–1495.

Haldar, K., and Holder, A. A. (1993). Export of parasite protein to the erythrocyte in *Plasmodium falciparum*-infected cells. *Semin. Cell Biol.* **4,** 345–353.

Haldar, K., and Uyetake, L. (1992). The movement of fluorescent endocytic tracers in *Plasmodium falciparum*-infected erythrocytes. *Mol. Biochem. Parasitol.* **50,** 161–178.

Haldar, K., Ferguson, M. A. J., and Cross, G. A. M. (1985). Acylation of *Plasmodium falciparum* merozoite surface antigen via *sn* 1,2-diacylglycerol. *J. Biol. Chem.* **260,** 4969–4974.

Haldar, K., Uyetake, L., Ghori, N., Elmendorf, H. G., and Li, W-1. (1991). The accumulation and metabolism of a fluorescent ceramide derivative in *Plasmodium falciparum*-infected erythrocytes. Mol. Biochem. Parasitol. **49,** 143–156.

Howard, R. J., Uni, S., Lyon, J. A., Taylor, D. W., Daniel, W., and Aikawa, M. (1987). Export of *Plasmodium falciparum* proteins to the host erythrocyte membrane: Special problems of protein trafficking and topogenesis. *NATO ASI Ser., H* **11,** 281–296.

Kanaani, J., and Ginsburg, H. (1989). Metabolic interconnection between the human malarial parasite *Plasmodium falciparum* and its host erythrocyte. *J. Biol. Chem.* **264,** 3194–3199.

Kirk, K., Horner, H. A., Elford, B. C., Clive Ellory, J. and Newbold, C. I. (1994). Transport of diverse substrates into malaria infected erythrocytes via a pathway showing functional characteristics of a chloride channel. *J. Biol. Chem.* **269,** 3339–3347.

Li, W-l., Das, A., Song, J-Y., Crary, J. L., and Haldar, K. (1991). Stage specific expression of plasmodial proteins containing an antigenic marker of the intraerythrocytic eisternae. *Mol. Biochem. Parasitol.* **49,** 157–168.

Pagano, R. E. (1989). A fluorescent derivative of ceramide: Physical properties and use in studying the Golgi apparatus of animal cells. *Methods Cell Biol.* **29,** 75–85.

Pagano, R. E., Martin, O. C., Kang, H. C., and Haugland, R. P (1991). A novel fluorescent ceramide analogue for studying membrane traffic in animal cells: Accumulation at the Golgi apparatus results in altered spectral properties of the sphingolipid precursor. *J. Cell Biol.* **113,** 1267–1279.

Perkins, M. E. (1992). Rhoptry organelles of apicomplexan parasites. *Parasitol. Today* **8,** 28–32.

Rothman, J. E., and Orci, L. (1992). Molecular dissection of the secretory pathway. *Nature (London)* **355,** 409–415.

Trager, W., Tershakovec, M., Lyandvert, L., Stanley, H., Lanners, N., and Gubert, E. (1981). Clones of the malaria parasite *Plasmodium falciparum* obtained by microscopic selection: Their characterization with regard to keroks, chloroquine sensitivity and formation of gametocytes. *Proc. Natl. Acad. Sci. U.S.A.* **78,** 6527–6530.

Trager, W., and Jensen, J. B. (1976). Human malaria in continuous culture. *Science* **193,** 673–675.

Trager, W., Jung, J., and Tershakovec, M. (1990). Initial extracellular development in vitro of erythrocytic stages of malaria parasites (*Plasmodium falciparum*). *Proc. Natl. Acad. Sci. U.S.A.* **87,** 5618–5622.

Ward, G. E., Miller, L. H., and Dvorak, J. A. (1993). The origin of the parasitophorous vacuole membrane lipids in malaria infected erythrocytes. *J. Cell Sci.* **106,** 237–248.

CHAPTER 13

Intracellular Survival by *Legionella*

Karen H. Berger* and Ralph R. Isberg*,†

*Department of Molecular Biology and Microbiology
†Howard Hughes Medical Institute
Tufts University School of Medicine
Boston, Massachusetts 02111

  I. Introduction
 II. Laboratory Cultivation of *L. pneumophila*
III. Tissue Culture of U937 Cell-Derived Macrophages
 IV. Intracellular Thymineless Death Enrichment
     A. Genetic Manipulation of Bacteria Prior to Reconstruction Studies
     B. Reconstruction Studies
     C. Enrichment of Mutants from a Mixed Bacterial Population
  V. Identification of Intracellular Growth Mutants from Enriched Bacterial Pools Using "Poke Plaque" Assays
 VI. Additional Remarks
     References

I. Introduction

In this chapter we describe a strategy that allows the isolation of bacterial mutants specifically defective for intracellular replication in cultured macrophages. This strategy is used to isolate mutants of the Gram-negative bacterium *Legionella pneumophila,* but should be applicable to other intracellular pathogens as well. The enrichment strategy is based on intracellular thymineless death of thymine-requiring but otherwise replication-competent microorganisms. Methods for cultivation of bacteria and tissue culture of the U937 macrophage cell line that is used are given (Sections II and III). We describe the reconstruction studies that are carried out to develop the enrichment strategy, and give specific details of the procedure that is used to enrich intracellular growth mutants from a mixed bacterial population via intracellular thymineless death (Section IV). Another procedure that is used to identify candidate mutants from enriched bacterial pools is described (Section V). Additional remarks on this enrichment are given (Section VI).

II. Laboratory Cultivation of *L. pneumophila*

L. pneumophila Philadelphila-1 is from the Center for Disease Control (CDC, Atlanta, GA). Avirulent mutant 25D is from Dr. Marcus A. Horwitz (1987). Solid bacteriological culture medium is charcoal yeast extract medium (CYE medium; Feeley *et al.*, 1979), buffered to pH 6.9 using *N*-(2-acetamido)-2-aminoethanesulphonic acid (ACES, Sigma). Bacteria are grown at 37°C in air. Under these conditions single colonies are typically visible in 3 days. Where noted, drugs are included in bacteriological media at the following concentrations: streptomycin, 100 μg/ml; kanamycin, 25 μg/ml; rifampicin, 5 μg/ml; trimethoprim, 10 μg/ml. Thymidine is included in bacteriological and cell culture media where noted at 100 μg/ml (10 mg/ml stock of thymidine is made up in distilled H_2O (dH_2O) and sterilized by filtration). Genetic manipulation of bacteria that is carried out prior to enrichment is described below (Section IV,A).

III. Tissue Culture of U937 Cell-Derived Macrophages

The U937 cell line is derived from a human histiocytic lymphoma (Sundström and Nilsson, 1976), and is from the American Tissue Culture Center (ATCC, Rockville, MD). All experiments are carried out using U937-derived macrophage monolayers after differentiation with phorbol esters (see below). Phorbol ester differentiation of U937 cells and incubation of differentiated U937 cell monolayers with *L. pneumophila* have been described (Pearlman *et al.*, 1988).

1. Medium for U937 cell culture is RPMI 1640 (Irvine Scientific), supplemented with 10% fetal calf serum (GIBCO) and 2 m*M* L-glutamine. Typically no antibiotics are included in cell culture medium. All U937 cell culture is at 37°C in 5% CO_2. Cells are maintained in their nonadherent, undifferentiated state in plastic tissue culture flasks (Falcon) and are passaged at a cell density of approximately 10^6 cells/ml. One-milliliter cell aliquots are stored frozen in 50% cell culture medium and 50% dimethyl sulfoxide (DMSO) at −80°C, thawed, and washed once in cell culture medium before culturing for experiments. After thawing, cells are serially passaged for no more than 15 times.

2. To differentiate U937 cells to an adherent macrophage-like state, transfer undifferentiated cells at a density of approximately 10^6 cells/ml to 50-ml plastic centrifuge tubes (Falcon) and pellet in a tabletop centrifuge (5 min at 1000 rpm, 25°C). Resuspend the pelleted cells in an equal volume of fresh cell culture

medium containing phorbol 12-myristate 13-acetate (PMA, Sigma) at a final concentration of 160 nM (PMA is stored at $-20°C$ in DMSO at $10,000\times$ final concentration and thawed just before use).

3. Transfer the resuspended cells to a fresh tissue culture flask (50 ml culture per 175-cm^2 flask) and incubate in the presence of PMA for 36–48 hr to allow differentiation.

4. To transfer PMA-differentiated U937 cells to tissue culture plates for incubation with bacteria, remove culture medium from differentiated U937 cell monolayers. Add 10 ml/flask of phosphate-buffered saline (PBS), pH 7.2, containing 2 mM EDTA. Incubate approximately 5 min with occasional agitation until cells have detached from plastic.

5. Transfer EDTA-detached cells to a 50-ml Falcon tube and add 0.5 ml of 100 mM MgCl$_2$ per 10 ml cell suspension (5 mM final concentration). Pellet cells as above (5 min at 1000 rpm, 25°C). Remove supernatant and resuspend cells gently but thoroughly in one-half the original culture volume (compared to that in step 3) of fresh cell culture medium lacking PMA. Transfer cells to 6-well (4 ml cell suspension/well) or 24-well (1 ml/well) tissue culture plates (Nunc, Falcon, or Costar) for reconstruction studies or enrichments (Section IV). One tissue culture flask containing a 50-ml culture of differentiated U937 cells typically yields approximately 2×10^7 cells, enough for confluent monolayers in one 24-well or 6-well plate. For poke plaque assays (Section V), cells are typically plated in 100-mm tissue culture dishes (Costar), with each 50-ml culture yielding one or two confluent plates.

6. The freshly plated differentiated U937 cells may be incubated for up to 2 days before addition of bacteria, replacing the spent cell culture medium with fresh medium as needed. Typically cells are plated approximately 12 hr before incubation with bacteria.

IV. Intracellular Thymineless Death Enrichment

It is generally useful to enrich for desired mutants in a mixed population, because it reduces the number of candidates that must be screened in order to identify mutants. The rationale for thymineless death enrichment of *L. pneumophila* intracellular growth mutants is analogous to the use of penicillin to enrich bacterial auxotrophs. A strategy similar to penicillin enrichment employs the β-lactam antibiotic methicillin to enrich intracellular growth mutants of the Gram-positive intracellular bacterial pathogen *Listeria monocytogenes* (Camilli *et al., 1989*). In the case of intracellular *L. pneumophila*, methicillin was determined to be bacteriostatic rather than bacteriocidal in reconstruction studies similar to those described below (data not shown). Therefore, neither methicillin nor other antibiotics that were tested (data not shown) were useful in our system to enrich intracellular growth mutants, and we adopted a different enrichment approach. Strains of *L. pneumophila* that are thymine auxotrophs (Thy⁻) require thymine for growth and cannot survive intracellularly in the absence of thymine, but can grow in cultured cells if cell culture medium is supplemented with thymine or thymidine (Mintz *et al.*, 1988). In contrast, we determined that a Thy⁻ derivative of a strain known to be defective for intracellular growth survives intracellularly in either the presence or the absence of added thymidine (Berger and Isberg, 1993). Therefore, starting with a mixed population of thymine auxotrophic bacteria, intracellular thymineless death selects against those bacteria that can grow intracellularly. Bacterial strains specifically defective for intracellular growth can be enriched simply by incubating a mixed bacterial population inside macrophages in the absence of added thymine or thymidine, as previously described (Berger and Isberg, 1993).

A. Genetic Manipulation of Bacteria Prior to Reconstruction Studies

We were fortunate to have available a known *L. pneumophila* mutant strain that is unable to grow intracellularly, but is still capable of intracellular survival (Horwitz, 1987). This allows us to perform reconstruction studies in which the intracellular survival of replication-competent and replication-defective *L. pneumophila* strains is compared in mixed incubations. Such studies allow us to evaluate the effectiveness of different enrichment protocols. Bacterial strains are distinguished in these reconstruction studies by different antibiotic resistances. Our studies use a replication-competent strain that is streptomycin-resistant (and rifampicin-sensitive; Str^R Rep⁺) and a rifampicin-resistant (and streptomycin-sensitive) isolate of the known replication-defective mutant (Rif^R Rep⁻), as described (Berger and Isberg, 1993).

Thymine auxotrophic derivatives of the Str^R Rep⁺ and Rif^R Rep⁻ strains are isolated by selecting for bacterial growth on CYE solid medium containing 100 μg/ml thymidine and 10 μg/ml trimethoprim. The antimetabolite drug trimethoprim selects against thymine prototrophs by inhibiting dihydrofolate reductase (Stacey and Simson, 1965), as previously described for isolation of *L. pneumophila* thymine auxotrophs (Mintz *et al.*, 1988). The thymine requirement of trimethoprim-resistant strains is confirmed by testing for colony formation on CYE solid medium lacking thymidine; Thy⁻ strains should grow only on medium supplemented with thymidine. Thy⁻ strains are additionally examined for whether they revert to thymine prototrophy at a high ($>10^{-6}$) frequency by plating bacteria on CYE medium lacking thymidine. The strains selected for use in reconstruction studies are not observed to undergo frequent reversion to Thy⁺; also, Thy⁺ bacteria were not found to represent an increased fraction of the population after a typical thymineless death enrichment. The Thy⁻ strain that is otherwise replication-competent remains able to multiply intracellularly when cell culture medium is supplemented with thymidine, as previously described (Mintz *et al.*, 1988). Details of a reconstruction study of the type used to develop the thymineless death enrichment procedure are given below.

B. Reconstruction Studies

1. For preliminary reconstruction studies, U937 cells are plated in 24-well plates, to allow a greater number of different conditions to be examined without increasing the number of cells used. Once optimal enrichment conditions are determined, the enrichment procedure is scaled up and 6-well plates are typically used.

2. Separately culture the Thy⁻Rep⁺ (Str^R) and Thy⁻Rep⁻ (Rif^R) strains described above from fresh single colonies onto CYE solid medium supplemented with thymidine and lacking antibiotics. Thinly spread bacteria to a uniform density on bacteriological culture plates using sterile cotton-tipped applicator swabs. Incubate plates until a fine lawn of bacterial growth is visible, typically overnight. It is important that bacterial growth on plates is not overly dense, so that bacteria of each strain are growing exponentially at the time of addition to macrophage monolayers.

3. Harvest bacteria separately into sterile test tubes containing 1–2 ml PBS at room temperature. Dilute bacterial suspensions into room-temperature coincubation medium: coincubation medium is RPMI containing 10% normal human serum, 0.4% bovine serum albumin (BSA), and 20 mM HEPES (pH 7), and is sterilized by filtration. Normal human serum is obtained from donors not known to have had Legionnaires' disease and is negative when examined for agglutination of *L. pneumophila* bacteria. No thymidine is added to this medium; however, if desired as a control, thymidine at 100 μg/ml may be added to additional samples that are tested in parallel. Dilute the individual suspensions of Thy⁻Rep⁺ (Str^R) and Thy⁻Rep⁻ (Rif^R) bacteria in coincubation medium to an approximate optical density at 600 nm (OD$_{600}$) of 0.2, estimated to be about 5×10^7 bacteria/ml, then dilute an additional 10-fold, so that the approximate bacterium : U937 cell multiplicity will be between 1 and 10. Mix the two diluted bacterial suspensions at different ratios (e.g., 1 : 1, 1 : 10, 1 : 100 Rep⁻ : Rep⁺), to stimulate an actual enrichment. Although mutants will probably not make up such a high percentage of an actual unenriched bacterial population, it is more convenient to determine the relative survival of these larger percentages of mutant bacteria in reconstruction studies.

4. Aspirate cell culture medium from PMA-differentiated U937 cell monolayers and add 0.5 ml of the mixed bacterial suspensions per well of a 24-well tissue culture plate. To separate wells, add comparable numbers of each bacterial strain in pure suspension; this allows determination of whether mixing the bacterial strains alters the survival of either. Reserve some of each suspension that is added to wells (see step 6). A plate used for a typical reconstruction study is diagrammed (Fig. 1A).

5. Allow U937 cell monolayers to take up bacteria. Incubate for 1–2 hr (all incubations of bacteria with U937-derived macrophages are at 37°C in 5% CO$_2$). Unlike many bacteria, *L. pneumophila* does not replicate in standard cell culture medium (Horwitz and Silverstein, 1980). Therefore, it is not necessary to include

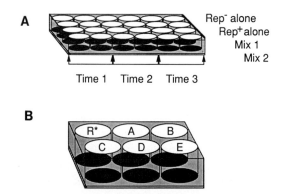

Fig. 1 (A) In a typical reconstruction study, U937 cell-derived macrophages are plated in 24-well tissue culture plates as described in the text. Thy⁻Rep⁻ and Thy⁻Rep⁺ bacterial strains that are distinguishable based on different drug resistances are cultured individually, and either pure or mixed bacterial suspensions are added to U937 cell monolayers. At different times, intracellular bacterial cfu of duplicate samples for each strain are determined as described (see text). (B) Enrichment is performed on separate pools of bacteria in individual wells of a 6-well tissue culture plate containing U937 cell monolayers. In parallel to pools undergoing enrichment (wells A–E), the identical procedure is performed on a mixture of the two strains that are used in reconstruction studies (well R*), and the survival of these strains is determined (see text).

an antibiotic such as gentamicin to prevent extracellular replication of bacteria during incubations with cells.

6. To determine the colony-forming units (cfu) added for each bacterial strain, serially dilute the remainders of bacterial suspensions that were added to U937 cell monolayers and plate aliquots in parallel on nonselective and differentially selective media (CYE Thy, CYE Thy Str, and CYE Thy Rif). This additionally determines the plating efficiencies of the drug-resistant bacterial strains on these media under the conditions of the reconstruction.

7. After an initial 1- to 2-hr incubation of bacteria with U937 cell monolayers, aspirate the medium, and wash monolayers twice with sterile PBS. Add fresh coincubation medium (lacking bacteria) and continue incubation.

8. At periodic intervals (e.g., 3, 8, and 24 hr; see below), aspirate coincubation medium and replace with fresh coincubation medium, prewarmed to 37°C, lacking NHS (see below) and containing gentamicin (diluted from a fresh 10 mg/ml stock in dH₂O, filter sterilized) at 50 μg/ml. Incubate 30 min to kill extracellular bacteria.

9. Aspirate medium from monolayers and wash three to four times with sterile PBS. (Note: it is important to wash out gentamicin so that bacteria are not killed once U937 cells are lysed. However, washing too vigorously will cause PMA-differentiated U937 cells to detach from plastic.) Add 0.5 ml/well

sterile dH$_2$O, incubate approximately 15–20 min, and lyse U937 cells by vigorous pipetting.

10. Plate serially diluted aliquots from each well of the reconstruction plate onto bacteriological culture media as above (step 6).

11. Incubate plates until single colonies are visible (3–5 days, depending on the ability of drug-resistant bacteria to grow in the presence of drug). Count colonies, and for each strain, compare that cfu initially added to monolayers with the cfu recovered following the trial enrichment procedure. A successful enrichment is defined by higher recovery of the intracellular growth mutant than recovery of the replication-competent strain when both are present in a mixed incubation.

To optimize the predicted success of the enrichment based on reconstruction studies before carrying out actual enrichments, a number of conditions may be varied. One condition that appears to affect the success of mutant enrichment in our reconstruction studies is the duration of incubation. Comparing total incubation times of 3, 8, and 24 hr (i.e., 2-hr initial incubation, wash, and 1-, 6-, or 22-hr additional incubation), we obtain the best result with 8 hr total coincubation, which results in approximately 10-fold enrichment, as described (Berger and Isberg, 1993).

In general, enrichment based on intracellular thymineless death is optimized when initial bacterial uptake is maximized. We did not attempt to increase initial bacterial-U937 cell contact by centrifuging U937 cell monolayers after adding bacteria, because this is somewhat toxic to the U937 cells. However, we wash the monolayers after allowing a limited time for initial bacterial uptake and then incubate for longer periods of time. This increases the percentage of total bacteria that are inside cells for the entire incubation period.

It is equally important that gentamicin killing of extracellular bacteria is effective. Because gentamicin killing of extracellular bacteria appears to be less effective when NHS is present based on reconstruction studies, NHS is omitted from coincubation medium during incubation with gentamicin. We also find that the presence of serum during incubation of bacteria with U937 cells appears to be necessary to obtain enrichment of the mutant. *L. pneumophila* enters monocytes less efficiently in the absence of a source of serum complement (Payne and Horwitz, 1987), so it is perhaps not surprising that reconstruction studies carried out in the absence of serum are less effective in enriching the intracellular growth mutant (data not shown).

Finally, it is important that bacteria are growing exponentially on plates just prior to harvesting for enrichment for intracellular thymineless death to be effective. The intracellular growth mutant is not enriched unless the replication-competent strain undergoes intracellular thymineless death, and this requires bacteria to be inside cells and not impeded or delayed from intracellular replication aside from their thymine requirement.

C. Enrichment of Mutants from a Mixed Bacterial Population

Briefly, pools of bacteria to be enriched are incubated with U937 macrophage monolayers in the absence of added thymine or thymidine. Enrichment protocol details are developed based on the results of reconstruction studies such as those described above. Specific steps of the thymineless death enrichment used to enrich *L. pneumophila* intracellular growth mutants are diagrammed (Fig. 2).

1. Prepare PMA-differentiated U937 cell monolayers in 6-well tissue culture plates, as described above (Section III).

2. Following bacterial mutagenesis, if applicable, grow individual bacterial pools on CYE solid medium supplemented with thymidine at 100 μg/ml. To maximize the number of unrelated mutants obtained, keep individual pools separate throughout the following steps. Monitor the effectiveness of enrichment by determining the relative survival of the known intracellular growth mutant (Rep$^-$) in a mixed incubation with the Rep$^+$ strain (i.e., a reconstruction), in parallel to enrichments. A typical enrichment plate is diagrammed (Fig. 1B).

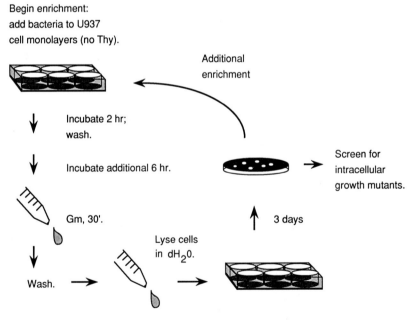

Fig. 2 Thymineless death enrichment procedure. Mixed (replication-competent and replication-defective) populations of Thy$^-$ bacteria are harvested from bacteriological culture plates and incubated in U937 cell-derived macrophages in the absence of thymine or thymidine in order to enrich replication-defective bacterial strains. Specific steps of the procedure are described in the text (Section IV).

3. Separately harvest bacteria from the individual pools that are to be enriched, and from the freshly plate-grown cultures of Rep$^+$ and Rep$^-$ strains to be used in the parallel reconstruction. As noted above, bacteria should be growing exponentially up to the point when they are harvested from bacteriological culture medium. They should therefore not be grown more than a few days on plates, nor stored in the cold prior to harvesting for enrichment. If necessary, plates of bacteria that have may been refrigerated may be replicaplated and grown up fresh for enrichment.

4. The essential details of the enrichment procedure are identical to those used to enrich the intracellular growth mutant in reconstruction studies and are diagrammed (Fig. 2). Briefly, harvest bacteria from plates into PBS and dilute in coincubation medium to the approximate bacterial cell density that is determined to be effective to enrich mutants in reconstruction studies. We find enrichment to be effective at an approximate multiplicity of 1–10 bacteria per U937 cell. At higher multiplicities, most bacteria will not be taken up by macrophages, increasing the proportion of extracellular bacteria that are not undergoing enrichment. Also, addition of large numbers of Rep$+$ bacteria is toxic to U937 cells.

5. Add 1 ml/well of individual bacterial suspensions to PMA-differentiated U937 cell monolayers as diagrammed (Figs. 1B and 2). To allow bacterial uptake, incubate for 1–2 hr, wash U937 cell monolayers, and continue incubation with fresh cell culture medium.

6. Incubate for the length of time that is determined to be effective to enrich the known intracellular growth mutant in reconstruction studies. Kill extracellular bacteria by gentamicin treatment, wash monolayers, and lyse U937 cells by pipetting with 0.2 ml/well of sterile dH$_2$O, as described above.

7. Plate bacteria from individually enriched pools separately onto solid bacteriological culture medium supplemented with thymidine. Bacteria from a single enrichment pool are typically plated onto several CYE Thy plates, and may be serially diluted to obtain growth of a dense lawn of individual bacterial colonies that are not quite confluent. Incubate to allow bacterial growth before either carrying out additional enrichment cycles or screening individual bacterial colonies to identify intracellular growth mutants (see Section V).

8. For additional enrichments, bacteria from the same pools that were plated separately are harvested and repooled. Enriched bacterial populations may be stored in 50% glycerol and 50% ACES-buffered yeast extract liquid culture medium (AYE; Gabay and Horwitz, 1985), supplemented with 100 μg/ml thymidine.

By requiring that bacterial strains that survive intracellular thymineless death grow on bacteriological culture media, the enrichment procedure selects mutants whose growth defect is specifically intracellular. Bacterial strains with intracellular growth defects that additionally exhibit more than a slight defect for growth on bacteriological culture medium relative to the starting (parental) strain will not be enriched by this procedure.

V. Identification of Intracellular Growth Mutants from Enriched Bacterial Pools Using ''Poke Plaque'' Assays

The poke plaque assay (Berger and Isberg, 1993) is a variation of the single plaque assay, which measures intracellular growth of bacteria in cultured cells (Oaks *et al.*, 1985) and was adapted for *L. pneumophila* as described (Marra *et al.*, 1992). The poke plaque assay allows relatively large numbers of bacterial strains to be examined for their ability to produce plaques on U937 cell monolayers, which is an indicator of intracellular growth competence.

1. Prepare confluent monolayers of PMA-differentiated U937 cells in 100-mm tissue culture dishes as described (Section II).

2. Aspirate cell culture medium and replace with 12 ml/dish of a mixture of 20% normal human serum, 0.7% molten Agar Noble (Difco) and $0.8 \times$ RPMI 1640 medium, supplemented with 100 μg/ml thymidine. Prepare Agar Noble at 1.8% and sterilize by autoclaving. Make up RPMI 1640 medium containing L-glutamine from powder (GIBCO) at $2 \times$ concentration, supplement with $NaHCO_3$ according to the manufacturer's instructions, and sterilize by filtration. Allow the overlay to solidify by cooling at room temperature before transferring overlaid monolayers to 37°C and 5% CO_2. Incubate monolayers at least 30 min before continuing to step 3.

3. Screen individual bacterial colonies that have been purified by restreaking twice on solid bacteriological culture medium by poking a sterile toothpick carrying a small portion of a single bacterial colony into the overlay. Bring the bacterial inoculum into contact with the monolayer without disrupting the monolayer integrity. Fifty to 100 bacterial strains may be conveniently screened on a single 100-mm dish in this manner, with the aid of a colony grid placed under the plate to mark the location of pokes.

4. In parallel to the poke plaque assay plate, patch each bacterial strain being tested onto the corresponding position on a CYE Thy plate. Label both plates in a manner that will allow candidate mutants on the CYE Thy plate to be identified subsequently. Incubate the assay plate at 37°C in 5% CO_2, and the CYE Thy plate at 37°C in air.

5. After 3 days, overlay the poke plaque assay plates with a second overlay consisting of 0.9% molten Agar Noble and $1 \times$ RPMI containing 0.01% (w/v) Neutral Red. Incubate the plate an additional few hours to overnight to allow the Neutral Red to diffuse through the first overlay and be taken up by the viable U937 cells. Nonviable U937 cells that are killed by replicating bacteria fail to take up the dye, producing large clear zones (Fig. 3). In contrast, candi-

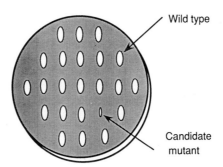

Fig. 3   The poke plaque assay is used to screen enriched populations of bacteria for intracellular growth mutants, as described (Section VI). A replication-competent bacterial strain produces a relatively large zone of clearing in a neutral red-strained U937 cell monolayer, due to intracellular bacterial growth and lysis of the U937 cells. In contrast, a candidate mutant strain causes either no localized U937 cell death, or produces a small zone of clearing relative to that produced by a replication-competent strain.

date intracellular growth mutants produce small zones of clearing relative to replication-competent strains on the U937 cell monolayer (Fig. 3).

6. Following identification of candidate mutants based on results of the initial poke plaque screen, these strains are examined using single plaque assays and by measurement of *L. pneumophila* intracellular growth in PMA-differentiated U937 cells, as previously described (Berger and Isberg, 1993). A relatively small number of candidate mutant bacterial strains are characterized using these more labor-intensive procedures.

VI. Additional Remarks

One significant consideration in the use of the thymineless death enrichment is whether intracellular growth of thymine auxotrophs can occur in the presence of extracellularly added thymine or thymidine, as is the case for *L. pneumophila* (Mintz *et al.*, 1988). *L. pneumophila* Thy⁻ strains may also be complemented genetically by plasmids containing the *tdΔi* gene, encoding thymidylate synthetase (Chu *et al.*, 1984; Berger and Isberg, 1993). If the intracellular growth defect caused by the thymine requirement cannot be effectively complemented or overcome by the addition of thymine or thymidine to the cell culture medium, this complicates both mutant identification and subsequent analysis. If candidate mutants can be identified, Thy⁺ revertants of candidate intracellular growth mutants may be isolated. The intracellular growth phenotype of the Thy⁺ revertants may then be described, provided the original Thy⁻ mutation undergoes reversion.

Intracellular thymineless death is a simple strategy to enrich intracellular growth mutants in a mixed population of microorganisms. Various measures such as those suggested here will help to optimize enrichment conditions. One caveat to the use of any enrichment is that it may select a limited class of mutant. Many of the mutants isolated using this thymineless death enrichment and identified by plaque assay screening were subsequently determined to be mutations in a single gene (*dot*A; Berger and Isberg, 1994). Mutants defective for synthesis of the *dot*A product are targeted to an intracellular compartment where they cannot grow, and this presumably allows them to survive in the absence of thymine. Equally important to bacterial survival of intracellular thymineless death, bacterial viability must not be appreciably reduced in the intracellular compartment to which microorgansims are targeted. We believe that this is the case for the *dot*A mutants, and that this is important for continued bacterial viability. Clearly, mutants killed intracellularly are not expected to be enriched by this procedure. To obtain a greater variety of candidate mutants using the thymineless death enrichment, we suggest that multiple enrichments be performed in which particular conditions are varied, such as the duration of incubation of bacteria in macrophage monolayers. We also suggest screening enriched populations for candidate mutants after each cycle of enrichment, to avoid overenriching the bacterial pools and possibly losing interesting mutants. Aside from these potential pitfalls, we determined enrichment by intracellular thymineless death to be both an effective and a straightforward method for enriching intracellular growth mutants from a mixed bacterial population.

Acknowledgments

We thank Dr. Marcus A. Horwitz for Mutant 25D, and Drs. Andrea Marra and Howard A. Shuman for advice on plaque assays. This work was supported by the Howard Hughes Medical Institute and an NSF Presidential Young Investigator Award to R.I., and by an NSF Graduate Fellowship to K.B. R.I. is an Associate Investigator of Howard Hughes Medical Institute.

References

Berger, K. H., and Isberg, R. R. (1993). Two distinct defects in intracellular growth complemented by a single genetic locus in *Legionella pneumophila*. *Mol. Microbiol.* **7,** 7–19.

Berger, K. H., Merriam, J. J., and Isberg, R. R. (1994). Altered intracellular targeting properties associated with mutations in the *Legionella pneumophila dotA* gene. *Mol. Microbiol.* In press.

Camilli, A., Paynton, C. R., and Portnoy, D. A. (1989). Intracellular methicillin selection of *Listeria monocytogenes* mutants unable to replicate in a macrophage cell line. *Proc. Natl. Acad. Sci. U.S.A.* **86,** 5522–5526.

Chu, F. K., Maley, G. F., Maley, F., and Belfort, M. (1984). Intervening sequence in the thymidylate synthase gene of bacteriophage T4. *Proc. Natl. Acad. Sci. U.S.A.* **81,** 3049–3053.

Feeley, J. C., Gibson, R. J., Gorman, G. W., Langford, N. C., Rasheed, J. K., Makel, D. C., and Baine, W. B. (1979). Charcoal-yeast extract agar: Primary isolation medium for *Legionella pneumophila*. *J. Clin. Microbiol.* **10,** 437–441.

Gabay, J. E., and Horwitz, M. A. (1985). Isolation and characterization of the cytoplasmic and outer membrane of the legionnaires' disease bacterium (*Legionella pneumophila*). *J. Exp. Med.* **161,** 409–422.

Horwitz, M. A. (1987). Characterization of avirulent mutant *Legionella pneumophila* that survive but do not multiply within human monocytes. *J. Exp. Med.* **166,** 1310–1328.

Horwitz, M. A., and Silverstein, S. C. (1980). Legionnaires' disease bacterium (*Legionella pneumophila*) multiplies intracellularly in human monocytes. *J. Clin. Invest.* **66,** 441–450.

Marra, A., and Shuman, H. A. (1992). Genetics of *Legionella pneumophila* virulence. *Annu. Rev. Genet.* **26,** 51–69.

Marra, A., Blander, S. J., Horwitz, M. A., and Shuman, H. A. (1992). Identification of a *Legionella pneumophila* locus required for intracellular multiplication in human macrophages. *Proc. Natl. Acad. Sci. U.S.A.* **89,** 9607–9611.

Mintz, C. S., Chen, J., and Shuman, H. A. (1988). Isolation and characterization of auxotrophic mutants of *Legionella pneumophila* that fail to multiply in human monocytes. *Infect. Immun.* **56,** 1449–1455.

Oaks, E. V., Wingfield, M. E., and Formal, S. B. (1985). Plaque formation by virulent *Shigella flexneri. Infect. Immun.* **48,** 124–129.

Payne, N. R., and Horwitz, M. A. (1987). Phagocytosis of *Legionella pneumophila* is mediated by human monocyte complement receptors. *J. Exp. Med.* **166,** 1377–1389.

Pearlman, E., Jiwa, A. H., Engleberg, N. C., and Eisenstein, B. I. (1988). Growth of *Legionella pneumophila* in a human macrophage-like (U937) cell line. *Microb. Pathol.* **5,** 87–95.

Stacey, K. A., and Simson, E. (1965). Improved method for the isolation of thymine-requiring mutants of *escherichia coli. J. Bacteriol.* **90,** 554–555.

Sundström, C., and Nilsson, K. (1976). Establishment and characterization of a human histiocytic lymphoma cell line (U937). *Int. J. Cancer.* **17,** 565–577.

CHAPTER 14

Isolation and Characterization of Pathogen-Containing Phagosomes

Prasanta Chakraborty, Sheila Sturgill-Koszycki, and David G. Russell

Department of Molecular Microbiology
Washington University School of Medicine
St. Louis, Missouri 63110

I. Introduction
II. Choice of Pathogens and Particles
    A. *Leishmania*
    B. *Mycobacterium*
    C. Ligand-Coated Beads: The "Model" System
III. Choice of Macrophage
IV. Particle Adherence and Internalization Conditions
V. Cell Lysis Conditions
VI. Isolation of Phagosomes
    A. *Leishmania* Phagosomes
    B. *Mycobacterium* Phagosomes
    C. IgG-Coated Bead Phagosomes
VII. Analysis of Phagosomal Constituents
VIII. Storage and Handling of Two-Dimensional SDS–PAGE Data
IX. Shortcomings
    References

I. Introduction

Different intracellular pathogens exploit different locations to sustain their infection. Despite the obvious perils of the host cell lysosomal system many pathogens remain inside their phagosome. Both the protozoan parasite *Leishmania* and the bacterial pathogen *Mycobacterium* enter their host macrophage by phagocytosis (Schlesinger and Horwitz, 1991; Schlesinger *et al.*, 1990; Alexander and Russell, 1992) and remain intravacuolar throughout their intracellular infection. Because the normal maturation pathway for phagosomes culminates in the lysosome, and subsequent degradation of the phagosomal contents, both these pathogens must subvert this progression. We have deliberately chosen these two pathogens because their strategies of phagosome modulation are diametrically opposite. *Leishmania* actively enhances the interchange between its vacuole and the host cell's endosomal network (Russell *et al.*, 1992), whereas *Mycobacterium* inhibits fusion of its vacuole with endosomal vesicles (Hart *et al.*, 1987; de Chastellier *et al.*, 1993). These functional differences facilitate a comparative approach to the identification of pathogen-specific modulation of their phagosome. In addition to overcoming these short-term problems, both pathogens must also procure an adequate supply of nutrients, and prevent their host cell from inducing a protective immune response. The resolution of both these infections is effected through T-cell mediated-activation of the host macrophage (Nathan and Hibbs, 1991). The biology of this intimate cell/cell interaction is obviously of interest to those wishing to tip the balance in favor of the host; however, the ability of these pathogens to modulate their intracellular compartment also offers a unique opportunity to unravel the sequence of events operating during phagosome formation and maturation.

The macrophage is usually promoted as the body's first bastion against microbial invaders; however, its very ability to phagocytose particles, and microbes, will have placed strong evolutionary pressure on the selection of intramacrophage survival mechanisms among potential pathogens. Macrophages internalize particles into phagosomes, which evolve or mature into acidic, hydrolase-rich lysosomes. Despite the obvious importance of the phagocytic vacuole in the entry and establishment of a range of pathogens, little progress has been made in delineating the parameters influencing formation and maturation of this compartment, and its subsequent subversion by invading microbes.

Phagocytosis is a highly localized, yet fairly extreme cellular event involving considerable rearrangement of the macrophage's cytoskeleton and the functional redefinition of an appreciable portion of the cell's plasmalemma. Analysis of the differentiation of the plasmalemma from surface membrane into intracellular vacuole membrane has produced differing results. The seminal studies from the laboratories of Steinman and Cohn (Steinman *et al.*, 1983) involving uptake of peroxidase-conjugated latex beads indicated that the protein profile of the phagosome membrane corresponded closely with that of the plasmalemma. However, in these studies, particle internalization was not through the signaling cascade of a recognized phagocytic receptor. Another study from the same laboratories on the internalization IgG-coated erythrocytes indicated the Fc receptor was preferentially enriched inside these phagosomes, suggesting that some measure of sorting of plasmalemma components was operating during phagosome formation (Mellman *et al.*, 1983). The more recent immunocytochemical analyses by Clemens and Horwitz (1992) on the sorting events concomitant with internalization of *Legionella* and *Escherichia coli* indicate that there are strong selective pressures on the identity of macrophage surface proteins internalized during phagosome formation.

We have developed a range of techniques to study the phagosome-differentiation mechanisms for model systems, such as IgG-coated latex particles, and for the pathogens *Leishmania* and *Mycobacterium*. Comparison of a range of phagosomes has already enabled us to demonstrate that the failure of the *Mycobacterium* phagosome to acidify is attributable to an absence of the vesicular proton–ATPase from the mycobacterial vacuole, despite its acquisition of other endosomal/lysosomal constituents such as LAMP 1 (Sturgill-Koszycki *et al.*, 1993).

II. Choice of Pathogens and Particles

A. *Leishmania*

Leishmania is a dimorphic pathogen that parasitizes the midgut of its sandfly vector as a uniflagellate, motile, promastigote form, and an aflagellate, amastigote form in the phagolysosome of its vertebrate host's macrophages (Alexander and Russell, 1992). *Leishmania* exists as a spectrum of species endemic to 86 countries in the world; all the species are categorized as biohazard level 2 pathogens. Most of the experiments in this laboratory have been conducted with the new world species *L. mexicana* (Strain No. MYNC/BZ/62/M379), which, although infective to humans, usually causes a self-limiting, localized, cutaneous infection. Both life-cycle stages can be maintained in culture, although it is debatable how "amastigote-like" the tissue-culture amastigotes (ALFs) really are. It is likely that they represent an intermediate stage, close to the intracellular amastigote. Previous research in this laboratory and that of Antoine at the Institut Pasteur has demonstrated that amastigotes of the mexicana complex reside within an acidic, hydrolase-rich compartment that is accessible to the macrophage's endosomal pathway (Antoine *et al.*, 1990; Prina *et al.*, 1990; Russell *et al.*, 1992).

Leishmania species attenuate in liquid culture, although this trend is considerably less marked in the new world species. However, to minimize this problem we store *L. mexicana* promastigotes as frozen aliquots of a culture established from parasites isolated from a mouse lesion. Fresh aliquots are thawed each month. Promastigotes are maintained in SDM 79 (Brun and Schonenberger, 1979) supplemented with either 5% heat-inactivated fetal calf serum (FCS), and 5% bovine embryonic fluid (Sigma, Catalog No. E1761). Amastigote cultures are established in UM54 modified from the original recipe to consist of 1L ME 199 plus 2.5 g glucose, 5 g Trypticase (BBL, Catalog No. 11921), 0.75 g glutamine, 20 mg hemin, 25 mM HEPES, and 20% FCS, pH 5.7. Cells are frozen at approximately 5×10^8 cells/ml in 50% serum and 5% glycerol.

B. *Mycobacterium*

Like *Leishmania,* the genus *Mycobacterium* includes a range of species that induce a spectrum of diseases of varying severity. This group of pathogens includes *M. tuberculosis, M. leprae, M. avium,* and BCG, the first two being biohazard level 3 pathogens, and the latter two biohazard level 2 pathogens. Given the manipulations required for isolation of phagosomes we were reluctant to work with *M. tuberculosis* and chose instead *M. avium* (Isolate No. 101). As discussed in the opening chapter, mycobacterial species must be treated with respect because of their resistance to desiccation and propensity to aerosolize. Rather than retain waste in biohazard bags we place all contaminated material directly into 5% sodium hypochlorite solution. Like *Leishmania, M. avium* attenuates in culture and its loss of virulence has been correlated with an alteration in colony formation on agar plates. To overcome this loss of virulence we isolated *M. avium* from the spleen of an infected mouse, then expanded and froze aliquots of the bacteria. Fresh aliquots are thawed every 2 weeks. These aliquots have a >85% smooth transparent (the virulent phenotype) colony morphology. The bacilli are maintained in Middlebrook medium (Difco, Catalog No. 0713-01-7) or on Middlebrook nutrient agar plates (Difco, Catalog No. 0627-01-2), both supplemented with OADC (Difco, Catalog No. 0722-64-0).

C. Ligand–Coated Beads: The "Model" System

Because phagosome maturation even for "inert" particles is very poorly understood, we developed a model system to assess the differentiation processes of phagosomes formed around inert particles. We opted for polystyrene beads that could be surface-derivatized with a ligand of choice to control the receptor(s) involved in particle internalization. The beads chosen were iron-loaded to facilitate purification of phagosomes from cell lysate by magnetic selection. Initial experiments were performed with 1- to 2-μm-diameter carboxylate beads (Polysciences) to which ligands were coupled with carbodiimide. We experienced some batch variation with this preparation and have subsequently switched to 3-μm-diameter tosylactivated polystyrene beads (Dynal, Catalog No. 142:03). In initial control experiments to verify that the binding and internalization of the bead was a function of the ligand, we cleaved IgG into Fab and Fc portions and demonstrated that at equivalent particle density the adherence and internalization of the Fc-coated particles was greater than 10- fold more than the Fab-coated or albumin-coated controls. Protein was coupled to the beads according to the manufacturer's instructions.

III. Choice of Macrophage

Macrophages are highly plastic cells that respond rapidly to alterations in their environment. This makes choosing the source of macrophages problematic, and of fundamental importance to the relevance of the experimental system. The commonly used murine macrophage-like cell lines J774 and P388D1 are poorly phagocytic with many particles (including *Leishmania* promastigotes); most of the clones from these lines lack key receptors such as the mannose receptor and the mannose 6-phosphate receptor. These macrophage-like lines also show a limited ability to respond to macrophage-activating cytokines such a IFN-γ. Although reports exist suggesting that IFN-g can induce microbicidal behavior in J774 (Rastogi and Blom, 1990), the capacity of this cell line to control both *L. mexicana* and *M. avium* does not compare well with primary murine macrophages.

Peritoneal macrophages are a more representative cell type; however, generating the numbers of cells needed for biochemical or electron microscopical analysis of isolated phagosomes is not feasible. We use murine bone marrow-derived macrophages as an alternative. The cells are eluted by passing cold medium through the center of isolated murine femurs. The macrophages are expanded for 7–10 days in DMEM containing 10% L929 cell-conditioned medium, which contains monocyte growth factors such as GMCSF and MCSF, and 10% FCS. These cells infect readily with both *Leishmania* and *Mycobacterium* and can be activated to kill both pathogens by stimulation with IFN-γ and LPS.

IV. Particle Adherence and Internalization Conditions

The various particles and pathogens used in this study differ in their relative affinities for macrophages and bind to receptors that exhibit differing temperature requirements. All these aspects have to be considered in tailoring the settling and adherence conditions of each particle.

Routine phagosome isolation is conducted in T75 tissue culture flasks. We normally use two flasks 70–90% confluent per preparation. Each flask contains $5–10 \times 10^6$ cells. We add the particles of choice at differing levels of excess, depending on the identity of the particle and the tightness of the time window required. Typically, we add to each flask 10^8 *L. mexicana* ALFs, 2.5×10^8 *M. avium* colony-forming units (cfu), or 10^8 IgG-coated beads. The protocols for "typical" phagosome isolation procedures for each of these particles are illustrated in Figs. 1–3. Particles are added in minimal volume of culture medium buffered with 25 m*M* HEPES, instead of bicarbonate, to minimize pH fluctuations. We have noticed that the efficiency of adherence and internalization of *M. avium* is markedly enhanced by the presence of heat-inactivated horse serum (5%) in the medium. Although this presumably affects the receptor involved it does not appear to influence the subsequent maturation of the vacuole nor the survival of the bacilli.

The particles are allowed to settle onto the cells by incubation on ice for 10 min. Because some receptors are not functional at 0°C, we place the flask at room temperature (rt) for 5 min to facilitate binding. In experiments requiring rigorous time points we remove the excess particles at this stage by gently

T75 Tissue culture flask of macrophages (X2).

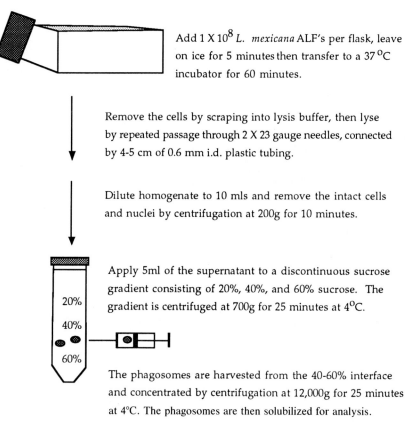

Add 1×10^8 L. *mexicana* ALF's per flask, leave on ice for 5 minutes then transfer to a 37 °C incubator for 60 minutes.

Remove the cells by scraping into lysis buffer, then lyse by repeated passage through 2 X 23 gauge needles, connected by 4-5 cm of 0.6 mm i.d. plastic tubing.

Dilute homogenate to 10 mls and remove the intact cells and nuclei by centrifugation at 200g for 10 minutes.

Apply 5ml of the supernatant to a discontinuous sucrose gradient consisting of 20%, 40%, and 60% sucrose. The gradient is centrifuged at 700g for 25 minutes at 4°C.

20%

40%

60%

The phagosomes are harvested from the 40-60% interface and concentrated by centrifugation at 12,000g for 25 minutes at 4°C. The phagosomes are then solubilized for analysis.

Fig. 1   Diagram showing the procedure used to isolate phagosomes containing the tissue-culture amastigote (ALF) form of *Leishmania mexicana*.

washing the cell monolayer $2\times$ with complete medium before placing the cells at 37°C. In experiments examining later time points, the cells are placed at 37°C for 10–15 min prior to removal of the excess particles by washing.

V. Cell Lysis Conditions

Prior to cell lysis, the cells are scraped from the culture flasks and resuspended in a minimal volume (2.5 ml/flask) of hypotonic homogenization buffer consisting of 20 mM HEPES, 0.5 mM EGTA, 0.25 M sucrose, and 0.1% gelatin (Sigma, Catalog No. G-9382), pH 7. The gelatin and sucrose help maintain the integrity of the vacuolar membrane during lysis and subsequent isolation steps.

T75 Tissue culture flask of macrophages (X2).

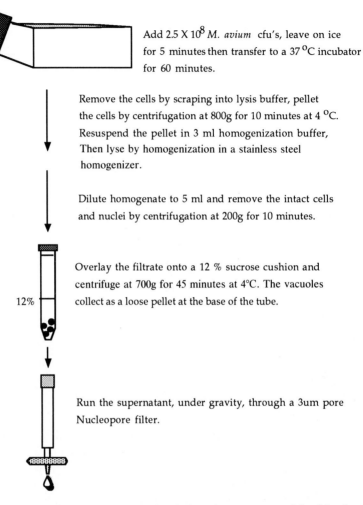

Add 2.5 X 10^8 *M. avium* cfu's, leave on ice
for 5 minutes then transfer to a 37 $^\circ$C incubator
for 60 minutes.

Remove the cells by scraping into lysis buffer, pellet
the cells by centrifugation at 800g for 10 minutes at 4 $^\circ$C.
Resuspend the pellet in 3 ml homogenization buffer,
Then lyse by homogenization in a stainless steel
homogenizer.

Dilute homogenate to 5 ml and remove the intact cells
and nuclei by centrifugation at 200g for 10 minutes.

Overlay the filtrate onto a 12 % sucrose cushion and
centrifuge at 700g for 45 minutes at 4°C. The vacuoles
collect as a loose pellet at the base of the tube.

12%

Run the supernatant, under gravity, through a 3um pore
Nucleopore filter.

Fig. 2   Diagram showing the procedure used to isolate phagosomes containing *Mycobacterium
avium* bacilli.

We routinely add a cocktail of protease inhibitors shown in Table I. In the case
of the *Leishmania* phagosomes, the cells are scraped directly into the lysis
solution and not centrifuged prior to lysis because we found that scraping itself
released a significant proportion of the phagosomes.

   Cell lysis is one of the most critical stages of isolating phagosomal compartments from macrophages. The success of this stage relies on one's ability to
lyse the macrophage without damaging the phagosomal membrane. This is

T75 Tissue culture flask of macrophages (X2).

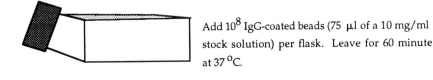

Add 10^8 IgG-coated beads (75 μl of a 10 mg/ml stock solution) per flask. Leave for 60 minute at 37 °C.

Remove the cells by scraping into lysis buffer, then lyse by repeated passage through 2 X 23 gauge needles, connected by 4-5 cm of 0.6 mm i.d. plastic tubing.

Transfer the cell lysate to a 5 ml borate glass tube and place in the Dynal MPC apparatus on ice. Leave for 6-10 minutes then pour off the buffer without disturbing the smear of bead phagosomes on the side of the tube. Repeat this procedure four times.

Collect the beads in 100-200 μl of solution and concentrate at the base of a microfuge tube by placing it in the MPC apparatus.

Fig. 3 Diagram showing the procedure used to isolate phagosomes containing iron-loaded, IgG-coated polystyrene beads.

obviously influenced by the "tightness" of the vacuole around the particle, a factor that has precluded us from successfully isolating the voluminous vacuoles that *L. mexicana* induce in a long-term infection. We have experimented with a range of buffers and lysis procedures before arriving at the following protocol. In brief, lysis by nitrogen cavitation caused breakage of the larger phagosomes, whereas methods relying on surface shear were more reproducible and reliable. We routinely lyse macrophages for isolation of bead and leishmanial phago-

Table I
Proteinase Inhibitors

Inhibitor (solvent for stock solution)	Source	Final concentration
Pepstatin (DMSO)	Sigma Cat. No. P-4265	50μg/ml
Leupeptin (H_2O)	Sigma Cat. No. L-2884	100 μg/ml
TLCK (DMSO)	Sigma Cat. No. T-7254	100 μg/ml
E64 (DMSO)	Sigma Cat. No. E-3132	50 μg/ml

somes by repeated passage (5–20 times) through two 1-ml syringes with 23-gauge needles connected by 4–5 cm of 0.6-mm-internal-diameter plastic tubing. This procedure is easily controlled and cell lysis is monitored by light microscopy to the point where approximately 90% of the cells are lysed. *M. avium*-containing macrophages may be lysed by this method; however, more recently, we have been lysing these macrophages in 3 ml of buffer with a 7-ml Dura-grind stainless-steel Dounce homogenizer (Wheaton, Catalog No. 357572). The smaller vacuole size of the bacilli allows a more extreme lysis procedure. The suspension is subjected to 3–10 passages in the homogenizer, and the efficiency of breakage and release monitored by light microscopy. Particular care is taken to avoid introduction of air bubbles during homogenization.

VI. Isolation of Phagosomes

The successful isolation of intact phagosomes is dependent on minimizing the interaction of the phagosomes with tube surfaces. One must avoid pelleting the phagosomes until the final stage in the isolation.

A. *Leishmania* Phagosomes

Following lysis by passage through the tubing, the suspension is diluted to 10 ml with chilled homogenization buffer and centrifuged at 200 g for 10 min to pellet intact cells and nuclei. The major contaminant of the final preparation is host cell nuclei that appear to have a density similar to that of the *Leishmania*-containing vacuoles. It is important to remove as many nuclei as possible at this stage.

The low-speed supernatant is then loaded onto a discontinuous sucrose gradient consisting of 3 ml 60% sucrose, 3 ml 40% sucrose, 3 ml 20% sucrose in HEPES saline (30 mM HEPES, 100 mM NaCl, 0.5 mM CaCl$_2$, 0.5 mM MgCl$_2$, pH 7) and centrifuged at 700 g for 25 min at 4°C. The vacuoles are harvested from the 40–60% interface and pelleted at 12,000 g for 25 min. The excess buffer is removed and the phagosomes are processed for analysis. An electron micrograph of an isolated, ALF-containing phagosome is shown in Fig. 4.

B. *Mycobacterium* Phagosomes

The infected cells are scraped into homogenization buffer and centrifuged at 800 g for 10 min at 4°C. The pellet is resuspended in 3 ml homogenization buffer with 0.5% gelatin (to reduce sticking during processing) and homogenized by 3–10 passages in a 7 ml Dura-Grind stainless steel homogenizer. Care must be taken to avoid introducing air bubbles into the homogenizer. The homogenate is diluted to 5 ml with homogenization buffer and centrifuged at 300 g for

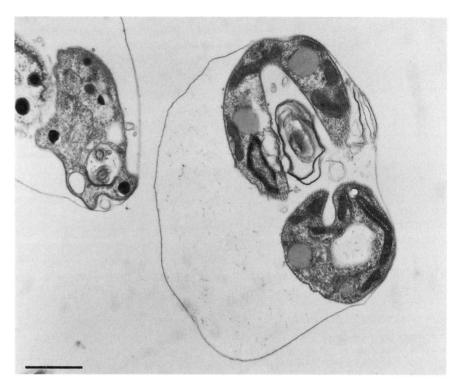

Fig. 4 An electron micrograph from a preparation of *Leishmania mexicana*-containing phagosomes isolated 60 min following internalization. The phagosomal membrane can clearly be seen
around the parasites. Scale bar: 1 μm.

10 min at 4°C to pellet intact cells and nuclei. This step may be repeated if too
many nuclei persist.

The supernatant is then layered onto a 12% sucrose cushion (4.0 ml) in a
15-ml Falcon tube, and centrifuged at 700 *g* for 45 minutes at 4°C. The bacterial
vacuoles are recovered as a loose pellet at the base of the tube. The vacuoles
are gently suspended in 5 ml of homogenization buffer and run, under gravity,
through either a 3 or 5-μm pore Nucleopore filter (Costar, Catalog Nos. 111112
and 111113), and washed with a further 3 ml. The flow-through is loaded onto
a second 4-ml 12% sucrose cushion in a 15-ml Falcon tube and centrifuged at
700 *g* for 45 minutes. The *M. avium* vacuoles are collected as a loose pellet in
the last 100 μl of the tube, while most of the remaining cellular debris and
cytoplasm are retained above the cushion. The *Mycobacterium*-containing
phagosomes are then processed for analysis. This procedure yields a very pure
preparation, although it is clear that some breakage of the vacuoles has occurred

during passage through the smaller Nucleopore filter. An electron micrograph of a *Mycobacterium* phagosome preparation is shown in Fig. 5.

C. IgG-Coated Bead Phagosomes

The iron-loaded beads-containing phagosomes are isolated by magnetic selection. Following lysis of the cells in 3 ml of homogenization buffer, the suspension is layered onto a 3 ml 12% sucrose cushion and centrifuged at 200 *g* for 10 min at 4°C to remove intact cells. The homogenate is then transferred to a 5-ml borate glass tube and placed in the Dynal magnetic particle concentrator (MPC) (Dynal, Catalog No. 120:01) in an ice bath for 6 min. The buffer is then carefully poured out of the tube leaving the beads as a brown smear associated with the side of the tube. Remove the tube from the MPC and gently resuspend the beads with 3 ml fresh, chilled homogenization buffer. The presence of gelatin in the buffer is critical, and may be augmented by bovine serum albumin (5 mg/ml), to reduce the association between the cytoplasmic face of the phagosomes and the tube wall. The BSA should be omitted from the final wash buffer to prevent it from interfering with migration of proteins analyzed by SDS–PAGE. The beads are washed a total of four times.

After the final wash, resuspend the beads in about 1 ml buffer and transfer to an microfuge tube. Set the MPC on its side in the ice bucket, magnet side up. Set the microfuge tube on top so that all the beads concentrate at the bottom. Remove the excess buffer and process the isolated fraction for analysis.

VII. Analysis of Phagosomal Constituents

To facilitate differentiation between macrophage and pathogen-derived proteins we labeled the macrophages metabolically by addition of 1 mCi [^{35}S]-methionine (Translabel from ICN, Catalog No. 51006) to each T75 flask of macrophages for 16–20 hr of continuous labeling prior to addition of the particles. The [^{35}S]methionine is added to methionine-free DMEM supplemented with sera dialyzed previously against PBS.

The macrophage constituents can be selectively solubilized from both the bead phagosomes and the *Mycobacterium* phagosomes, although in the latter case care has to be taken to ensure that the bacteria are not lysed. We usually solubilize the mycobacterial phagosomes in 0.03% Nonidet P-40 on ice for 15 min. The bacilli are removed from the extract by centrifugation at 6000 *g* for 10 min. We had previously determined that these conditions do not result in lysis of *M. avium*. The *Leishmania* phagosomes must be solubilized in total because the parasite is covered in a simple plasmalemma that is not resistant to detergent. The macrophage proteins of the phagosome are analyzed by two-dimensional (2D) SDS–PAGE and autoradiography, or by immunoblotting of SDS–PAGE-separated material. The 2D SDS–PAGE profiles of *Leishmania*-

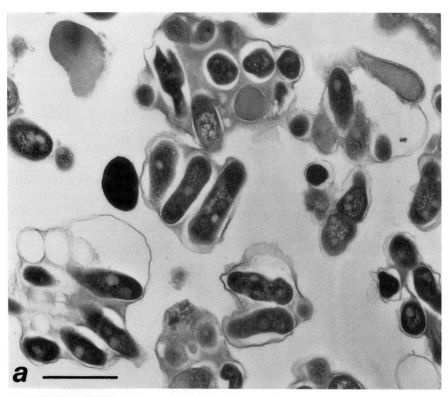

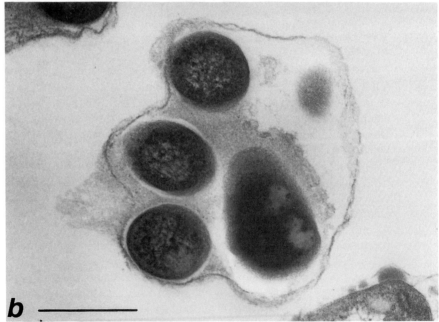

and *M. avium*-containing phagosomes isolated 60 min after internalization are shown in Fig. 6. The patterns generated by the two different isolates are highly comparable, and are very similar to the profile generated by the IgG-bead phagosomes (not shown).

VIII. Storage and Handling of Two-Dimensional SDS–PAGE Data

There are several computer/work station systems on the market, all of which are extremely expensive and are designed to import and process complex 2D gel patterns. These packages have the facility to "rubber-sheet" the patterns to compensate for run variability. However, our laboratory is interested in a reductionist approach and is trying to reduce the number of protein species of interest, therefore this type of sophisticated data processing is considered overkill.

As a cheap alternative we use a flat-bed scanner (Microtek Scanmaker 600ZS) to scan autoradiographs and import the data into a Quadra 700 Macintosh. The scanner software has an Adobe Photoshop plug-in. Photoshop has the capacity to stretch images along both dimensions which we find is sufficient to correct for minor run variation. The files are relatively large (approximately 1 MB at 300 dpi resolution), so we use a rewritable optical drive (Fujitsu DynaMO) and 128 MB floppy discs for data storage.

IX. Shortcomings

The phagosome represents a vesicle in transit. Initially, it is derived from plasmalemma constituents, but with time it acquires more and more intracellular proteins as it evolves into a lysosomal compartment. Given that the phagosome is in a constant state of flux, what markers should be used for biochemical confirmation that our isolated phagosomal preparations are pure? To date we have used electron microscopy to monitor our preparations. This procedure is adequate for contaminants of defined-structural organization, such as nuclei and mitochondria, but how do we verify that small vesicles associated with isolated phagosomes are not an artifact of the isolation procedure? At present we cannot definitively dismiss this possibility. However, the comparative similarity between the 2D SDS–PAGE profiles from phagosomes isolated by markedly

Fig. 5 Electron micrographs from a preparation of *Mycobacterium avium*-containing phagosomes isolated 60 min following internalization. These vacuoles were isolated without using the Nucleopore filter. The phagosomal membranes are intact and the vacuolar preparation shows little indication of contamination with other cellular components. Scale bars: (*a*) 1 μm; (*b*) 0.5 μm.

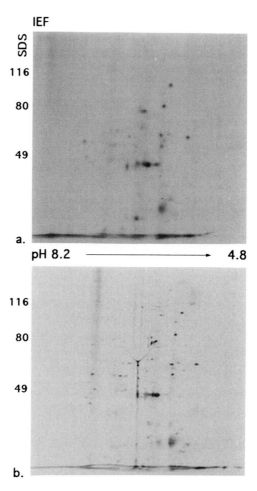

Fig. 6 Autoradiographs of 2D SDS–PAGE gels run with (a) *Leishmania* and (b) *Mycobacterium*-containing phagosomes isolated from ^{35}S-methionine-labeled macrophages 60 min following internalization of the particles. The major protein constituents of the vacuoles are common to both preparations.

different protocols, coupled with our emerging data concerning the differential exclusion of functionally defined host proteins such as the proton–ATPase (Sturgill-Koszycki *et al.*, 1993), argues for the relevance of our results.

We are currently developing control experiments that should allow us to evaluate the degree of contamination, and to discriminate between non-specific contamination and selective contamination. We prepare 4 comparable flasks of bone-marrow macrophages, two of which are metabolically-labeled as described. Particles are added to one labeled and one unlabeled flask and left to be internalized. The labeled flask with particles is scraped, and the labeled cells mixed with cells removed from an unlabeled flask with no particles. In parallel, the unlabeled flask with particles is scraped, and the unlabeled cells mixed with labeled cells removed from a labeled flask with no particles. The two mixed cell populations are processed in parallel. The amount of radiolabel recovered from the two preparations is compared to indicate the % total contamination, and 2D SDS–PAGE gels are run to determine the identity of the protein contaminants. These procedures are under development, however, we felt it important to include an outline in this chapter. In preliminary experiments with the iron-loaded IgG beads the total levels of contamination vary from 6–10%.

In an effort to avoid any spurious interpretation of the differential distribution of proteins, we take pains to confirm our analysis of isolated phagosomes through the use of immunoelectron microscopy on the pathogen-containing vacuoles *in situ*. The methods for immunoelectron microscopy of infected macrophages are detailed in the following Chapter 15.

References

Alexander, J., and Russell, D. G. (1992). The interaction of *Leishmania* species with macrophages. *Adv. Parasitol.* **31**, 175–254.

Antoine, J. C., Prina, E., Jouanne, C., and Bongrand, P. (1990). Parasitophorous vacuoles of *Leishmania amazonensis*-infected macrophages maintain an acidic pH. *Infect. Immun.* **58**, 779–787.

Brun, R., and Schonenberger, M. (1979). Cultivation and in vitro cloning of procyclic culture forms of *Trypanosoma brucei* in a semi-defined medium. *Acta Trop.* **36**, 289–292.

Clemens, D., and Horwitz, M. (1992). Membrane sorting during phagocytosis: Exclusion of major histocompatibility complex molecules but not complement receptor CR3 during conventional and coiling phagocytosis. *J. Exp. Med.* **175**, 1317–1326.

de Chastellier, C., Frehel, C., Offredo, C., and Skamene, E. (1993). Implication of phagosome-lysosome fusion in restriction of *Mycobacterium avium* growth in bone marrow macrophages from genetically resistant mice. *Infect. Immun.* **61**, 3775–3784.

Hart, P. D., Young, M. R., Gordon, A. H., and Sullivan, K. H. (1987). Inhibition of phagosome-lysosome fusion in macrophages by certain mycobacteria can be explained by inhibition of lysosomal movements observed after phagocytosis. *J. Exp. Med.* **166**, 933–946.

Mellman, I., Plutner, H., Steinman, R. M., Unkless, J. C., and Cohn, Z. (1983). Internalization and degradation of macrophage Fc receptors during receptor-mediated phagocytosis. *J. Cell Biol.* **96**, 887–895.

Nathan, C., and Hibbs, J. (1991). Role of nitric oxide synthesis in macrophage antimicrobial activity. *Curr. Opin. Immunol.* **3**, 65–70.

Prina, E., Antoine, J. C., Wiederanders, B., and Kirschke, H. (1990). Localization and activity of various lysosomal proteases in *Leishmania amazonensis*-infected macrophages. *Infect Immun.* **58**, 1730–1737.

Rastogi, N., and Blom, P. M. (1990). A comparative study on the activation of J-774 macrophage-like cells by gamma-interferon, 1,25-dihydroxyvitamin D3 and lipopeptide RP-56142: Ability to kill intracellularly multiplying *Mycobacterium tuberculosis* and *Mycobacterium avium*. *Int. J. Med. Microbiol.* **273**, 344–361.

Russell, D. G., Xu, S., and Chakraborty, P. (1992). Intracellular trafficking and the parasitophorous vacuole of *Leishmania mexicana*-infected macrophages. *J. Cell Sci.* **103**, 1193–1210.

Schlesinger, L. S., and Horwitz, M. A. (1991). Phagocytosis of *Mycobacterium leprae* by human monocyte- derived macrophages is mediated by complement receptors CR1 (CD35), CR3 (CD11b/CD18), and CR4 (CD11c/CD18) and IFN-g activation inhibits complement receptor function and phagocytosis of this bacterium. *J. Immunol.* **147,** 1983–1994.

Schlesinger, L. S., Bellinger, K. C., Payne, N. R., and Horwitz, M. A. (1990). Phagocytosis of *Mycobacterium tuberculosis* is mediated by human monocyte complement receptors and complement component C3. *J. Immunol.* **144,** 2771–2780.

Steinman, R., Muller, W., and Cohn, Z. (1983). Endocytosis and the recycling of the plasma membrane. *J. Cell Biol.* **96,** 1–27.

Sturgill-Koszycki, S., Schlesinger, P., Chakraborty, P., Haddix, P. L., Collins, H. L.,Fok, A. K., Allen, R. D., Gluck, S. L., Heuser, J., and Russell, D. G. (1993). Lack of acidification of in *Mycobacterium* phagosomes produced by exclusion of the vesicular proton-ATPase. *Science* **263,** 678–681.

CHAPTER 15

Immunoelectron Microscopy of Endosomal Trafficking in Macrophages Infected with Microbial Pathogens

David G. Russell

Department of Molecular Microbiology
Washington University School of Medicine
St. Louis, Missouri 63110

I. Introduction
II. The Host–Pathogen Interplay
III. Intersection with the Endosomal Pathway
    A. Ligands for Endocytic Receptors
    B. Fluid-Phase Markers
    C. Surface Biotinylation of Exposed Macrophage Proteins
IV. Processing of Infected Macrophages for Immunoelectron Microscopy
V. Blocking Cryosections and Incubation with Primary Antiserum
    A. General Points
    B. Detection of Biotinylated Markers
VI. Gold-Conjugated Second Antibodies
VII. Controls
VIII. Final Preparation of the Grids
IX. Routine Protocol for Analysis of Fluid-Phase Trafficking
    References

I. Introduction

Analysis of the intracellular compartment of infected cells *in situ* is a necessary component to any project studying the biology of intracellular parasitism. In the Chapter 14 we outlined procedures for the analysis of isolated phagosomes and parasite-containing vacuoles. However, any results obtained from such analyses should be confirmed by immunoelectron microscopy of the respective *in situ* infection. The endocytic network of mammalian cells has been mapped through the differential distribution of certain membrane proteins. This distribution has been used to assess the functional position of an endosome or phagosome relative to early recycling compartments, or established lysosomes (Kornfeld and Mellman, 1989; Russell *et al.*, 1992). The lysosomal glycoproteins LAMPs, or lgp's, 1 and 2, are present in late endosomes and in lysosomes (Chen *et al.*, 1986,1988), whereas the mannose 6-phosphate receptor (M6PR), which is responsible for the delivery of lysosomal hydrolases, is present in late endosomes and absent from the lysosomes themselves (Kornfeld and Mellman, 1989). The vesicular proton–ATPase appears in the early phagosome (Sturgill-Koszycki *et al.*, 1993) and is responsible for the pH decrease as the phagosome differentiates through the endosomal/lysosomal continuum. Ligands and markers endocytosed by mammalian cells segregate differentially following internalization. Some ligands, such as transferrin dissociate early from their receptor and do not penetrate deeply into the endosomal network (Goldenthal *et al.*, 1988); in contrast, fluid-phase markers such as dextran can permiate throughout the endosomal system.

Immunoelectron microscopy provides a powerful tool for the elucidation of intracellular trafficking pathways that can either deliver or remove material from the microbe-containing vacuole. These results provide additional information about the relative position of the compartment within the endosomal network, and the extent to which the pathogen has induced "aberrant" behavior in its host vacuole.

In this chapter we outline some approaches that have been used in this laboratory for the analysis of murine bone marrow macrophages infected with either *Leishmania* or *Mycobacterium*. This chapter is not intended as an exhaustive description of immunoelectron microscopy, for that I would urge the reader to consult Griffiths (1993).

II. The Host–Pathogen Interplay

The relative pathogenicity of the microbe to its host cell must be considered in the design of these experiments. Some microbes, such as *Salmonella,* are highly "toxic" to their host macrophage and will kill the cell if left on the cells for too long or if added in too large an excess. In contrast, both *Leishmania* and *Mycobacterium* form extremely stable interactions with the host cell, so it is feasible to establish longer-term experiments examining infections from the point of initial infection up to 2 weeks later (Russell *et al.*, 1992; Xu *et al.*, 1994).

We routinely establish our infections in T25 tissue culture flasks with 60–80% confluent bone marrow-derived macrophage monolayers. Estimating the macrophage density in a confluent flask to be around 1.5–3.0×10^6 cells, we infect the cells with a 5- to 10-fold excess of *Mycobacterium avium* or *Leishmania mexicana* promastigotes. We use a greater excess of microbes for shorter term cultures where little replication is expected. Antibiotic is omitted from the medium used with cultures infected with *Mycobacterium*. The microbes are left on the macrophages for 2–4 hr, after which time the flasks are washed with warm medium and returned to culture. For longer-term infections with *L. mexicana* (>2 days) we incubate the cultures at 35°C because in the vertebrate host the infection is restricted to the skin, which has a lower temperature.

III. Intersection with the Endosomal Pathway

We have exploited the biotin–streptavidin interaction to detect the trafficking of a range of different ligands or fluid phase markers in infected macrophages (Russell *et al.*, 1992). The sections were colabeled with antibodies against known endosomal constituents, most notably LAMP 1, M6PR, and the proton–ATPase, which are discussed in more detail in Section V,B.

A. Ligands for Endocytic Receptors

Ligands were biotinylated with 10–100 molar excess of *N*-hydroxysuccinimide (NHS) sulfo-biotin (Pierce Catalog No. 21217) at pH 7.6–8, conditions varying with the sensitivity of the protein to loss of function. For example, we found that overlabeling of transferrin with biotin prevented us from inhibiting its uptake with unlabeled protein, and rerouted the protein to lysosomes. In contrast, both mannosylated BSA and β-glucuronidase (affinity-purified on M6PR-sepharose), which were internalized by the mannose receptor and M6PR, respectively, exhibited no loss of function (Russell *et al.*, 1992). These ligands were added to the infected macrophages at 10–50 μg/ml in a minimal volume of prewarmed medium.

We routinely control for function by verifying with FITC–streptavidin that an excess of unlabeled protein can competitively inhibit uptake of the labeled ligand. In our original experiments on *Leishmania*-infected macrophages we confirmed this result by inhibiting internalization of NHS–carboxyfluorescein-labeled ligand (Boeringher Catalog No. 1055 089) with its biotinylated counterpart (Russell *et al.*, 1992).

B. Fluid-Phase Markers

Fluid-phase markers such as biotinylated dextran (10 kDa) (Molecular Probes, Catalog No. D-1856) or biotinylated albumin were added at 0.5–1 mg/ml to a minimal volume of prewarmed, equilibrated medium. In routine experiments, the ligand or fluid-phase marker was incubated with the cells for 15 min and the cells were washed and returned to warm, equilibrated medium for 45 min prior to fixing the cell monolayer. This period of time allowed appreciable accumulation of the dextran within the *Leishmania*-containing vacuole.

C. Surface Biotinylation of Exposed Macrophage Proteins

One additional approach that we have found useful in assessing the accessibility of intracellular compartments to endocytosed material is based on the rationale described by Thilo and colleagues, who developed a method for the "non-selective" labeling of surface glycoproteins with [³H]galactose via galactosyltransferase (Thilo, 1983; Lang *et al.*, 1988). This procedure employed autoradiography to visualize the redistribution of labeled plasmalemma components and "score" bulk internalization. We have used surface-labeling with NHS–sulfobiotin to similar ends (Russell *et al.*, 1992). Routinely, infected macrophages are washed in PBS, chilled on ice, and biotinylated in ice-cold PBS (pH 7.8) with 0.5 mg/ml NHS–sulfobiotin for 20 min. This approach employs suboptimal labeling conditions and an excess of label which, in combination, produces a satisfactory level of surface-labeling without stressing the cells unduly. The cells were then washed in three changes of warm, complete medium to remove unreacted label and returned to culture temperature. The cells are then left at 37°C for a determined period of time, usually 45 min prior to fixation of the monolayer.

IV. Processing of Infected Macrophages for Immunoelectron Microscopy

Monolayers of infected macrophages were fixed in a range of different fixatives depending on the sensitivity of the epitope of interest to aldehyde fixation. We use three different basic recipes, all in HEPES saline (30 mM HEPES, 100 mM NaCl, 0.5 mM CaCl$_2$, 0.5 mM MgCl$_2$, pH 7). Biotin is resistant to aldehyde, so we normally fix our cell monolayers with 1% glutaraldehyde. For epitopes that show an intermediate level of sensitivity we use 0.2% glutaraldehyde and 4% formaldehyde, prepared fresh by dissociation of paraformaldehyde at 100°C cooled and filtered prior to use. And for the most sensitive epitopes we use 4% formaldehyde alone.

Monolayers are fixed for 30–60 min, and removed by gentle scraping with a rubber policeman (Sarstedt cell scraper, Catalog No. 83.1830), which causes minimal damage to the fixed cells. The cell suspension is washed in two changes of HEPES saline and resuspended in a microfuge tube in 50–100 μl 10% warm gelatin. The cells are then pelleted and placed on ice to allow the gelatin to set. The tip of the tube is cut off and placed in a fresh microfuge with 1 ml of the relevant fixative for 30 min on a rocking table. The microfuge tube tip is transferred to another tube containing 1 ml of 2.3 M sucrose and 20% polyvinyl pyrrolidone (10 kDa) (Sigma Catalog PVP-10) as cryopreservant and plasticizer.

The sample is left on a Nutator rocking table for 2 hr at room temperature (rt) or overnight at 4°C. The gelatin-embedded cell pellet has usually detached from the tube tip by this time and can be trimmed and mounted on an aluminum stud ready for freezing in liquid nitrogen. Frozen sections are then cut using an RMC MT7 with CR 21 cryochamber and transferred to carbon-coated Formvar grids that have been freshly glow-discharged.

V. Blocking Cryosections and Incubation with Primary Antiserum

A. General Points

Grids are blocked in either 10% fetal calf serum in PBS, or 5% goat serum and 5% fetal calf serum in PBS. Both solutions are filtered prior to use, and the sections are floated, face downward, on 2.5 ml of blocking solution in the wells of a 24-well plate. All washes are carried out in this manner, and all antibody incubations are conducted on 15 μl of solution in the wells of a Terasaki microwell plate (Nunc Catalog No. 439225). A strip of moist blotting paper is placed at the bottom of the microwell plate to maintain humidity.

The choice and source of primary antibody for analysis of microbial infections have afforded us hours of pointless amusement which, I am sure the "regular" cell biologist has not, to date, appreciated. First, if you work on *Mycobacterium* as we do, all polyclonal antiserum generated with Freund's complete adjuvant is useless because BCG, the attenuated strain of *M. bovis,* is what makes the adjuvant complete. Affinity purification of the antibody against the original immunogen will overcome much of the background; however, we also add 1 mg/ml of filtered BCG sonicate to the primary and secondary antibody incubation solutions. The second problem encountered by microbiologists involves antibody responses to normal gut fauna or chance infections. This problem is particularly prevalent in rabbits, which tend not to be maintained in pathogen-free facilities. Antibodies against *Escherichia coli* and *Salmonella* LPSs appear an almost unavoidable constitutent of rabbit serum. Again, these difficulties can be overcome by with affinity purification or addition of a sonicated extract. We have also found that reactivity can be removed by incubation of the antiserum with boiled bacterial sonicates coupled directly to activated CH–Sepharose.

Finally, even if the immunizing antigen is not pure it is feasible to affinity-purify small amounts of antibody for immunoelectron microscopy. In brief, the antigen fraction is separated by SDS–PAGE and transferred to nitrocellulose. The membrane is stained with 0.1% Ponceau S (available as a 20× stock from Sigma, Catalog No. P-7767), and the relevant bands are excised from the gel and washed in several changes of PBS. The nitrocellulose pieces are blocked in PBS with 10% FCS, and incubated in immune serum. The nitrocellulose strips are washed in PBS with 10% FCS and the antibody is eluted with 500 μl 100 m*M* glycine, pH 2.5. The solution is neutralized with 1.0 *M* Tris, pH 8, and dialyzed against PBS.

B. Detection of Biotinylated Markers

In initial experiments we found that streptavidin–gold was somewhat inefficient in the detection of biotinylated proteins in cryosections. We believe that this is due to soluble streptavidin that has desorbed from the gold particles. To avoid this problem we developed an antibody-sandwich technique using rabbit antistreptavidin (Sigma, Catalog No. S-6390). We affinity-purified the antibody on a streptavidin–agarose column (Sigma, Catalog No. S-1638), eluted with 100 m*M* glycine, pH 2.5, and stored it at 1 mg/ml in PBS with 0.02% Na azide. Sections are blocked, incubated in 1 μg/ml streptavidin (Sigma, Catalog No. S-4762) in blocking buffer for 30 minutes at rt, then washed for 10 min in a 24-well plate as described above. The sections are incubated in 1 μg/ml rabbit antistreptavidin in blocking buffer for 60 min rt, or overnight at 4°C.

We routinely use a double-labeling procedure and add a second antibody to the antistreptavidin antibody solution. We have found that the anti-LAMP 1 rat monoclonal antibody 1D4B, from Dr. Tom August's laboratory is an invaluable marker. LAMP 1 is present predominantly in late endosomal/lysosomal compartments (Chen *et al.*, 1986,1988). The monoclonal antibody 1D4B recognizes a glutaraldehyde-resistant epitope on murine LAMP 1, and the hybridoma line is available from the Developmental Studies Hybridoma Bank (tel 410-955-3985/ 319-335-3826). Double labeling with a well-characterized antibody provides a measure of internal control. The differential distribution of LAMP 1 (1D4B) versus rabbit antilipoarabinomannan (LAM), a cell wall constituent of *Mycobacterium* species (Chan *et al.*, 1991; Chatterjee *et al.*, 1992), is shown in the immunoelectron micrograph in Fig. 1. The rabbit anti-LAM was raised by Dr. Delphi Chatterjee, Colorado State University, by inoculating LAM emulsified in Freund's incomplete adjuvant (no BCG).

VI. Gold-Conjugated Second Antibodies

We prefer to use species-specific anti-immunoglobulin antibodies conjugated to gold, rather than protein A– or protein G–gold because it facilitates easier double-labeling with antibodies from different sources and, in our hands, shows less variation between monoclonal antibodies. We use two different suppliers; our anti-rabbit IgG (15 nm gold, Catalog No. RPN 422) and anti-rat IgG (5 nm gold, Catalog No. RPN 434) came from Amersham, and our anti-mouse IgG (12 nm, Catalog No. 115-205-071, and 18 nm gold, Catalog No. 115-215-146), anti mouse IgM (12 nm gold, Catalog No. 115-205-075) from Jackson Immunoresearch Laboratories. The anti-IgG and anti-IgM antibodies from Jackson have been minimized for cross-reactivity between antibody classes, so one can double

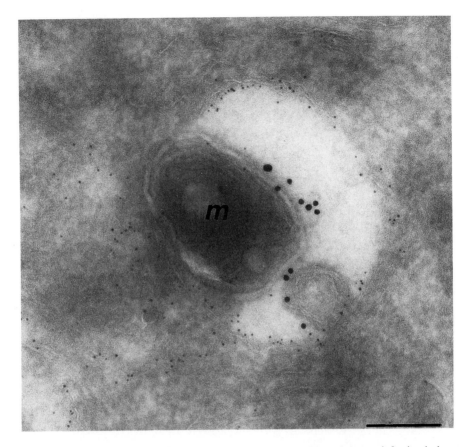

Fig. 1 An immunoelectron micrograph of *M. avium* (strain 101), 4 days postinfection in bone marrow-derived murine macrophages. The bacteria were always found enclosed in LAMP 1-positive vacuoles, revealed by the rat monoclonal antibody 1D4B and 5-nm-gold anti-rat IgG. Sections were costained with rabbit anti-lipoarabinomannan (LAM) and 15-nm gold antirabbit IgG. Scale bar: 0.2μm.

label with mouse IgG and IgM antibodies. All these antibodies are used at between 20 and 50 : 1 dilution in the relevant blocking buffer, with 1 mg/ml boiled bacterial sonicate for *Mycobacterium, Salmonella,* and *E. coli* preparations.

VII. Controls

The relevant controls necessary to ensure that the labeling is indeed restricted to the antigen of interest varies considerably with the experimental conditions. A list of controls that we have used under various experimental conditions is given below.

1. Some controls, such as ensuring that the gold-conjugated second antibody does not bind to your section in the absence of primary antibody, are obvious and straightforward. These should be conducted at the start of any series of studies on a particular sample, or when a new batch of second antibody is used.

2. Labeling with preimmune serum is useful if complete, immune, antiserum is used. This can aid in identification of antimicrobial antibodies due to infection or the normal gut flora and fauna of the immunized animal. In the case of raising antibodies to *Toxoplasma, Histoplasma, Candida, Salmonella, Helicobacter,* or *E. coli* antigens in particular, I advise that the rabbit be screened before immunization, unless you intend to affinity-purify the final antibody.

3. In our trafficking studies with biotinylated ligands and markers, the omission of the biotinylated compound controlled for the level of background in the experiments (Russell *et al.,* 1992). Furthermore, the distribution of the label varied, correlating with the identity of the ligand or marker.

4. Antipeptide antibodies can be competitively inhibited by addition of free peptide to the primary antibody solution, as described (Wallis *et al.,* 1994).

5. The use of the expressor versus nonexpressor phenotype may be possible in transformed or transfected cells, in some culture cell lines, and in activated versus resting macrophages. We have used transformed versus wild-type *E. coli* to localize cytosolic proteins from both *E. coli* (Arnqvist *et al.,* 1992; Pfeifer *et al.,* 1992) and *Helicobacter pylori* (Frazier *et al.,* 1993). We have also used *Leishmania*-infected P388D1 cells, a macrophage-like cell line that lacks the mannose 6-phosphate receptor, to control for the unusual distribution of the receptor in *Leishmania*-infected bone marrow-derived macrophages (Russell *et al.,* 1992).

VIII. Final Preparation of the Grids

The grids are washed sequentially on 2.5 ml of blocking buffer, PBS, and ddH$_2$O in a 24-well plate on a rocking platform for 10 min in each solution. The grids are placed on individual droplets of 10% polyvinyl alcohol with 0.3% uranyl acetate on Parafilm on glass on top of ice for 10 min. The grids are removed with nickel–chrome wire loops, the excess stain is withdrawn with filter paper triangles, and the grids are left in the loops to air-dry. It is preferable to leave the grids for 2 hr, or overnight, in a desiccator before putting them in the electron microscope, to minimize evaporation of stain. Micrographs from both *Leishmania*- and *Mycobacterium*-infected macrophage incubated with biotinylated markers are illustrated in Figs. 2 and 3.

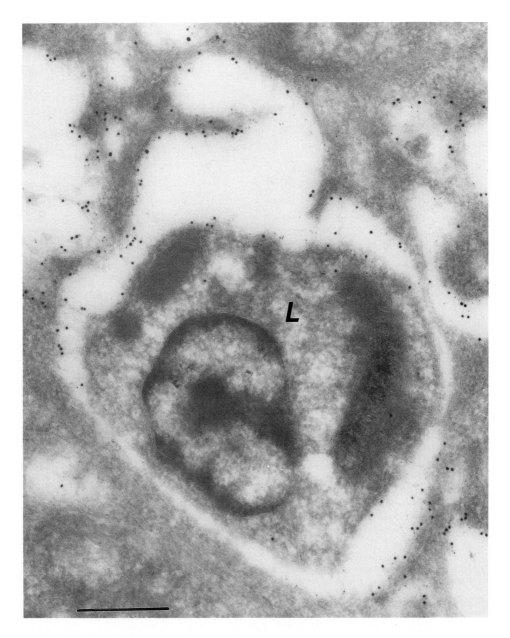

Fig. 2 A cryoimmunoelectron micrograph from a bone marrow macrophage infected 5 days previously, and incubated with biotin–NHS-derivatized dextran (0.5 mg/ml) for 20 min followed by a 45-min chase period. The section was probed with 1D4B (rat anti-LAMP 1)/5-nm gold anti rat IgG and streptavidin/rabbit anti-streptavidin/15-nm gold anti-rabbit IgG. The fluid phase marker freely enters the parasitophorous vacuole. Reproduced from Russell *et al.* (1992) with permission. Scale bar: 0.5 μm.

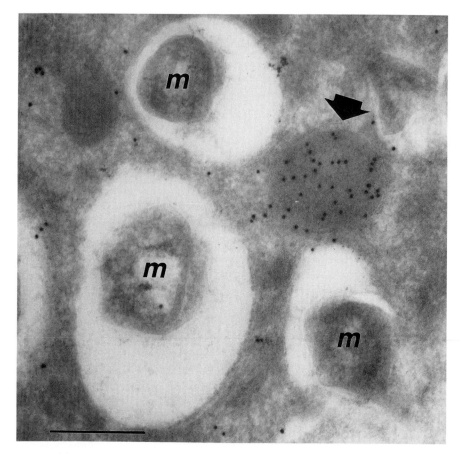

Fig. 3 A cryoimmunoelectron micrograph from a bone marrow macrophage infected 4 days previously with *Mycobacterium avium*, and incubated with biotin–NHS-derivatized mannosylated BSA (25 μg/ml) for 15 min followed by a 45-min chase period. The section was probed with 1D4B (rat anti-LAMP 1)/5-nm gold anti rat IgG and streptavidin/rabbit antistreptavidin/15-nm gold anti-rabbit IgG. The ligand appears to be excluded from the bacteria-containing vacuoles and concentrates in dense lysosome-like compartments. Scale bar: 0.5μm.

IX. Routine Protocol for Analysis of Fluid-Phase Trafficking

1. Take the infected macrophage monolayer in a T25 flask, remove all but 3 ml of the medium, and add 150 μl of a 10 mg/ml stock of biotinylated-dextran (10 kDa). Replace the flask in the incubator for 15 min, then wash with fresh, warm, equilbrated medium and return the flask to the incubator for 45 min.

2. Pour off the medium and add 5 ml of 1% gluteraldehyde in HEPES saline.

Leave the flask on a rocking platform for 45 min. Pour off the fixative and wash the cells with two changes of HEPES saline.

3. Scrape the cells from the flask into 1 ml of HEPES saline and transfer to a microfuge tube; gently pellet the cell suspension in a microfuge. Remove most of the saline and carefully suspend the cells in approximately 20 μl of saline. Add 50 μl of warm, 10% gelatin and pipette the suspension up and down to disperse the cells. Centrifuge to pellet the cells, and place the microfuge tube on ice for 10 min.

4. Cut off the tip of the microfuge tube and place in 1% glutaraldehyde in HEPES saline on the rocking platform for 30–60 min. Transfer the microfuge tube tip directly to a fresh microfuge tube containing 2.3 M sucrose/20% PVP and place on a Nutator for 2 hr rt or overnight at 4°C.

5. Remove the gelatin-embedded cell pellet from the tube tip and sucrose/PVP. Place under a dissecting microscope and trim the block. The block is then transferred to an aluminum stud and frozen by plunging in liquid nitrogen.

6. Sections are cut on a cryoultramicrotome and placed on carbon-coated Formvar grids that have been glow-discharged.

7. Grids are blocked by placing face-down on 10% FCS in HEPES saline for 10 min. The grids are then placed on individual droplets of 15μl blocking solution with 1 μg/ml streptavidin in the wells of a Terasaki multiwell plate and left for 30 mins rt.

8. Grids are washed for 10 min by floating face-down on 2.5 ml blocking solution in a 24-well plate on a racking platform. The grids are incubated in primary antibody (1 μg/ml rabbit anti-streptavidin + 1D4B culture supernatant diluted 1 : 5 in blocking buffer) overnight at 4°C or 1 hr at rt. To mycobacterial samples we routinely add 1 mg/ml of boiled, filtered BCG sonicate.

9. Grids are washed as described above and placed on 15 μl of second antibody solution containing 15 nm gold anti-rabbit IgG (30 : 1) and 5 nm gold anti-rat IgG (25 : 1) for 1 hr rt.

10. The grids are washed sequentially with blocking buffer, PBS, and ddH$_2$O for 10 min each and stained with 10% PVA and 0.3% uranyl acetate.

References

Arnqvist, A., Olsen, A., Pfeifer, J., Russell, D. G., and Normark, S. J. (1992). The Crl protein activates cryptic genes for curli formation and fibronectin binding in *Escherichia coli* HB101. *Mol. Microbiol.* **6,** 2443–2452.

Chan, J., Fan, X. D., Hunter, S. W., Brennan, P. J., and Bloom, B. R. (1991). Lipoarabinomannan, a possible virulence factor involved in persistence of *Mycobacterium tuberculosis* within macrophages. *Infect Immun.* **59,** 1755–1761.

Chatterjee, D., Hunter, S. W., McNeil, M., and Brennan, P. J. (1992). Lipoarabinomannan. Multiglycosylated form of the mycobacterial mannosylphosphatidylinositols. *J. Biol. Chem.* **267,** 6228–6233.

Chen, J. W., Chen, G. L., D'Souza, M. P., Murphy, T. L., and August, J. T. (1986). Lysosomal membrane glycoproteins: Properties of LAMP-1 and LAMP-2. *Biochem. Soc.Symph.* **51,** 97–112.

Chen, J. W., Cha, Y., Yuksel, K. U., Gracy, R. W., and August, J. T. (1988). Isolation and sequencing of a cDNA clone encoding lysosomal membrane glycoprotein mouse LAMP-1. Sequence similarity to proteins bearing onco-differentiation antigens. *J. Biol. Chem.* **263,** 8754–8758.

Frazier, B., Pfeifer, J., Russell, D., Falk, P., Olsen, A., Hammar, M., Westblom, T., and Normark, S. J. (1993). Paracrystalline inclusions of a novel ferritin containing nonheme iron, produced by the human gastric pathogen Helicobacter pylori: Evidence for a third class of ferritins. *J. Bacteriol.* **175,** 966–972.

Goldenthal, K. L., Hedman, K. W., Chen, J. W., August, J. T., Vihko, P., Pastan, I., and Willingham, M. C. (1988). Pre-lysosomal divergence of alpha 2-macroglobulin and transferrin: A kinetic study using a monoclonal antibody against a lysosomal membrane glycoprotein (LAMP-1). *J. Histochem. Cytochem.* **36,** 391–400.

Griffiths, G. (1993). "Fine Structure Immunocytochemistry." Springer-Verlag, Heidelberg.

Kornfeld, S., and Mellman, I. (1989). The biogenesis of lysosomes. *Annu. Rev. Cell Biol.* **5,** 483–525.

Lang, T., de Chastellier, C., Ryter, A., and Thilo, L. (1988). Endocytic membrane traffic with respect to phagosomes in macrophages infected with non-pathogenic bacteria: Phagosomal membrane acquires the same composition as lysosomal membrane. *Eur. J. Cell Biol.* **46,** 39–50.

Pfeifer, J., Wick, M., Russell, D., Normark, S., and Harding, C. V. (1992). Recombinant *E. coli* express a defined, cytoplasmic epitope that is efficiently processed by macrophages for Class II MHC presentation to T lymphocytes. *J. Immunol.* **149,** 2576–2584.

Russell, D. G., Xu, S., and Chakraborty, P. (1992). Intracellular trafficking and the parasitophorous vacuole of *Leishmania mexicana*-infected macrophages. *J.Cell Sci.* **103,** 1193–1210.

Sturgill-Koszycki, S., Schlesinger, P., Chakraborty, P., Haddix, P. L., Collins, H. L., Fok, A. K., Allen, R. D., Gluck, S. L., Heuser, J., and Russell, D. G. (1993). Lack of acidification of in *Mycobacterium* phagosomes produced by exclusion of the vesicular proton-ATPase. *Science* **263,** 678–681.

Thilo, L. (1983). Labeling of plasma membrane glycoconjugates by terminal glycosylation (Galactosyltransferase and Glycosidase). *In* "Methods in Enzymology" (S. Fleischer and B. Fleischer, eds). Vol. **98,** 415–421. Academic Press, New York.

Wallis, A., Russell, D. G., and McMaster, W. R. (1994). *Leishmania major*: Organization and conservation of genes encoding repetitive peptides and subcellular localization of the corresponding proteins. *Exp. Parasitol.* **78,** 161–174.

Xu, S., Sturgill-Koszycki, S., van Heyningen, T., Cooper, A., Chatterjee, D., Orme, I., Allen, P., and Russell, D. G. (1994). Intracellular trafficking and the *Mycobacterium*-infected macrophage. *J. Immunol.* (in press).

CHAPTER 16

Measuring the pH of Pathogen-Containing Phagosomes

Paul H. Schlesinger

Department of Cell Biology and Physiology
Washington University School of Medicine
St. Louis, Missouri 63110

I. Introduction
    A. Why Does the Phagosome pH Matter?
    B. Using the Pathogen to Probe Phagosomal pH
    C. The System of Intracellular Vesicles through Which the Parasite Enters the Host Cell
    D. The Generation of Intravesicular Acid pH
II. Materials
    A. Labeling of the Particle for Uptake and the Experimental System for the Measurement of Phagosomal pH
    B. Biological Fidelity of the Experimental Method
III. Procedures
    A. Fluorescent Determination of pH
    B. Local Environmental Factors That Affect pH Measurement
    C. *Histoplasma*: An Example of Some Difficulties in the Measurement of Phagosomal pH
    D. Calibrate, Calibrate, Calibrate . . .
IV. When Things Are Not Perfect, or Even Very Close
V. Conclusion
    References

I. Introduction

A. Why Does the Phagosome pH Matter?

It has long been realized that the environmental pH will exert a strong or even controlling effect upon the survival and growth of microorganisms, both free living and intracellular. This reflects the importance of pH, both intracellular and environmental, to all organisms (Roos and Boron, 1981). Thus the acid pH of the phagosome is a major contributor in the hostile environment presented to potential pathogens. Now there is increasing and direct experimental evidence that several organisms that survive for extended periods and replicate within eukaryotic cells are found in phagosomes with an elevated pH during the infection (Black *et al.*, 1986; Sibley *et al.*, 1985; Horwitz and Maxfield, 1984). Using the methods described herein, the mechanisms whereby two such organisms reduce phagosomal acidification have been explored (Eissenberg *et al.*, 1993; Sturgill-Koszycki *et al.*, 1994). The fact that these two organisms have found different ways to reduce acidification would seem to indicate that a number of approaches to this issue might exist. In addition, an equally important question and one that has yet to be carefully considered is the manner in which a particular intraphagosomal pH plays a significant role in the intracellular survival and disease-producing potential of microorganisms. The study of these questions forms the basis of a practical interest in the accurate *in situ* study of phagosomal pH and the mechanism by which it is achieved.

The study of phagosomal acidification has progressed more slowly than that of other intracellular vesicles. The small numbers of this organelle, which is not constituitively expressed, have certainly made their study more difficult. In addition the presence of functionally different phagosomes (having distinct internal pHs) in one cell (Eissenberg *et al.*, 1993) has made it critical to know what functional group is being observed. Consequently our model for the molecular mechanism of phagosome acidification comes almost entirely by analogy with the endosomal system. The details of their differences and similarities becomes especially relevant as we attempt to develop approached to the treatment of intracellular pathogens, which must be targeted at the phagosomal system. At present our understanding of the biochemical relationships between host and parasite with regard to the phagosomal environment are rudimentary but at least one example of a molecular relationship provides some important insight. The intracellular parasite, *Legionella pneumophila* (Byrd and Horwitz, 1991), becomes dependent upon an acid intravesicular pH for the transferrin-mediated delivery of required iron to the organism. This observation makes it clear that the pathogen can come to depend upon acid intracellular vesicles whereas other acid vesicles pose the greatest threat to its survival. We should be reminded to avoid overly simplistic manipulation of the vesicle/phagosome environment. The use of pH and ion gradients in the transport of nutrients, osmotic regulation, and establishment of voltage gradients can be employed by either the host or the pathogen. As we proceed to study the characteristics of intracellular vesicles containing pathogens, the intricate relationship between the host cell and its internal guest will become more apparent.

In addition the recent demonstration of *in vitro* fusion between phagosomes and endosomes has added new pathways to phagosome maturation (Mayorga *et al.*, 1991). A common assumption has always been that acidification was triggered by the addition of the molecular engines of acidification to the phagosome by fusion with lysosomes. The complementary view that inhibition of acidification resulted from either avoidance or inactivation of these transport systems has also dominated our thinking. This paradigm should now be modified to include both the addition and the release of molecules from the phagosomal compartment and interaction with the endosomal compartments as well as lysosomes. Our natural assumption has been that the selection of such components for retention or release would be under the regulation of the host cell. Now this assumption must also be revised since it is clear that in at least one case the selective release of membrane components in the phagosome is under the direction of the intracellular organism (Sturgill-Koszycki *et al.*, 1994). Therefore it appears that phagosome acidification involves delivery and retrieval, as well as the activity of the acidification machinery. Therefore simplistic mechanisms of control should viewed very carefully. In order to consider the effects of pH upon metabolite transport, ion gradients, or membrane potential it will be crucial to determine quantitatively the actual pH value. Finally both from analogy with the endocytosis pathway (Forgac, 1989) and from studies on the phagosomal system (Geisow *et al.*, 1981; Jensen and Bainton, 1973) we now are well aware that these vesicles are dynamic and that the ability to follow the pH in living cells as a function of time is essential. We need details of the transport systems that support the phagosomal environment and the manner in which that environment is created, regulated, or eliminated. In this chapter I discuss a system in which these parameters can be studied and which has provided information that is physiologically relevant.

B. Using the Pathogen to Probe Phagosomal pH

We have used pathogens that survive intracellularly to study the environment of the phagosome that occurs during the normal infection. This requires that both the pathogen and the host cell survive for the course of the experiment. This is a minimum requirement for the experiment to achieve its desired result and allow the study of the host–parasite relationship. Since we will be discussing the association of a mammalian cell and another organism where both parts of the complex are viable entities I would like to establish a standard set of terms for this system. I shall refer to the intracellular organism as a "parasite" and the eukaryotic cell as the "host."

Studies of intracellular organisms have tended to emphasize the host cell's abilities and activities. Although the complexity of the host and its apparent "ownership" of the contested turf support this view, the fact that not all organisms cause disease and that even fewer survive in intracellular vesicles should give us cause to consider the organism as both a participant in the outcome and a potential tool in the study of this important process. In the present situation we will make most use of the pathogen for its ability to confer specificity and biologic veracity to our studies. We will use the parasite labeled with reporting molecules to study the phagosomal pH during a cellular infection.

However, the interaction of the parasite with the host has many aspects and it would be profitable to pursue other characteristics using different reporting molecules to quantitatively study additional parameters.

Using the parasite to introduce our pH probe provides biologic specificity for its introduction into the parasite–phagosome space (that part of the intracellular vesicle compartment to which the parasite has access) and thereby eliminates spurious signals from uninfected cells and from nearby but uninvolved vesicles in the infected cell. This is a critical advantage of the method and is not easily duplicated with other techniques. We have used this approach with spectrofluorimetry, which provides information on the population of parasite-containing intracellular phagosomes; however, microspectrofluorimetry has been applied to single vesicles (Black et al., 1986; Sibley et al., 1985; Horwitz and Maxfield, 1984). At present we see real advantages to averaging the pH over the vesicles of a biologically selected class. From studies on the endosomal system, it appears that the pH of intracellular vesicles varies over a rather wide range but in a manner that is related their functional classification (Forgac, 1989; Ymamshiro and Maxfield, 1987; Lukacs et al., 1992). Furthermore the dynamic nature of the endosomal system of vesicles means that intravesicular pH can vary with time as well as an anatomical-type of vesicle. Thus far the pH classification of phagosomes has not been reported but they are dynamic structures having interactions with other vesicles. The kinetic aspects of phagosome maturation and the acidification of newly forming phagosomes have previously been noted (Geisow et al., 1981; Jensen and Bainton, 1973). By using a pH probe-labeled parasite we are functionally defining a class of phagosomes as those that contain the labeled organism. It is possible by this technique to study the pH of that functional group of phagosomes. By having directly determined the average pH of a functional class of vesicles there is no need to extrapolate from a necessarily small number of measurements done by microspectrofluorimetry to the entire class of vesicles.

C. The System of Intracellular Vesicles through Which the Parasite Enters the Host Cell

To gain access to the eukaryotic cell it now appears that potential parasites must interact with the host cell endocytic system. Although the mechanism of this entry is incompletely understood it appears that in most cases both the host cell and the parasite can participate. Classically molecules on the parasite would bind with molecules on the host and then the parasite would be enclosed by plasma membrane of the host creating a phagosome. There may be exceptions to the phagocytic mechanism of entry (e.g., that the parasite engineers its own entry without the consent or participation of the host) but these do not appear to be common and would arguably lead to the death of the host cell, hardly an advantage to the parasite. Upon formation of the phagosome the parasite is met by an array of host-defensive measures that are intended to kill, dismember, or otherwise inactivate the parasite. Perhaps one of the better supporting evidences for the importance a common phagocytic route of entry is the array of defenses that the host has strewn before the parasite. Some parasites avoid these defenses by exiting from the phagocytic vesicles as soon as possible. Of more interest currently are those that remain and grow inside the cellular phagosome, since they have found ways to neutralize the cellular defenses. Both the host and the parasite play important roles in phagocytosis; it is not acceptable to employ just any particle labeled with a pH-sensitive fluorophore and conclude that the results are a model for a particular pathogen. Our reason for using a viable organism as a probe is to mimic the normal infection as closely as possible and to the extent that any particular parasite does influence events we would like this influence to be intact. Therefore it will be important in each case to make comparisons between the infection with the native and the labeled parasite in as many ways as are possible. These comparisons have included the viability of the host, protein synthesis by the host, tryphan blue exclusion by the host, viability of the parasite, rates of infection, and multiplication of the parasite (Eissenberg *et al.*, 1988; Sturgill-Koszycki *et al.*, 1994). At its most simple this could be accomplished by a few hours over a microscope but it can also become much more complex.

The easiest way to characterize what we understand of the intracellular vesicle/membrane system of the typical phagocytic cell is "complex." Fortunately in this chapter I will deal primarily with the regulation of vesicle pH and focus the discussion on phagosomes. Although this reduces the scope of the subject, there is less detailed knowledge concerning phagosomes and I will often resort to analogy with the endocytic pathway. A large number of observations support the longstanding dogma that lysosome fusion with phagosomes occurs. There is now direct evidence that phagosomes and lysosomes exchange membrane components and contents (Eissenberg *et al.*, 1988; Hart *et al.*, 1987; Wang and Goren, 1987; Bizal *et al.*, 1991; Sturgill-Koszycki *et al.*, 1994). Therefore the molecular complexes involved in phagosome acidification could be arriving from lysosomal vesicles. However, by using *in vitro* assays to study the fusion of intracellular vesicles it has now been shown that endosomes and phagosomes can also fuse (Pitt *et al.*, 1992; Mayorga *et al.*, 1991). This is more consistent with the kinetics of phagosome acidification shown in Fig. 1, which is rapid like that of endosomes (also shown). This comparison associates phagosome acidification with that of endosomes by their similar time course and final pH and not with the later fusion with lysosomes that occurs. However, a recurring theme is that phagosomes are very heterogeneous in their properties and perhaps we should forgo the idea that there is a "prototypical" phagosome pathway. We know that the program of phagosomes containing mycobacteria or histoplasma are very different than that of zymosan phagosomes (Eissenberg *et al.*, 1988; Sturgill-Koszycki *et al.*, 1994). Furthermore several intracellular parasites are known to prevent acidification of the vesicle in which they reside (Black *et al.*, 1986; Sibley *et al.*, 1985; Horwitz and Maxfield, 1984), and there

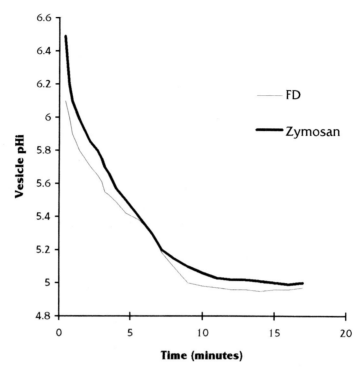

Fig. 1 Time course of vacuole acidification. Mouse peritoneal macrophages were loaded with FITC-dextran or with FITC-zymosan and then the time course of acidification was followed by determination of the excitation ratios at 450 and 497 nm as described in the text.

are examples of cells that are unable to form acidifying endosomes (Merion *et al.*, 1983). Although variation is found in both endosome and phagosome acidification there are vesicle-specific characteristics that account for some differences. As an example, the size of the resulting vesicles can produce differences in their properties. The vesicles produced in the fluid phase endocytosis of FITC-dextran (FD) are initially quite small, whereas the phagocytosis of a particle produces vesicles at least the size of the particle. Because of the low internal pH and the semipermeable surrounding membrane, it is possible to increase the vesicle pH with NH$_4$Cl added to the medium around the cell (Krogstad and Schlesinger, 1986). Furthermore the continuing activity of the proton pump, partial permeability to the protonated form of the weak base, and ionic and osmotic adjustments that occur lead to a characteristic time course of the endosomal pH during a cycle of NH$_4$Cl addition, equilibration, and removal, as shown in Fig. 2. In large part due to the difference in vesicle size and the slower rate of intravesicular pH change in the larger phagosomal vesicles, the pattern of pH change is very different in very large phagosomes

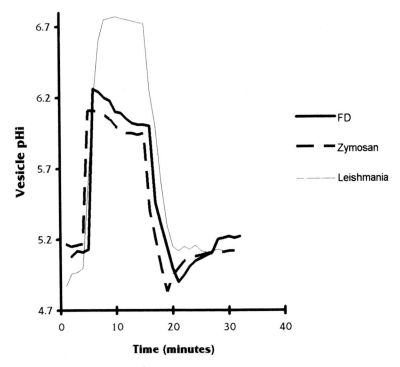

Fig. 2   The effects of a weak base. The time course of NH$_4$Cl alkalinization of acid intracellular vacuoles containing FITC-dextran, zymosan, and *Leishmania mexicana*.

as the vesicles containing *Leishmania mexicana*. Variation in intracellular vesicle pH has also been correlated with functional classes of vesicles. Early endosomes and Golgi have been proposed to have higher pH values whereas late endosomes and lysosomes are felt to have a pH close to 5 (Forgac, 1989). For those vesicles occupied with intracellular receptor-mediated transport and targeting the role of acid pH has been clearly related to the binding and dissociation of receptor–ligand complexes and the pH dependence of those reactions determine occupancy of the receptors. In vesicles that are involved in degradation of their contents, the role of pH has been identified with that activation of the degradative enzymes and the denaturation of the substrates. Clearly phagosomes may incorporate one or both of these functions at different times in their life cycle. The measurements of zymosan-phagosome pH indicate that usually the pH drops to approximately 5 within 20 min of formation and then stays at a low pH for an extended period of time. In those cases where an intracellular parasite has been reported to prevent acidification of the phagosome, the observations have tended to support an inhibition of the initial acidification and then a maintenance of the elevated pH. Since we have now studied zymosan uptake in several macrophage cell lines and in resident peritoneal macrophage, aveolar macrophages, and bone monocytes it seems that for this particle the pattern of acidification is consistent. However, our recent work also indicates that the nature of the labeled particle can result in phagosomes of different pH in the same cell (Eissenberg *et al.*, 1993).

What has not yet been studied are the existence and consequences of ion gradients or membrane potential that could result from the pH gradient or can be coupled to that gradient. Using these forces and gradients the invading parasite could presumably accumulate required substances from the cytoplasm of the host while the cell might attempt to use the same forces to separate the parasite from required substances. Thus far we have identified a number of intracellular parasites that apparently assist their survival by blunting the acidification of the phagosome but we do not know what other concentration gradients might be important to this host–parasite relationship.

D. The Generation of Intravesicular Acid pH

With examples of intracellular vesicle acidification from mammals to plants all appearing to use the same mechanism it is overwhelmingly assumed that the phagosome will continue the trend and employ the same electrogenic vacuolar proton pump to generate an intravesicular acid pH. It catalyzes the energy driven transport of protons to a concentration approximately 100-fold that of the surrounding cytoplasm requiring Mg^{2+}-ATP (see Fig. 3). In coated vesicles (Xie and Stone, 1988), endosomes (Forgac, 1989), Golgi (Glickman *et al.*, 1983), and the osteoclast ruffled membrane (Mattsson *et al.*, 1994), this proton pump is in parallel with an anion conductive pathway, which reduces the membrane potential created by the electrogenic proton pump. These two transporters are independent and can be isolated separately without dramatically altering either of their properties (Blair and Schlesinger, 1990; Forgac, 1989; Xie and Stone, 1988). It has been shown that inhibition of the anion (chloride) conductance will reduce the pump-mediated acidification of endosomes, coated vesicles, and the osteoclast bone resorption compartment (Forgac, 1989; Glickman *et al.*, 1983; Lukacs *et al.*, 1992). Therefore it is possible to introduce a second site at which the parasite might influence phagosomal acidification although there is no reported instance of this occurring. Other transport pathways may contribute to the acidification of phagosomal vesicles but these two components appear to be central to acidification.

1. The Phagosomal Proton Pump

At present the nature of the phagosomal proton pump is assumed to be similar to that of other intracellular vesicles but it has not been isolated or studied biochemically. Several recent studies have identified pump subunits on phago-

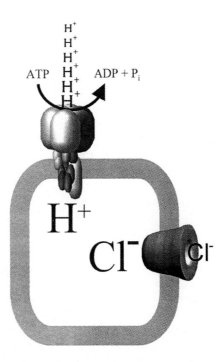

Fig. 3 Vacuolar acidification machinery. By analogy with the endosomal vesicular system the phagosome is presumed to contain a vacuolar-type proton pump that is a MgATP-requiring electro-genic proton transporter that introduces protons into the vacuole interior and a passive anion conductance that is required to reduce membrane potential in the vesicles to levels that produce vesicle acidification.

somes using antibodies raised against the endosomal pump (Pitt *et al.*, 1992; Sturgill-Koszycki *et al.*, 1994), and therefore it seems that the similarity holds up at the molecular level. Studies in our laboratories have indicated that when normal acidification is occurring the 116-kDa glycoprotein is present in the pump complex of phagosomes although this subunit has been reported absent in some vacuolar proton ATPases (Gluck and Caldwell, 1987). Although the vacuolar proton pumps have been biochemically characterized, their regulation is not understood. If the acidification system for phagosomes is indeed "bor-rowed" from endosomes then there will be no need to regulate its activity separate from that of the endosome pump. When the need for phagosome acidification is concluded, the pump could merely be returned to the endosome compartment. It has also been proposed that assembly of the vacuolar proton pump might control its activity *in vivo* (Brown *et al.*, 1991). The phagosome might be an excellent candidate for such a regulatory mechanism since it is not a constituitively expressed organelle in all cell types. This organelle is only required to be active after a cell internalizes a particle and the ability to turn its acidification on and off would conserve energy. One mechanism by which this could operate would be for the hydrophilic sector of the proton pump to dissociate from the membrane regions under appropriate conditions. The remaining hydrophobic sector might be a proton channel, but lacking the ATPase subunits of the pump it would not actively transport protons. Thus far all mechanisms of molecular regulation of the vacuolar proton pump directly are speculation since studies of the mechanism of phagosomal acidification are just beginning. It is clear that the absence of pump subunits is correlated with the lack of acidification, which is consistant with the first proposal (Sturgill-Koszycki *et al.*, 1994).

2. The Vacuolar Compensating Current: A Proposed Chloride Channel

Figure 3 also indicates a chloride channel in the phagosome membrane. Indications are that an anion conductance (most likely a chloride channel) is present in parallel with the vacuolar proton pump in almost all circumstances. Golgi, endosomes, and coated vesicles have been shown to require charge compensation to acidify (Forgac, 1989; Glickman *et al.*, 1983; Lukacs *et al.*, 1992). In addition a chloride channel has been purified from the osteoclast ruffled membrane, which also contains a vacuolar proton pump (Blair *et al.*, 1989; Mattsson *et al.*, 1994; Chatterjee *et al.*, 1992). Very little is known about this channel but a candidate protein has been isolated from the osteoclast ruffled membrane and reconstituted into artificial lipid membrane vesicles producing a chloride conductance (Blair and Schlesinger, 1990). An immunologically homologous protein has been isolated and cloned from bovine renal tissue (Redhead *et al.*, 1992; Landry *et al.*, 1993). These proteins are reported to be localized to intracellular organelles and recent reports indicate that the channel proteins are phosphorylated (Forgac, 1989; Mulberg *et al.*, 1991; Landry *et al.*, 1978; Bae and Verkman, 1990; Tilly *et al.*, 1992). Furthermore the phosphorylation has been linked to the regulation of the chloride conductance in the endosomal vesicles and therefore to their acidification. This provides a mechanism for regulation of endosomal acidification, which if the analogy is extended would provide for regulation of phagosomal acidification. However, one must quickly point out that the putative chloride channel has not been specifically identified in phagosomes and evidence for its role in phagosomal acidification is by analogy with other intracellular vesicles. Despite this it is probably appropriate to re-state the hypothesis: "The chloride channel conductance is critical to acidification of the endosome and lysosome and as such provides an indirect way of regulating their pH." A similar mechanism is possibly active in the phagosome.

II. Materials

A. Labeling of the Particle for Uptake and the Experimental System for the Measurement of Phagosomal pH

1. Nonliving Particles

For any cell that is able to take up particles it will be possible to prepare zymosan that is labeled with fluorescein and to measure the pH of the resulting vacuoles. Zymosan is the autoclaved cell walls of yeast particles and possesses sufficient amino groups for reaction with FITC or other forms of fluorescent probes. This cell wall is degraded very slowly in macrophages and can be used to measure phagosomal pH for extended periods of time, including up to 24 hr in culture. Consequently I have used this particle as a primary standard and probe for testing the uptake and pH measurement when beginning studies in a new experimental system or whenever starting a series of experiments or if I want to test the method. There are some variations in the quality of zymosan particles but the preparations are typically quite usable (Eissenberg *et al.*, 1988; Sturgill-Koszycki *et al.*, 1994). After labeling I have found the labeled particles to be chemically stable for some time (at least a week when kept refrigerated and following the usual injunctions about sterility and preventing bleaching by strong persistent light).

Other particles that we have used because they can be modified chemically with additional proteins and ligands include primarily latex beads (Sturgill-Koszycki *et al.*, 1994). These have been more difficult because they seem to vary considerably in their preparation. It has been possible to prepare latex beads that are well labeled, but success in this regard has been dependent upon the quality of the bead preparation. Unfortunately I cannot identify the factors that result in this variability and therefore the use of latex beads is still very empirical. Most important are the maneuvers that are described subsequently in the detailed discussion of calibration. However, just as important is the determination that the modified beads are stable to the biological conditions to which they will be exposed. One must determine that over the time that phagosomal pH is to be measured the particle and its fluorescence must remain within the compartment of interest. This must be determined as objectively and critically as possible and therefore in addition to the purely morphological studies, which can be difficult to interpret, it is also desirable to define biochemical and functional criteria. The effect of ammonium chloride on the fluorescence of a pH-sensitive dye in an acidified compartment surrounded by a semipermeable biological membrane is characteristic and I find it very useful in validating the preparations. The most useful attitude is to be skeptical of any situation that produces a significantly different response than those shown in Fig. 2. Another criteria that is useful to us is the response of cell-associated fluorescence to changes in the external pH. An immediate response to increasing or decreasing the external pH by 0.5 units indicates that the particle and probe are still exposed to the external medium. A slower (5 to 10 min) and incomplete response (0.1–0.3 pH units) to the same stimulus indicates that the particle and the fluorescent probe are inside a closed vesicle.

The rationale for using latex beads is primarily to allow the uptake of beads after specific binding to particular receptors (for example, the Fc receptors). Since these ligands have to date all been proteins or peptides there is the additional complication of whether the ligand is stable and how much difference that makes after internalization. Whether the fate of a phagosome or its contents will change depending upon the binding of contents to receptors or the occupancy of included receptors is not clear. However, as a partial compromise we have adopted the approach of first modifying the particle with the ligand of interest and then labeling the attached ligand with the fluorescent probe. Therefore if the ligand separates from the bead it should be apparent morphologically or because the fluorescence is released from the cells. It is hoped that we will not unknowingly be following the pH of a particle that has lost its specificity, although admittedly we do not understand the importance of the specific binding after the initial internalization of the particle. A further complication of this technique results from the consequence of actually labeling the protein ligand. As discussed shortly the pH dependence of fluorescence can be strongly affected by the environment of the fluorphore (see Fig. 5). Since proteins can dramatically effect the local environment of peptide chains the pH-dependent fluorescence of a protein-substituted bead can be quite surprising. This effect is discussed in the section on calibration of the pH response.

In summary it is perfectly possible to label inert particles with pH-sensitive fluorescent probes and then to use these probes to study phagosomal pH. Since these particles can be modified with receptor-specific ligands, they afford a possibility to study the role of receptor binding specificity in the acidification of the phagosome and to explore aspects of the phagosomal acidification. The degree to which changing the surface compliment of protein will affect the fate of the particles after internalization is not clear at this point.

2. Labeling of Living Particles

a. Maintaining Viability

In order to use a probe for determination of the phagosomal pH during an entry and establishment of the invading organism it goes without comment that using the living organism as a probe is preferable. Although it is possible to employ the proteins and structural elements of the microorganism for the same purpose, the result will be less satisfying since there will be no guarantee that some critical activity associated with the living organism has not been lost. In addition it is quite apparent that many intracellular microorganisms contribute macromolecules and metabolically to the function of the host–parasite complex (Moulder, 1985). In our experience it has not been terribly difficult to achieve at least moderate viability. Microorganisms almost always have a outer covering in addition to the plasma membrane and if the labeling can occur on whatever capsule is there, the immediate and direct influence upon the cell is small. Using fluorescein and its derivatives the labeling can be done under mild conditions, neutral pH, and low temperatures. This results in high viability immediately after the labeling, which we have judged typically by the exclusion of dyes and the morphological appearance of the organism. Longer-term survival and multiplication in culture are critical for confirmation of viability. The final test, however, will be the ability of the organism to grow and divide intracellularly in the host cell (Eissenberg *et al.*, 1988; Sturgill-Koszycki *et al.*, 1994). Finally, for the model to be complete it is important that the microorganism cause the same effects in the host after labeling as it did before the probe was added.

In any particular experimental system these criteria will be achieved to a degree that will determine the meaningfulness of the experiment. For example, if the labeled organism does not survive in the host cell and no inhibition of acidification is detected, it might be that the labeling interfered with defenses of the parasite to alter host cell acidification or it might mean that the labeling made the parasite nonviable. We have been able to label successfully three types of organisms and to get them to produce typical infections in host cells. Therefore I do not anticipate that this will in a general way prevent the study of the phagosome acidification and the potential of this method is substantial.

B. Biological Fidelity of the Experimental Method

In the study of biological systems by any method there are questions associated with perturbations required to obtain data. This typically arises in the present method when in order to increase the fluorescent signal we over inoculated the cells with parasite. Since parasite killing the of host cell will probably produce and altered phagosomal acidification, the results are not illuminating. The issue of overinoculation is a particularly difficult one since it is natural to attempt to have the maximum signal and one of the greatest difficulties we have with the measurement of phagosomal populations is the signal-to-noise ratio. Furthermore, since the parasite will disable the host cells in some respect this issue becomes almost a philosophical one. It is reasonable to attempt to study the same level of host cell damage that is routinely found in natural infections. Perhaps a general rule that could be useful is that if the phagosomal pH depends too directly on an experimental parameter (such as multiplicity of inoculation) then one should consider the possibility that the experimental perturbation is controlling the system. There may also be guidelines from the natural situation, such as the number of organisms per cell that is normally found.

Other factors that can call biologic fidelity into question include the identity of the host cells. Frequently it has been difficult to decide on the appropriate host cell for studies. For intracellular parasites the natural host cell is often the macrophage and the most convenient, but perhaps not the best, choice is to use macrophage cell lines. These cell lines are probably not equivalent to *in vivo* macrophages in ways that are important to the acidification and processing of phagosomes. This would be especially relevant if one is studying the effects that might be modulated by humoral factors. In cases where the effects on acidification are direct, such as the release of an inhibitor, we have successfully used the macrophage cell lines to study the inhibition of acidification. In the end it would seem most useful to focus on testable molecular hypotheses and then attempt to test them under as wide a range of cellular conditions as possible.

III. Procedures

A. Fluorescent Determination of pH

The measurement of intravesicular pH has been relatively easily to accomplish for approximately 10 years. The method derives from the original use of pH-sensitive fluorescent dyes to measure intracellular pH by Thomas and its first applications to vesicles by Poole and Okhuma (Thomas, 1986; Thomas *et al.*, 1979; Ohkuma and Poole, 1978). In order to be accurate it is necessary to determine simultaneously the total amount of fluorescent dye and the distribution of that dye among its pH-dependent forms. Several approaches have evolved to accomplish that task but one that appears to have most general usefullness employs a dye whose structure permits an isosbestic point in its fluorescent spectra. Then by using the ratio of the pH-independent fluorescence (isosbestic point) and the pH-dependent fluorescence one has normalized the emission for the amount of fluorescent dye and the result can be used to determine pH. In this regard, fluorescein has many suitable properties with an isosbestic point and a pK $\sim$5.5 in aqueous solution. The excitation spectra at pH values from 3 to 9 for 6-carboxyfluorescein in dilute aqueous solution are shown in Fig. 4. Using the excitation that is pH-insensitive (450 nm) and the pH-sensitive peak (495 nm), one can form a ratio that is corrected for the concentration of fluorescein, and this can be used to determine the pH in unknown samples. In well-defined samples with a large signal this is a very accurate method of determining pH and unaffected by fluorescein concentration up to micromolar concentrations, where autoquenching become significant.

B. Local Environmental Factors That Affect pH Measurement

Clearly there are other factors that must also be considered and most of these result from the well-known influence of environment on fluorescence emission. Because chemical fluorescence results from the generation and decay of an excited molecule in solution, the interaction of that molecular excited state with the local environment is critical in determining the nature and strength of the emission. Therefore any solvent or solute molecule or species present in

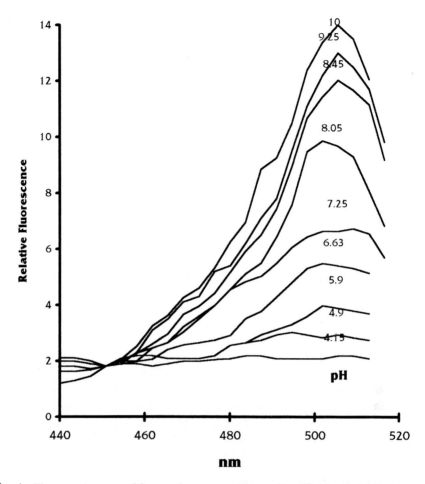

Fig. 4 Fluorescent spectra of fluorescein-zymosan. The spectra of fluorescein-labeled zymosan particles in 140 mM KCl at the indicated pH values buffered with a mixture of acetate, HEPES, and ammonium buffers at 10 mM. The excitation wavelength is indicated on the x-axis and the emission was at 513 nm. The zymosan was labeled with fluorescein as described (Eissenberg *et al.*, 1988; Sturgill-Koszycki *et al.*, 1994).

the local environment can potentially interact with the pH-sensitive fluorophore and alter the emission. The most useful antidote to these sorts of technical difficulties is the performance of *in situ* calibration and a healthy skepticism that calibration in one set of conditions will carry over to another set of experimental conditions.

In this regard the experimentor is aided significantly by the system under investigation. The intracellular environment is carefully monitored by the host cell. Therefore within rather narrow limits, the concentrations of ions are constant within the cytoplasm. Although we might suspect that a similar situation occurs in the phagosome, that conclusion is not warranted. This is particularly relevant when comparing the conditions in a phagosome formed around a pathogen and those of some easily obtainable (and easily labeled) particles such as zymosan. The calibration may be clear with zymosan, which is inert, but it must also be applicable or repeated when comparing virulent and avirulent forms of the relevant organisms. In this regard it becomes important to calibrate pH under conditions that are close to those in which the intracellular organism can survive and grow.

C. *Histoplasma:* An Example of Some Difficulties in the Measurement of Phagosomal pH

For the last several years I have been involved in the study of phagosomal pH of macrophage cell lines infected by *Histoplasma capsulatum*. We were concerned that the cell wall of this organism would cause problems with the use of fluorophores attached to its outer surface to measure the intravesicular pH (Eissenberg and Goldman, 1991). As previously discussed the local environment can directly affect fluorescence and the reporting of pH. We initially had several concerns: (i) the cell wall and capsule of the organism might act as a barrier, preventing exposure to the phagosomal pH; (ii) the difference between the rough (pathogen) and smooth capsule might lead to a difference in the pH response of the attached fluorophore; (iii) the closeness of phagosome–vesicle membrane approximation will influence the pH measurement by labeled organisms. Because of these concerns, we performed repeated calibrations of the labeled organism both free and *in situ* after uptake by cells. In addition we employed several methods and treatments of the cells during and after labeling in attempts to reduce these effects.

D. Calibrate, Calibrate, Calibrate . . .

Many potential mistakes can be avoided by frequent and careful study of the relationship of fluorescence to pH under controlled conditions. Our experience with *Histoplasma* makes the potential errors abundantly clear. Figure 5 shows the ratio verses pH curves for the rough and smooth histoplasma and zymosan. Although not shown, the calibration curve of fluorescein linked to dextran is very similar to that for fluorescein-labeled zymosan. All of these calibrations have a large enough total change of ratio with pH that they can be used to determine pH from an intracellular particle. The problem becomes apparent when one considers the same ratio obtained from a zymosan particle indicating pH 5.5 would reflect a pH of 6–6.5 surrounding a histoplasma particle. Therefore without the multiple calibrations we would not have discerned that the phagosomal pH is elevated in the histoplasma phagosome. Looking at this figure it should be obvious that to interpret experimental results successfully, one needs to be familiar with the response of the probe under conditions that are as close

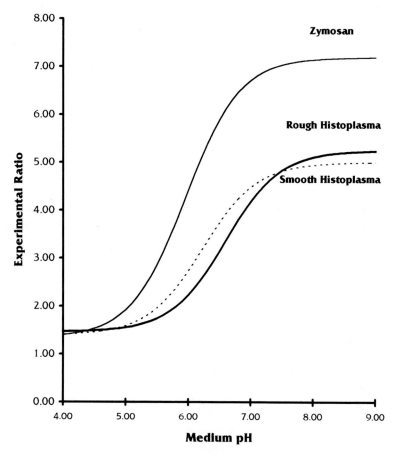

Fig. 5 Calibration of the pH response of intracellular fluorescent probes. Three particles loaded into a macrophages cell line were placed into a medium of 120 mM KCl; 10 mm each of Na acetate, NH₄Cl, and HEPES; and 11 μM nigericin. The medium pH was then adjusted to the indicated values and the ratio of fluorescence (497 nm/ 450 nm) was allowed to stabilize and was determined. The data were fitted using nonlinear regression, and the best fit values for upper and lower plateaus and for the curve midpoint were determined.

as possible to actual conditions within the infected cell. An important fact to remember here is that the fluorophore and the paricle are not totally interchangeable. The local environment, which determines the fluorescence, is dependent on the chemistry of the particle as well as the pH of the vacuole and the two combine to determine the pH calibration curve. This dependence on the particle surface might be reduced by attaching the fluorophore to a chemical arm and extending it from the particle surface into the surrounding solution. However, this increases the chemical manipulation of the particle and has been detrimental to parasite viability in our hands.

In almost all cases we have been successful with reactive derivatives of fluorescein that are readily available. These will react with protein amino groups at close to neutral pH and the incubation during the reactions are mild. The particles are then thoroughly washed to remove unreacted fluorescence and the relationship of fluorescence to pH is determined in suspension. The fluorescein derivatives have spectra as shown in Fig. 4 and the ratio versus pH for the labeled organism in suspension is determined. The plots obtained should be sigmoidal, resembling those in Fig. 5. Two criteria are necessary for a useful pH probe. First, the curves should have a reasonable midpoint of 5.6–6, indicating that the fluorophore has access to the aqueous environment around the particle and the midpoint of the ratio approximates the pK of the fluorescein fluorophore. Second, the plateaus need to be separated enough to allow for reasonable accuracy of pH determination. This usually requires at least 2 ratio units difference for high and low pH conditions. If the probe satisfies these criteria in suspension, it is recalibrated after uptake by cells.

As a second stage of verification, the probe is allowed to enter the cells under conditions resembling those of a natural infection, and after a period of at least 2 hr, but frequently overnight, the cells are placed in isosmotic KCl medium in an attempt to equalize intracellular and extracellular K^+ concentrations. The nigericin is added to the medium (we have used 10 μM) and the pH of the medium is adjusted to various calibration values. Nigericin is a $K^+:H^+$ ionophore and will equilibrate intra- and extracellular pH (Pressman et al., 1967). This equilibration happens quickly to a monolayer of cells in culture and can be monitored by watching the ratio as a function of time. Since the cells are cultured on the small glass chips that can be transferred to media of different pH values, a small number of chips can be used to develop complete calibration curves as shown in Fig. 5. If the calibration curve in suspension is the same as that from particles in cells, it is then appropriate to proceed with studies to determine the phagosomal pH using the in situ calibration curves developed in this fashion. If they do not correspond, it is then necessary to explain the lack of similarity or to modify the method to get good correspondence.

As mentioned above, the most common difficulty is low fluorescence relative to the endogenous cellular background. If this occurs the pH dependence of the fluorescence will be slight and the problem will be apparent. Using a different chemical form of the dye increased labeling and corrected this problem for us in the case of mycobacteria. Another difficulty that has been easily solved is the noncovalent fluorescent labeling of particles or cells. This can be a problem early in the experiment if the particles are not washed well. The method of washing that we have employed is centrifugation, but it also helps to change the medium during culture of the cells after uptake. Successful washing of the particles will decrease the occurrence of excessive "bleeding" of fluorescence during culturing.

IV. When Things Are Not Perfect, or Even Very Close

More often than one wishes the results of an experiment are disappointing and the calculated pH values are not as expected. The most common cause of this has been a low signal level from the infected cells. Fluorescence is always measured on a relative scale and therefore a one- or twofold smaller change in the signal can be obscured by attenuation settings on the spectrofluorimeter. This level of change can be important when working close to the limits of detectability. Frequently in order to maintain viability of the parasite and to reduce the infectious burden to the host cells, the experimental signal is maintained at as low a level as will provide data. In order to obtain meaningful data it is necessary that the endogenous fluorescence of the cell be 10% or less of the total measured fluorescence of the sample. Typically the endogenous fluorescence of cultured cells is not pH sensitive and gives a low ratio value when determined as described below. This produces an artifactually lower apparent pH and in most cases is welcomed by the investigator as evidence of viable cells and a stable fluorescent probe. However, if when 5–10 mM NH$_4$Cl is added to the medium the fluorescent signal does not respond in a manner similar to that shown in Fig. 2, one is most likely studying cellular background fluorescence. If the signal is not responsive to NH$_4$Cl it is imperative to test the endogenous fluorescence of identical samples with and without loading of the fluorescent probe. Increasing the labeling of the probe, its loading into cells, the viability of the host cells, the stability of the probe, and cleaning of the equipment are areas where improvements can be made.

We have typically used glass slides cut into 9 × 25-mm slices to support the cells. They are cheap, thick enough to be relatively easy to handle, have good optical properties for visible (>400 nm) fluorescence, and tolerate cleaning with fuming nitric acid. Before each use these glass "chips" are immersed for a few minutes in fuming nitric acid, which removes fluorescein and other endogenous fluorophores and produces a very low background. Since fluorescein readily binds to glassware on contact it is important to clean the cuvettes with fuming nitric acid also. A second important contaminate is phenol red, which will be in many culture media and will affect the fluorescent measurements. Fortunately this washes out of the cells readily. Using these techniques we have routinely obtained a background from the cuvettes and chips that is less than 10% of the endogenous background of the cells and most importantly this background is not pH dependent. Without cleaning in fuming nitric acid the fluorescence background of the cuvettes and chips rapidly becomes greater than the fluorescence of cells loaded with probe.

Most cells will have some characteristic or property requiring a specialized culture technique, but these can frequently be accommodated in this method. If the cells are dividing in culture they should be plated so that on the day of measurement they are not confluent. A good density for us has been approximately a 50% covering of the surface area. We routinely sterilize the glass chips with alcohol, flaming, and cooling just before adding the cells and then place the cells directly upon the chip in a small volume of suitable medium. This is incubated in a small humidified culture dish until the cells are attached, when the dish is flooded with medium. The uptake of particles can be done in small volumes added directly to the surface of the chip and washed away rapidly by dilution. Temperature can be rapidly changed by addition of preequilibrated medium. For fluorometric measurements the chip is placed into a lucite square that fits into the top of a standard cuvette and has a groove in it that is slightly wider than the thickness of the glass chip. Using a small rectangle of rubber glove as a gasket, the chip can be wedged into the lucite square and held firmly in place. With a small screw on top of the lucite square as a handle, the cells slide into the cuvette, which contains the appropriate medium. The groove is cut diagonally across the lucite but actually at 49° to the excitation light to reduce scattering and refraction from the glass surface. Changing medium or incubation conditions is rapidly accomplished by sliding the chip out and placing it in another cuvette. If necessary a gas mixture can be perfused through a tube placed in the lucite cap to provide for CO_2-buffered medium. This method is not technically complicated and we have found it to be flexible and useful for studying several types of cells. In order to get accurate and reliable results, the cell culture must be healthy and the most helpful change we have made toward this end is to keep the culture and measurement system simple.

In almost all cases where we have not been able to get accurate measurements of vacuolar pH, the difficulty has been resolved with improvement of the health of the cultured cells. Cells that have been stressed in culture do not take up particles vigorously, do not acidify intracellular compartments well, and display increased levels of endogenous fluorescence (probably a personal subjective observation). One of the ways of accidentally stressing the cells is hyperosmolar stress due to evaporation of medium when transferring the chips from one solution to another. At 37°C the rate of evaporation from a 50- to 200-μm layer of fluid on the surface of a glass chip can be rapid and since the salts remain behind, the osmolarity of this layer can increase rapidly. My impression is that all these transfers must be made with less than 20 sec exposure out of the bulk solution and then some recovery time might be required. Each transfer will also be accompanied by some loss of cells but for most macrophages this will be vey small and probably presents dead or injured cells. However, soon after inoculation with a pathogen when many of the cells are stressed, cell loss can be greater and it is wise to be gentle in washing and transferring the chips. Cells also tend to come off the chip when they overgrow or become very confluent. Chips that are not clean can result in cell loss, but this rarely is a problem if the fuming nitric acid cleaning procedure is used.

As mentioned at the beginning of this section each cell type and experimental goal will have its own requirements, but by keeping the technical level of the system simple we have been able to modify it easily and accommodate a number of different circumstances.

══ V. Conclusion

In closing it seems appropriate to emphasize that *in situ* calibration of the pH response is a powerful, if demanding, way to validate results. It has often been tempting to proceed with measurements using a pathogen after an initial calibration with zymosan or to move to a new cell type without recalibration. Almost every time I have succumbed to this my reward has been to repeat the experiments. On the other hand, I have not had occasion to revise a series of experiments that have been bracketed by careful calibrations. Furthermore, each time we have increased the complexity of the system it has seemed to impair its utility. The cells are sensitive to osmolar stress, which can occur during transfers, and more transfers will increase the possibility of this occurrence. The particles are frequently pathogens and therefore any increase in manipulation greatly increases the precautions that are necessary. Therefore, trying to load with more than one particle is difficult. Finally, the measurements are close to the sensitivity limits of fluorescence and therefore care and cleanliness are important. This approach has been successful in each instance that we have undertaken. Therefore, the method seems to have merit and would appear to be capable of extension to other organisms and to other phagosomal parameters for which there are suitable probes.

Acknowledgment

This work was supported by USPHS Grant AR42370.

References

Bae, H.-R., and Verkman, A. S. (1990). Protein kinase A regulates chloride conductance in endocytic vesicles from proximal tubule. *Nature* (*London*) **348,** 637–639.

Bizal, C. L., Butler, J. P., and Feldman, H. A. (1991). Kinetics of phagocytosis and phagosome-lysosome fusion in hamster lung and peritoneal macrophages. *J. Leukocyte Biol.* **50,** 229–239.

Black, C. M., Paliescheskey, M., Beaman, B. L., Donovan, R. M., and Goldstein, E. (1986). Acidification of phagosomes in murine macrophages: Blockage by nocardia asteroides. *J. Infect. Dis.* **154,** 952–958.

Blair, H. C., and Schlesinger, P. H. (1990). Purification of a stilbene sensitive chloride channel and reconstitution of chloride conductivity into phospholipid vesicles. *Biochem. Biophys. Res. Commun.* **171,** 920–925.

Blair, H. C., Teitelbaum, S. L., Ghiselli, R., and Gluck, S. (1989). Osteoclastic bone resorption by a polarized vacuolar proton pump. *Science* **245,** 855–857.

Brown, D., Sabolic, I., and Gluck, S. (1991). Cholchicin-induced redistribution of proton pumps in kidney epithelial cells. *Kidney Internat.* **40,** 579–583.

Byrd, T. F., and Horwitz, M. A. (1991). Chloroquine inhibits the intracellular multiplication of *Legionella pneumophila* by limiting the availability of iron. A potential new mechanism for the therapeutic effect of chloroquine against intracellular pathogens. *J. Clin. Invest.* **88,** 351–357.

Chatterjee, D., Chakraborty, M., Leit, M., Jamsa-Kellokumpu, S., Fuchs, R., Bartkiewicz, M., Hernando, N., and Baron, R. (1992). The osteoclasts proton pump differs in its pharmacologic and catalytic subunits from other vacuolar H + -ATPases. *J. Exp. Biol.* **172,** 193–204.

Eissenberg, L. G., and Goldman, W. E. (1991). *Histoplasma capsulatum* variation and adaptive strategies for parasitism: New perspectives on histoplasmosis. *Clin. Microbiol. Rev.* **4**, 411–421.

Eissenberg, L. G., Schlesinger, P. H., and Goldman, W. E. (1988). Phagosome-lysosome fusion in P388D1 macrophages infected with histoplasma capsulatum. *J. Leukocyte Biol.* **43**, 483–491.

Eissenberg, L. G., Goldman, W. E., and Schlesinger, P. H. (1993). Histoplasma capsulatum modulates the acidification of phagolysosomes. *J. Exp. Med.* **177**, 1605–1611.

Forgac, M. (1989). Structure and function of vacuolar class of ATP-driven proton pumps. *Physiol. Rev.* **69**, 765–796.

Geisow, M. J., D'Arcy Hart, P., and Young, M. R. (1981). Temporal changes of lysosome and phagosome PH during phagolysosome formation in macrophages: Studies by fluorescence spectroscopy. *J. Cell Biol.* **89**, 645–652.

Glickman, J., Croen, K., Kelly, S., and Al-Awqati, Q. (1983). Golgi membranes contain an electrogenic proton pump in parallel to a chloride conductance. *J. Cell Biol.* **97**, 1303–1308.

Gluck, S., and Caldwell, J. (1987). Immunoaffinit purification and characterization of H + -ATPase from bovine kidney. *J. Biol. Chem.* **262**, 15770–15780.

Hart, P. D., Young, M. R., Gordon, A. H., and Sullivan, K. H. (1987). Inhibition of phagosome-lysosome fusion in macrophages by certain mycobacteria can be explained by inhibition of lysosomal movements observed after phagocytosis. *J. Exp. Med.* **166**, 933–946.

Horwitz, M. A., and Maxfield, F. R. (1984). *Legionella pneumophila* inhibits acidification of its phagosome in human monocytes. *J. Cell Biol.* **99**, 1963–1943.

Jensen, M. S., and Bainton, D. F. (1973). Temporal changes in PH within the phagocytic vacuole of the polymorphonuclear neutrophilic leukocyte. *J. Cell Biol.* **56**, 379–388.

Krogstad, D. J., and Schlesinger, P. H. (1986). A perspective on antimalarial action: Effects of weak bases on *Plasmodium falicparum*. *Biochem. Pharmacol.* **35**, 547–552.

Landry, D. W., Reitman, M., Cragoe, E. J., and Al-Awqati, Q. (1987). Epithelial chloride channel. Development of inhibitory ligands. *J. Gen. Physiol.* **90**, 779–798.

Landry, D. W., Sullivan, S., Nicolaides, M., Redhead, C., Edelman, A., Field, M., Al-Awqati, Q., and Edwards, J. (1993). Molecular cloning and characterization of P64, a chloride channel protein from kidney microsomes. *J. Biol. Chem.* **268**, 14948–14955.

Lukacs, G. L., Chang, X.-B., Kartner, N., Rotstein, O. D., Riordan, J. R., and Grinstein, S. (1992). The cystic fibrosis transmembrane regulator is present and functional in endosomes. Role as a determinant of endosomal PH. *J. Biol. Chem.* **267**, 14568–14572.

Mattsson, J. P., Schlesinger, P. H., Keeling, D. J., Teitelbaum, S. L., Stone, D. K., and Xie, X.-S. (1994). Isolation and reconstitution of a vacuolar-type H + -ATPase from the osteoclast ruffled membrane. *J. Biol. Chem.* (In press).

Mayorga, L. S., Bertini, F., and Stahl, P. D. (1991). Fusion of newly formed phagosomes with endosomes in intact cells and in a cell-free system. *J. Biol. Chem.* **266**, 6511–6517.

Merion, M., Schlesinger, P. H., Brooks, R. M., Moehring, J. M., Moehring, T. J., and Sly, W. S. (1983). Defective acidification of endosomes in Chinese hamster ovary cells: Mutants 'cross-resistant' to toxins and viruses. *Proc. Natl. Acad. Sci. U.S.A.* **80**, 3334–3338.

Moulder, J. W. (1985. Comparative biology of intracellular parasitism. *Microbiol. Rev.* **49**, 298–337.

Mulberg, A. E., Tulk, B. M., and Forgac, M. (1991). Modulation of coated vesicle chloride channel activity and acidification by reversible protein kinase A-dependent phosphorylation. *J. Biol. Chem.* **266**, 20590–20593.

Nilsen, A., Nyberg, K., and Camner, P. (1988). Intraphagosomal PH in alveolar macrophages after phagocytosis in vivo and in vitro of fluorescein-labeled yeast particles. *Exp. Lung Res.* **14**, 197–207.

Ohkuma, S., and Poole, B. (1978). Fluorescence probe measurement of the intralysosomal pH in living cells and the perturbation of pH by various agents. *Proc. Natl. Acad. Sci. U.S.A.* **75**, 3327–3331.

Pitt, A., Mayorga, L. S., Stahl, P. D., and Schwartz, A. L. (1992). Alterations in the protein composition of maturing phagosomes. *J. Clin. Invest.* **90**, 1978–1983.

Pressman, B. C., Harns, E. J., Jagger, W. S., and Johns, J. H. (1967). Antibiotic-mediated transport of alkali ions across lipid barriers. *Proc. Natl. Acad. Sci. U.S.A.* **58,** 1949–1956.

Redhead, C. R., Edelman, A., Brown, D., Landry, D. W., and Al-Awqati, Q. (1992). A ubiquitous 64 KDa protein Ia a component of a chloride channel of plasma membrane and intracellular membranes. *Proc. Natl. Acad. Sci. U.S.A.* **89,** 3716–3720.

Roos, A., and Boron, W. F. (1981). Intracellular pH. *Physiol. Rev.* **61,** 296–434.

Sibley, L. D., Weidner, E., and Krahenbuhl, J. L. (1985). Phagosome acidification blocked by intracellular toxoplasma *Gondii. Nature (London)* **315,** 416–419.

Sturgill-Koszycki, S., Schlesinger, P. H., Chakraborty, P., Haddix, P. L., Collins, H. L., Fok, A. K., Allen, R. D., Gluck, S. L., Heuser, J., and Russell, D. G. (1994). Lack of acidification in *Mycobacterium* phagosomes produced by exclusion of the vesicular proton-ATPase. *Science* **263,** 678–681.

Thomas, J. A. (1986). Intracellularly trapped PH indicators. *Soc. Gen. Physiol. Ser.* **40,** 311–325.

Thomas, J. A., Buchsbaum, R. N., Simmiek, A., and Racker, E. (1979). Intracellular pH measurements to Ehrlich ascites tumor cells utilizing spectroscopic probes generated in situ. *Biochemistry* **18,** 2210–2218.

Tilly, B. C., Mancini, G. M. S., Bijman, J., van Gageldonk, P. G. M., Beeren, C. E. M. T., Bridges, R. J., de Jonge, H. R., and Verheijen, F. W. (1992). Nucleotide activated chloride channels in lysosomal membranes. *Biochem. Biophys. Res. Commun.* **187,** 254–260.

Wang, Y. L., and Goren, M. B. (1987). Differential and sequential delivery of fluorescent lysosomal probes into phagosomes in mouse peritoneal macrophages. *J. Cell Biol.* **106,** 1749–1754.

Xie, X.-S., and Stone, D. K. (1988). Partial resolution and reconstitution of the subunits of the clathrin-coated vesicle proton ATPase responsible for Ca^{2+}-activated hydrolysis. *J. Biol. Chem.* **263,** 9859–9867.

Ymamshiro, D. J., and Maxfield, F. R. (1987). Acidification of morphologically distinct endosomes in mutant and wild-type chinese hamster ovary cells. *J. Cell Biol.* **105,** 2723–2633.

CHAPTER 17

Techniques for Studying Phagocytic Processing of Bacteria for Class I or II MHC-Restricted Antigen Recognition by T Lymphocytes

Clifford V. Harding

Institute of Pathology
Case Western Reserve University
Cleveland, Ohio 44106

I. Introduction
    A. Antigen Processing Pathways
    B. Compartmentalization of Microbes in Host Cells
II. Generating MHC-I- and MHC-II-Restricted T Cells to Detect Antigen Processing
    A. MHC-II-Restricted T Cell Hybridomas
    B. MHC-I-Restricted T Cell Hybridomas
III. Antigen Presentation Assays
    A. Basic Assay for T Cell Recognition of Antigen
    B. Modifications for Testing Presentation of Bacterial Antigens
IV. Observations and Implications
    References

I. Introduction

A. Antigen Processing Pathways

T lymphocytes play a critical role in immunity to most intracellular pathogens. Recognition of foreign antigens by T cells is determined by the clonotypic T cell receptor, which recognizes immunogenic peptides bound to or "presented by" major histocompatibility (MHC) molecules expressed on antigen presenting cells (T cells generally do not directly recognize native protein antigens). There are two different classes of MHC molecules, which follow different intracellular processing pathways to acquire immunogenic peptides derived from protein antigens for presentation (Harding and Geuze, 1993a). Despite many differences (below), both pathways involve the intracellular cleavage of native protein antigens to produce immunogenic peptides for presentation. These antigen processing mechanisms can be bypassed by adding exogenous peptides (produced synthetically or by *in vitro* proteolysis of the native antigen), which are in a sense "preprocessed" antigens that can bind directly to MHC molecules on the cell surface.

Class I MHC (MHC-I) molecules primarily present peptides that are derived from the cytoplasm of the antigen presenting cell, such as viral nucleoproteins, although there are exceptions to this rule that we will discuss below. Antigen catabolism in the host cell cytoplasm produces peptides that are transported across the ER membrane by the TAP peptide transporter and bind to MHC-I molecules in the lumen of the ER prior to their export to the plasma membrane.

In contrast, class II MHC (MHC-II) molecules bind invariant chain in the ER, which prevents the association of peptides until the cleavage of invariant chain, which occurs after the MHC-II molecules are transported to an endocytic compartment. Thus, MHC-II molecules bind peptides derived from antigens that are catabolized in vesicular endocytic or phagocytic compartments of the antigen presenting cell (Harding, 1993).

MHC-I and MHC-II molecules are also recognized by the accessory molecules CD4 and CD8, respectively, such that CD4+ T cells are generally MHC-II restricted and CD8+ T cells are generally MHC-I restricted. CD4+ T cells are referred to as T helper (T_H) cells; they secrete cytokines that promote both cellular and humoral immune reactions. CD8+ cells are often cytolytic; i.e., they can kill cells that present specific peptide-MHC-I complexes (e.g., infected cells presenting a viral antigen); these cells are termed cytolytic T cells (CTL).

Antigen processing can be defined as the mechanisms that convert native protein antigens into peptides bound to MHC molecules, i.e., peptide-MHC complexes, which can be recognized by T cells. In addition to providing the foundation for T cell recognition of microbial antigens, antigen processing represents a fascinating system to study the functions and interplay of intracellular organelles, including various endocytic compartments. For example, processing of antigens for presentation by MHC-II molecules appears to involve functions of late, lysosome-like endocytic compartments, yet peptides are recycled from these compartments to the plasma membrane. This provides a unique tool to examine recycling pathways from late endocytic compartments, which are poorly appreciated by other approaches.

B. Compartmentalization of Microbes in Host Cells

The differential sampling of peptides from the cytoplasm and vesicular compartments by MHC-I and MHC-II, respectively, influences the nature of the pathogenic microbes that are detected by CD4 and CD8 T cells (Pamer, 1993). After invading host cells, many pathogens remain within vacuolar compartments, which they often functionally modify, whereas others escape from vacuolar compartments into the cytoplasm (Table I). For example, viable *Listeria monocytogenes* organisms express listeriolysin O (LLO), which allows them to escape from vacuolar compartments into the cytoplasm after invading host cells and elicit MHC-I-restricted CD8 T cell responses, while nonviable or nonvirulent *Listeria* organisms that do not escape from vacuolar compartments also do not generate MHC-I-restricted T cell responses (Brunt *et al.*, 1990). Thus, a large number of intravacuolar pathogens may be able to escape detection by CD8 cytolytic T cells due to sequestration of their antigens from cytosolic processing mechanisms. However, recent results suggest that an accessory pathway may exist in some host cells to allow MHC-I presentation of antigens from vacuolar organisms (Pfeifer *et al.*, 1993). Evidence for this pathway comes from both *in vitro* studies that utilized techniques described below and from *in vivo* studies (Flynn *et al.*, 1990; Aggarwal *et al.*, 1990).

II. Generating MHC-I- and MHC-II-Restricted T Cells to Detect Antigen Processing

A. MHC-II-Restricted T Cell Hybridomas

The response of antigen-specific T cells provides a sensitive bioassay for the production of peptide–MHC complexes by antigen processing. Antigen-specific, untransformed T cell lines or clones can be isolated following immunization with an antigen and used for antigen processing experiments, but the generation of T cell hybridomas by fusion of such T cells with a transformed cell line provides a much more convenient source of clonal T cells with defined

Table I
Compartmentalization of Some Intracellular Bacteria in Host Cells

Species	Compartmentalization	MHC restriction
Listeria monocytogenes (w.t., LLO+)	Cytoplasm (initially vacuolar)	I, II[a]
Listeria monocytogenes (LLO−)	Vacuolar	II[a]
Salmonella typhimurium	Vacuolar[b]	I[c], II[d]
Mycobacterium tuberculosis	Vacuolar (some cytoplasmic escape may occur)[e]	I, II

[a] Brunt *et al.* (1990).
[b] Pfeifer *et al.* (1992, 1993).
[c] Pfeifer *et al.* (1993); Aggarwal *et al.* (1990); Flynn *et al.* (1990).
[d] Pfeifer *et al.* (1992).
[e] McDonough *et al.* (1993); Myrvik *et al.* (1984).

specificity for this purpose. Furthermore, the responses of untransformed T cells are more highly dependent on "costimulator" molecules in addition to peptide–MHC complexes, and costimulator activity may be destroyed by manipulations such as fixation that are employed in many microbial antigen processing protocols (below), making T cell hybridomas of great advantage for this application. Since the proliferation of T hybridoma cells is no longer antigen-dependent, recognition of peptide–MHC complexes is determined by measuring the increase in interleukin-2 (IL-2) secretion that occurs upon recognition of antigen by these cells.

The generation of MHC-II-restricted murine T cell hybridomas is now well established and described (Kruisbeek, 1992) and will be only summarized here. This procedure involves immunization of a mouse with the protein antigen of interest, subsequent harvesting of regional lymph nodes, and expansion of CD4+ T cells *in vitro* by addition of the protein antigen. An activated antigen-specific bulk T cell line generated in this manner can be fused with BW5147.G1.4 cells (ATCC), an established HAT-sensitive fusion partner derived from a Thy 1.1$^+$ murine lymphoma cell line, which has been extensively used to generate murine T cell hybridomas. Generation of human T cell hybridomas remains problematic due to lack of an established effective fusion partner.

B. MHC-I-Restricted T Cell Hybridomas

The generation of MHC-I-restricted CD8 T cell hybridomas requires the use of additional techniques that have been developed recently. Since soluble antigens do not generally generate MHC-I restricted responses *in vivo* (Collins *et al.*, 1992), alternative approaches are necessary for the initial *in vivo* immunization (Table II). In addition, CD8-transfected BW5147 cells (Burgert *et al.*, 1989) should be used as the fusion partner to ensure effective expression of CD8.

Protocol 1. Generation of MHC-I-restricted T hybridomas specific for ovalbumin.

This summarizes the protocol used for generating T hybridomas specific for the ovalbumin 257-264 epitope presented by the H-2K^b murine MHC-I molecule (Pfeifer *et al.*, 1993). It utilizes liposome-encapsulated antigen to elicit MHC-I-restricted T cells.

1. Inject two or three mice (e.g., C57BL/6, H-2^b) iv with 0.1 mg liposome-encapsulated antigen (e.g., ovalbumin). See Protocol 2 for preparation of liposome-encapsulated antigen.

2. After 7–10 days sacrifice the mice, remove their spleens, and prepare a suspension of splenocytes (by teasing with forceps, lightly grinding between sterile glass slides or light homogenization in a sterile loose Douncc homogenizer).

Table II
Approaches for Generating MHC-I-Restricted CD8 T Cell Lines

Approach	Reference
Live virus	
Live bacteria[a]	Flynn *et al.* (1990); Aggarwal *et al.* (1990); Brunt *et al.* (1990); Pfeifer *et al.* (1993); Wick *et al.* (1993)
Liposome-encapsulated antigen	Harding *et al.* (1991a); Collins *et al.* (1992); Reddy *et al.* (1991); Zhou *et al.* (1992); Nair *et al.* (1992)
ISCOMS[b]-associated antigen	Takahashi *et al.* (1990); Mowat and Reid (1992)
Cell-associated antigen	Carbone and Bevan (1990)
Immunization with peptide or lipopeptide	Deres *et al.* (1989); Carbone and Bevan (1989)

[a] May be effective with vacuolar bacteria, e.g., *Salmonella typhimurium*, as well as those that escape into the cytosol, e.g., *Listeria monocytogenes*.
[b] Immune stimulatory complexes.

3. Incubate approximately 10^7 splenocytes with irradiated stimulator cells that will present the appropriate antigen on MHC-I molecules at a 20:1 ratio (splenocyte:stimulator cell). Effective means for generating MHC-I restricted presentation *in vitro* are summarized in Table III. For example, the stimulator cells may be cells transfected to express antigen (Moore *et al.*, 1988), cells with antigen introduced for MHC-I processing and presentation by electroporation (Harding, 1992; Chen *et al.*, 1993) or osmotic lysis of endosomes (Moore *et al.*, 1988), or cells exposed to immunogenic peptide that can bind to surface MHC-I molecules. This and subsequent steps should be done with appropriate medium, such as DMEM (GIBCO) 12100-061), which we supplement with 10%

Table III
Approaches for *in Vitro* Analysis or Expansion of CD8 T Cells[a]

Approach	Reference
Osmotic lysis of endosomes	Moore *et al.* (1988)
Acid-sensitive liposomes	Harding *et al.* (1991a); Reddy *et al.* (1991); Nair *et al.* (1992)
Transfection of stimulator cells	Moore *et al.* (1988)
Stimulator cells primed with synthetic peptide	
Bacterial antigen	Pfeifer *et al.* (1993)
Particle-conjugated antigen	Kovacsovics-Bankowski *et al.* (1993)

[a] Approaches listed above for *in vivo* use are also generally effective *in vitro*. Acid-sensitive liposomes should be used for efficient *in vitro* processing of liposome-encapsulated antigens.

FCS (Hyclone), $5 \times 10^{-5} M$ 2-ME, L-arginine HCl (116 mg/liter), L-asparagine (36 mg/liter), NaHCO$_3$ (2 g/liter), sodium pyruvate (1 mM), and antibiotics.

4. Replenish medium as necessary over the next 5 days. At this point the cells can be stimulated with IL-2 and fused with CD8-transfected BW5147 cells (technique as for MHC-II-restricted hybridomas), or they can be further passed on stimulator cells as indicated.

Protocol 2. A simple approach for preparation of liposome-encapsulated antigens using the dehydration–rehydration technique.

A number of approaches can be used for generating liposome-encapsulated antigens (Gregoriadis, 1990). We have used the detergent dialysis method (Harding *et al.,* 1991b) and the dehydration–rehydration method (Collins *et al.,* 1992). The following protocol is a very simple version of the latter, requiring a minimum of specialized equipment.

This protocol is for the preparation of liposomes composed of dioleoylphosphatidylcholine (DOPC) and dioleoylphosphatidylserine (DOPS), which form liposomes that are largely insensitive to varying pH within physiological ranges (i.e., "acid-resistant"). Acid-sensitive liposomes can be similarly prepared, formulated using phosphatidylethanolamine, which by itself does not form bilayer membranes under physiological conditions, combined with a proton-titratable stabilizing lipid, e.g., palmitoylhomocysteine (PHC) or cholesteryl hemisuccinate (CHEMS) (Table IV). Most of the above lipids are commercially available (DOPC, DOPS, and DOPE from Avanti Polar Lipids, CHEMS from Sigma). Acid-sensitive liposomes have advantages for *in vitro* delivery of antigen for MHC-I presentation (Harding *et al.,* 1991a). Both acid-sensitive and acid-resistant liposomes can be used *in vivo* to elicit MHC-I-restricted CD8 T cell responses (Collins *et al.,* 1992).

Table IV
Properties of Liposomes with Different Membrane Compositions

	Liposome membrane composition[a]	
Property	PC/PS	PE/CHEMS[b]
pH sensitivity	Acid-resistant	Acid-sensitive
Site of antigen delivery	Lysosomes	Early endosomes, cytosol[c]
In vitro antigen processing	MHC-II	MHC-II, MHC-I
In vivo antigen processing	MHC-II, MHC-I	MHC-II, MHC-1

[a] PC, phosphatidylcholine; PS, phosphatidylserine; PE, phosphatidylethanolamine; CHEMS, cholesteryl hemisuccinate.

[b] Acid-sensitive liposome formulations include PE/CHEMS (Ellens *et al.,* 1984) and PE/succinylglycerol (Collins *et al.,* 1990).

[c] For PE/CHEMS liposomes, cytosolic delivery is estimated to be 0.01–10% (Chu *et al.,* 1990).

1. Combine in a glass tube 8 μmol DOPS in chloroform (628 μl at 10 mg/ml) with 2 μmol DOPS in chloroform (166 μl at 10 mg/ml).

2. Dry under nitrogen gas.

3. Centrifuge under vacuum for 20–30 min in Speedvac (Savant) or similar apparatus.

4. Add 180 μl of 0.2× PBS/pH 7.9/EDTA (20% PBS in water, pH 7.9, 0.2 mM EDTA).

5. Sonicate well to obtain a milky suspension.

6. Add the antigen to be entrapped in distilled water (e.g., 300 μl at 10 mg/ml) and vortex. In order to monitor trapping efficiency a trace of radioactive antigen should be employed; this should be thoroughly mixed with the unlabeled antigen prior to addition to the liposome suspension.

7. Freeze in a dry ice/ethanol bath.

8. Lyophilize overnight.

9. Add 46 μl distilled water (the volume here must be 20% of the volume of 0.2× PBS/pH 7.9/EDTA used above to restore isotonicity) and vortex thoroughly. Trapping will occur upon rehydration, and trapping efficiency is increased by minimizing the volume at this step.

10. Add 400 μl PBS, vortex.

11. Measure total volume and remove 10 μl to determine radioactivity, calculate total cpm in tube.

12. Filter through presterilized 200-nm polycarbonate Nuclepore filters, 13-mm diameter (Costar) (the filters and the syringe holder can be sterilized by autoclaving).

13. Pellet in a Beckman airfuge (30 min at 30 psi).

14. Remove supernatant and resuspend in PBS (e.g., 300 μl).

15. Count an aliquot and determine total liposomal radioactivity. Compare with step 11 to determine trapping efficiency and calculate the amount of antigen in the liposome preparation. Trapping efficiency varies with different antigens and preparations, but is often 5–10%.

III. Antigen Presentation Assays

A. Basic Assay for T Cell Recognition of Antigen

Antigen recognition by untransformed CD4+ T cell clones results in activation with proliferation and the secretion of cytokines, either of which can serve as the basis of an assay for class II antigen processing and presentation by MHC-II molecules. Antigen recognition by CD8+ T cell clones is often determined by a cytotoxicity assay, which measures lysis of antigen presenting "target" cells. Thus, cytotoxicity assays are often used as a measure of class I antigen processing and presentation by MHC-I molecules. As mentioned above, the use of T cell hybridomas instead of untransformed T cell clones provides many technical advantages. Since T cell hybridomas proliferate constitutively and often exhibit poor cytolytic activity (even if CD8 +), antigen recognition must be determined by cytokine secretion when these cells are used, as described below. T cell hybridomas, including CD8 + MHC-I-restricted hybridomas, usually secrete interleukin-2 (IL-2) upon antigen recognition, and this cytokine secretion can be quantitated as a measure of antigen presentation.

IL-2 secretion can be quantitated by two techniques. A commercially available kit (Genzyme) allows quantitative determination IL-2 levels by ELISA. A frequently used alternative is an IL-2 bioassay, which depends on the proliferation and incorporation of [^{3}H]thymidine by IL-2-dependent cells, e.g., CTLL-2 cells (ATCC). The ELISA assay is quicker, but more expensive, especially if large numbers of samples must be processed. The ELISA assay also easily provides specificity for the cytokine (IL-2). Cytokine specificity can be achieved in the bioassay system as well by the use of an appropriate cell line maintained under proper conditions and the use of appropriate controls (Bottomly et al., 1991). For the purposes of antigen processing assays, the limitations of the CTLL-2 bioassay are of little consequence, and it allows the processing of large numbers of samples.

We will first describe a basic assay that can be used to determine the processing of protein antigens or nonviable microbes, as well as the presentation of immunogenic peptides (this assay does not require fixation of the antigen-presenting cells). This protocol represents two nested bioassays: (1) A T cell bioassay for the level of peptide–MHC complexes that generates samples with varying concentrations of IL-2. (2) The CTLL-2 bioassay for IL-2, which is described elsewhere (Bottomly et al., 1991; Miller, 1991). The following section will demonstrate how this assay can be adapted for other purposes, including the quantitation of the processing of viable microbes.

Protocol 3. A continuous processing or nonfixation assay for antigen processing.

This is the simplest antigen processing and presentation assay. It involves just the combination of antigen-presenting cells, antigen (e.g., protein or heat-killed microbes), and antigen-specific T hybridoma cells. If particulate antigens, such as bacteria, are used, macrophages should be used as antigen-presenting cells (see below). Otherwise, transformed B cell lines are generally the most convenient source of antigen-presenting cells. In any case, the antigen-presenting cell must express the appropriate MHC allele that restricts the response of the T cell hybridoma employed.

1. In flat-bottom 96-well plates (e.g., Costar) combine the following in a total of 0.2 ml/well. Antigen-presenting cells: 5×10^4–2×10^5/well (use fewer cells with rapidly proliferating cells than with more quiescent cells). T hybridoma

cells: 10^5/well. Antigen: generally a range of doses is employed; include appropriate negative (no antigen) and positive (e.g., synthetic immunogenic peptide, if available) controls.

2. Incubate 20–24 hr at 37°C.

3. Harvest an aliquot (e.g., 0.1 ml) of the supernatants for the CTLL-2 bioassay (CTLL-2 cells may be set up with multiple dilutions of the supernatants for quantitative determination of IL-2 levels). Freeze the supernatant to kill any viable cells therein before the CTLL-2 assay (supernatants can also be stored at -20°C until the CTLL assay is performed).

4. Run CTLL-2 bioassay as described (Bottomly *et al.*, 1991), adding 100 μl standard medium containing CTLL cells at 5×10^4/ml (5×10^3/well). Include negative (CTLL-2 cells plus normal medium) and positive (CTLL-2 cells plus medium containing IL-2) controls. Incubate 24 hr. Add 20 μl of medium containing [methyl-^{3}H]thymidine (New England Nuclear, NET-027A) at 20 μCi/ ml. Incubate overnight and then harvest plate (e.g., with a Skatron harvester) or freeze at -20°C until harvesting. Count samples in a scintillation counter.

B. Modifications for Testing Presentation of Bacterial Antigens

Protocol 4. Fixation assay for the study of microbial antigen processing.

The fixation assay involves the exposure of antigen-processing cells to antigen in one incubation and the subsequent fixation of these cells, which halts all further processing and leaves only the cell surface peptide–MHC complexes that were previously generated available for recognition by T cells. Thus, the antigen-processing incubation and the T cell bioassay are separate steps in this protocol. This allows the investigator to control the kinetics of antigen processing independent of the 20- to 24-hr T cell bioassay. It also allows the use of pharmacologic agents to perturb antigen-processing events, followed by their removal after fixation to avoid artifacts of their toxicity on the T cell bioassay. An additional advantage for microbial processing experiments is that the fixation step kills microbes, sterilizing the well, so that contamination during the subsequent T cell and CTLL incubations is not a problem. In this case an appropriate positive control is still immunogenic peptide, which can be added after fixation during the T cell incubation. For the fixation assay it is advantageous to use adherent antigen-presenting cells (e.g., macrophages), since the extensive washing steps can then be simply executed. The use of nonadherent antigen-presenting cells is also possible, but this necessitates the centrifugation of the cells at each wash step.

1. Add antigen (bacteria) to adherent antigen presenting cells in 96-well plates (for macrophages adhere 2×10^5 cells/well for 2 hr and then wash) in a volume of 0.1 ml or to nonadherent antigen-presenting cells in sterile tubes. Bacteria may be centrifuged onto the cells to enhance the efficiency of uptake (1000 $\times$

g for 5–10 min) (Pfeifer *et al.*, 1992,1993). It may be important to use an antibiotic-free medium during all incubations with viable microbes.

2. Incubate at 37°C (time may vary, 1–2 hr is often sufficient).

3. Wash cells in standard medium, e.g., DMEM without serum (most extracellular bacteria may be largely removed from nonadherent antigen-presenting cells by washing with centrifugation at lower speeds (e.g., 300–400 $\times$ g for 5–7 min).

4. Fix in 1% paraformaldehyde in PBS (in practice, many people add an equal volume of 2% paraformaldehyde in PBS to medium left in the well at the end of the last wash. Caution must be employed to avoid spattering fixative or generating drops of fixative clinging to the top of the wells, since this fixative may not be removed with later washing and will kill the T cells.

5. Incubate with fixative for 10–15 min at room temperature.

6. Wash once with standard medium. It may be advisable to perform this and subsequent washing steps with a larger volume than employed during the fixation step, to remove thoroughly all fixative from the upper portions of the wells.

7. Add lysine wash solution composed of 1 part 0.4 M lysine, pH 7.4, and 1 part standard medium (lysine provides a convenient source of amino groups to react with remaining paraformaldehyde). Incubate 20 min at room temperature.

8. Wash three to four times more with standard medium (for nonadherent cells in tubes this may be decreased to two to three times, since each wash may be more complete).

9. Replace medium with standard medium with serum.

10. Add T cells and proceed as indicated in the preceding protocol.

Comment: Generating antigen-presenting cells

The protocols described above have been developed primarily for use with adherent macrophages, but they may also be used with other adherent cells or nonadherent cells with the indicated modifications. It is important to consider the conditions to achieve adequate expression of the appropriate MHC molecules. Although B cell lines constitutively express MHC-II molecules, MHC-II levels are regulated on macrophages. Thus, macrophages must be appropriately activated prior to MHC-II antigen-processing experiments. MHC-I is constitutively expressed by most cells, including macrophages and B cells, but treatment with interferon-γ (below) increases its level of expression. Activation, e.g., of macrophages, also alters a myriad of functions important to microbial defense mechanisms, which may also affect antigen-processing mechanisms.

For studies with macrophages elicited peritoneal macrophages (Harding and Geuze, 1992) or bone marrow macrophages (Brunt *et al.*, 1990) may be used. Activated peritoneal macrophages can be elicited by ip injection of *Listeria monocytogenes* (Harding and Geuze, 1992) or concanavalin A (Sigma Type IV,

C2010, 0.1 mg/mouse 4 days prior to harvesting). Prior to plating for adherence the macrophages should be kept on ice to prevent losses due to adherence to the walls of tubes). In addition, macrophages elicited by these or other methods can be activated *in vitro* (Harding and Geuze, 1992) by overnight incubation with 10 ng/ml (100 U/ml) recombinant murine interferon-γ (Genzyme).

Comment: Methods for testing presentation of bacterial antigens

The previous approaches can be used to study the processing of microbes when T cell lines that respond to microbial antigens are available or can be generated (above). It is not always necessary to generate T cells to natural microbial antigens. For some purposes, one can study the processing of a recombinant antigen expressed in a microbial strain. This approach has been used to study both MHC-I processing (Pfeifer *et al.*, 1993; Wick *et al.*, 1993) and MHC-II processing (Pfeifer *et al.*, 1992; Wick *et al.*, 1993) *in vitro* using the techniques described above. An example is given in Fig. 1, which shows processing and presentation of a Crl-OVA fusion protein expressed in *Escherichia coli*. Recombinant antigens are also important for *in vivo* studies of bacterial vaccine vectors (Flynn *et al.*, 1990; Aggarwal *et al.*, 1990).

The preparation of microbes for antigen-processing experiments must be

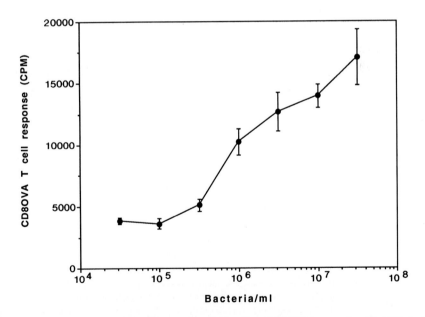

Fig. 1 A dose–response study of processing of *E. coli* strain HB101 expressing Crl-OVA fusion protein. The Crl-OVA fusion protein contains the 257–264 OVA epitope that is recognized by CD8OVA T hybridoma cells (Pfeifer *et al.*, 1993). This experiment was performed as described in Protocol 4, using different bacterial titers in a well volume of 0.1 ml.

planned to generate a suspension of viable organism. Bacteria can be grown on agar plates or in liquid medium, but in any case fresh cultures should be employed. Nonviable organisms can also be employed, such as heat-killed organisms (Harding and Geuze, 1992). Organisms should be resuspended in medium that is devoid of antibiotics and that is compatible with both the microbes and the host cell.

IV. Observations and Implications

The methods described above can be adapted to study the processing of various microbes by different antigen-presenting cells. It is often useful to complement these studies with other approaches, some of which are detailed in this volume, that can further define the biochemical or cell biological mechanisms involved in the processing pathway. For example, our studies (Harding and Geuze, 1992,1993b) have combined antigen-processing experiments with immunoelectron microscopy to define of the distribution of MHC-II molecules and the role of phagolysosomal compartments in macrophages during the processing of *Listeria monocytogenes* for MHC-II presentation (these studies employed heat-killed *Listeria* organisms). We have also employed conventional transmission electron microscopy to define the compartments in macrophages that are targeted by bacteria after phagocytic uptake during the processing of *E. coli* and *Salmonella typhimurium* for both MHC-I (Pfeifer *et al.*, 1993) and MHC-II (Pfeifer *et al.*, 1992) presentation. This combination of approaches allows the investigation of roles of subcellular compartments in different processing mechanisms. One interesting result has been the definition of a novel phagocytic pathway for processing of vacuolar bacteria (i.e., exogenous bacteria that do not reach the cytoplasm, such as *E. coli* or *Salmonella*) for MHC-I presentation, which may employ alternative processing mechanisms distinct from the conventional MHC-I processing pathway. This pathway is likely to be of practical therapeutic importance, since it appears to provide the mechanism whereby *Salmonella* vaccine vectors can induce protective CD8+ T cell responses.

References

Aggarwal, A., Kumar, S., Jaffe, R., Hone, D., Gross, M., and Sadoff, J. (1990). Oral *Salmonella*: Malaria circumsporozoite recombinants induce specific CD8+ cytotoxic T cells. *J. Exp. Med.* **172,** 1083–1090.

Bottomly, K., Davis, L. S., and Lipsky, P. E. (1991). Measurement of human and murine interleukin 2 and interleukin 4. *In* "Current Protocols in Immunology" (J. E. Coligan, A. M. Kruisbeek, D. H. Margulis, E. M. Shevach, and W. Strober, eds.), pp. 6.3.1–6.3.12. Wiley, New York.

Brunt, L. M., Portnoy, D. A., and Unanue, E. R. (1990). Presentation of Listeria monocytogenes to CD8+ T cells requires secretion of hemolysin and intracellular bacterial growth. *J. Immunol.* **145,** 3540–3546.

Burgert, H.-G., White, J., Weltzien, H.-U., Marrack, P., and Kappler, J. W. (1989). Reactivity of Vβ17a$^+$ CD8$^+$ T cell hybrids. Analysis using a new CD8$^+$ T cell fusion partner. *J. Exp. Med.* **170**, 1887–1904.

Carbone, F. R., and Bevan, M. J. (1989). Induction of ovalbumin-specific cytotoxic T cells by in vivo peptide immunization. *J. Exp. Med.* **169**, 603–612.

Carbone, F. R., and Bevan, M. J. (1990). Class I-restricted processing and presentation of exogenous cell-associated antigen in vivo. *J. Exp. Med.* **171**, 377–387.

Chen, W., Carbone, F. R., and McCluskey, J. (1993). Electroporation and commercial liposomes efficiently deliver soluble protein into the MHC class I presentation pathway. Priming in vitro and in vivo for class I-restricted recognition of soluble antigen. *J. Immunol. Methods* **160**, 49–57.

Chu, C.-J., Dijkstra, J., Lai, M.-Z., Hong, K., and Szoka, F. C. (1990). Efficiency of cytoplasmic delivery by pH-sensitive liposomes to cells in culture. *Pharm. Res.* **7**, 824–834.

Collins, D. S., Litzinger, D. C., and Huang, L. (1990). Structural and functional comparisons of pH-sensitive liposomes composed of phosphatidyl-ethanolamine and three different diacylsuccinylglycerols. *Biochim. Biophys. Acta* **1025**, 234–242.

Collins, D. S., Findlay, K., and Harding, C. V. (1992). Processing of exogenous liposome-encapsulated antigens in vivo generates class I MHC-restricted T cell responses. *J. Immunol.* **148**, 3336–3341.

Deres, K., Schild, H., Wiesmüller, K.-H., and Rammensee, H.-G. (1989). In vivo priming of virus-specific cytotoxic T lymphocytes with synthetic lipopeptide vaccine. *Nature (London)* **343**, 561.

Ellens, H., Bentz, J., and Szoka, F. C. (1984). Acid-induced destabilization of phosphatidylethanolamine containing liposomes: Role of biolayer contact. *Biochemistry* **23**, 1532–1538.

Flynn, J. L., Weiss, W. R., Norris, K. A., Siefert, H. S., Kumar, S., and So, M. (1990). Generation of a cytotoxic T-lymphocyte response using a *Salmonella* antigen-delivery system. *Mol. Microbiol.* **4**, 2111–2118.

Gregoriadis, G. (1990). Immunological adjuvants: A role for liposomes. *Immunol. Today* **11**, 89–97.

Harding, C. V. (1992). Electroporation of exogenous antigen into the cytosol for antigen processing and class I MHC presentation. *Eur. J. Immunol.* **22**, 1865–1869.

Harding, C. V. (1993). Cellular and molecular aspects of antigen processing and the function of class II MHC molecules. *Am. J. Respir. Cell Mol. Biol.* **8**, 461–467.

Harding, C. V., and Geuze, H. J. (1992). Class II MHC molecules are present in macrophage lysosomes and phagolysosomes that function in the phagocytic processing of Listeria monocytogenes for presentation to T cells. *J. Cell Biol.* **119**, 531–542.

Harding, C. V., and Geuze, H. J. (1993a). Antigen processing and intracellular traffic of antigens and MHC molecules. *Curr. Opin. Cell Biol.* **5**, 596–605.

Harding, C. V., and Geuze, H. J. (1993b). Immunogenic peptides bind to class II MHC molecules in an early lysosomal compartment. *J. Immunol.* **151**, 3988–3998.

Harding, C. V., Collins, D. S., Kanagawa, O., and Unanue, E. R. (1991a). Lysosomes process liposome-encapsulated antigens for class II MHC presentation, while cytosolic antigen delivery engenders class I MHC presentation. *J. Immunol.* **147**, 2860–2863.

Harding, C. V., Collins, D. S., Slot, J. W., Geuze, H. J., and Unanue, E. R. (1991b). Liposome-encapsulated antigens are processed in lysosomes, recycled and presented to T cells. *Cell (Cambridge, Mass.)* **64**, 393–401.

Kovacsovics-Bankowski, M., Clark, D., Benacerraf, B., and Rock, K. L. (1993). Efficient major histocompatibility complex class I presentation of exogenous antigen upon phagocytosis by macrophages. *Proc. Natl. Acad. Sci. U.S.A.* **90**, 4942–4946.

Kruisbeek, A. M. (1992). Production of mouse T cell hybridomas. *In* "Current Protocols in Immunology" (J. E. Coligan, A. M. Kruisbeek, D. H. Margulis, E. M. Shevach, and W. Strober, eds.), pp. 3.14.1–3.14.11. Wiley, New York.

McDonough, K. A., Kress, Y., and Bloom, B. R. (1993). Pathogenesis of tuberculosis: Interaction of *Mycobacterium tuberculosis* with macrophages. *Infect. Immun.* **61**, 2763–2773.

Miller, R. A. (1991). Quantitation of functional T cells by limiting dilution. *In* "Current Protocols in Immunology" (J. E. Coligan, A. M. Kruisbeek, D. H. Margulis, E. M. Shevach, and W. Strober, eds.), pp. 3.15.5–3.15.6. Wiley, New York.

Moore, M. W., Carbone, F. R., and Bevan, M. J. (1988). Introduction of soluble protein into the class I pathway of antigen processing and presentation. *Cell (Cambridge, Mass.)* **54**, 777–785.

Mowat, A. McI., and Reid, G. (1992). Preparation of immune stimulating complexes (ISCOMS) as adjuvants. *In* "Current Protocols in Immunology" (J. E. Coligan, A. M. Kruisbeek, D. H. Margulis, E. M. Shevach, and W. Strober, eds.), pp. 2.11.1–2.11.12. Wiley, New York.

Myrvik, Q. N., Leake, E. S., and Wright, M. J. (1984). Disruption of phagosomal membranes of normal alveolar macrophages by the H37Rv strain of *Mycobacterium tuberculosis*. A correlate of virulence. *Am. Rev. Respir. Dis.* **129**, 332–328.

Nair, S., Zhou, F., Reddy, R., Huang, L., and Rouse, B. T. (1992). Soluble proteins delivered to dendritic cells via pH-sensitive liposomes induce primary cytotoxic T lymphocyte responses in vitro. *J. Exp. Med.* **175**, 609–612.

Pamer, E. G. (1993). Cellular immunity to intracellular bacteria. *Curr. Opin. Immunol.* **5**, 492–496.

Pfeifer, J. D., Wick, M. J., Russell, D., Normark, S. J., and Harding, C. V. (1992). Recombinant E. coli express a defined, cytoplasmic epitope that is efficiently processed in macrophage phagolysosomes for class II MHC presentation to T lymphocytes. *J. Immunol.* **149**, 2576–2584.

Pfeifer, J. D., Wick, M. J., Roberts, R. L., Findlay, K., Normark, S. J., and Harding, C. V. (1993). Phagocytic processing of bacterial antigens for class I MHC presentation to T cells. *Nature (London)* **361**, 359–362.

Reddy, R., Zhou, F., Huang, L., Carbone, F., Bevan, M., and Rouse, B. T. (1991). pH sensitive liposomes provide an efficient means of sensitizing target cells to class I restricted CTL recognition of a soluble protein. *J. Immunol. Methods* **141**, 157–163.

Reddy, R., Zhou, F., Nair, S., Huang, L., and Rouse, B. T. (1992). In vivo cytotoxic T lymphocyte induction with soluble proteins administered in liposomes. *J. Immunol.* **148**, 1585–1589.

Takahashi, H., Takeshita, T., Morein, B., Putney, S., Germain, R. N., and Berzofsky, J. A. (1990). Induction of CD8$^+$ cytotoxic T cells by immunization with purified HIV-1 envelope protein in ISCOMs. *Nature (London)* **344**, 873–875.

Wick, M. J., Pfeifer, J. D., Findlay, K. F., Harding, C. V., and Normark, S. J. (1993). Compartmentalization of epitopes expressed in *E. coli* has only a minor influence on the efficiency of phagocytic processing for class I and class II presentation to T cells. *Infect. Immun.* **61**, 4848–4856.

Zhou, F., Rouse, B. T., and Huang, L. (1992). Induction of cytotoxic T lymphocytes in vivo with protein antigen entrapped in membranous vehicles. *J. Immunol.* **149**, 1599–1604.

INDEX

A

ADC enrichment recipe, 109
Antigens
  processing
    assays, 319–324
    generating T cells to detect, 315–319
      MHC-I-restricted, 316–319
      MHC-II-restricted, 315–316
    pathways, 313–314
    bacterial, testing presentation of, 321–324

B

Bacteria
  adhesion, 165–188
    to cells in culture, 169–172
      assay for, 170–172
    to glycoproteins adsorbed onto
      hydroxyapatite, 180
    quantitation in solid phase using ELISA,
      179–180, *181*
    in vitro assays, 175–185
  binding
    to immobilized receptors in solid phase,
      175–180
    to receptors in solution, 181–184
      binding of soluble proteins to bacteria,
        182–183
      determination of relative binding
        efficiency of soluble
        glycoconjugates, 183–184
  colonization of cultured cells by, assay,
    170–172
  host receptor distribution for, *in situ*
    screening, 166–169
    labeling procedure, 166–168
    tissue section overlay assay, 168–169
  inhibition experiments *in situ*, 174–175
  phagocytic processing for class I/II MHC-
    restricted antigen recognition by T
    lymphocytes, 313–324
  receptors for, *in situ,* biochemical
    characterization of molecular nature of,
    172–174

Bacterial antigens, testing presentation of,
  321–324
Bacterial glycolipid receptors, detection of, by
  HPTLC overlay, 176–178
Bacterial protein/glycoprotein receptors,
  detection of, in Western blots, 176
Biosafety considerations, for microbial
  pathogens, 1–2
Bradyzoites, *T. gondii,* 30

C

CD36, as ligand for *Plasmodium falciparum*-
  infected erythrocytes, 196
Cell lysis, in isolating phagosomes from
  macrophages, 266–269
Chemical mutagenesis, of mycobacterium,
  118–119
Chloride channel, in phagosome membrane,
  298
Cloning
  of adhesin as translational fusion, 186–187
  by complementation of *Toxoplasma gondii,*
    57–58
Conjugation, in mycobacteria genetics, 115
Culture media
  for growth of bone marrow-derived
    macrophages, 132–133
  for murine macrophage, 131–133
  for mycobacterium, 109–111
Cytoadherence
  Plasmodium falciparum-infected
    erythrocyte, *see Plasmodium*
    falciparum-infected erythrocyte,
    cytoadherence
  as *in vitro* model of sequestration, 197–207
    assay conditions, 200–204
    example, 204–207
    incubation buffer composition, 200
    parasite lines, 197–198
    physical agitation, 200–201
    quantitation of cytoadherence, 201–204
    target cells, 199
    temperature, 200

Cytometry, flow, quantitation of MCH class II antigen expression in murine macrophage, 145

D

Dipeptides, diverse, hemolytic effects, 244–245
DNA plasmid transformation, in mycobacterium genetics, 115–117

E

ELAM-1, as ligand for *Plasmodium falciparum*-infected erythrocytes, 195
Electroporation
in *Leishmania* transfection experiments, 71–73
of plasmid DNS into mycobacteria, 116–117
Endosomal trafficking
in macrophages infected with microbial pathogens, immunoelectron microscopy, 277–288
pathways, intersection with, 279–280
Environmental factors, local, phagosomal pH and, 302–304
Erythrocytes
in culture of malaria parasites, 10
Plasmodium falciparum-infected cytoadherence, *see Plasmodium falciparum*-infected erythrocyte, cytoadherence
sheep, opsonized, phagocytosis, 151–152
Eukaryotic cell glycoconjugates, probing, with purified bacterial adhesins, 185–187

F

Flexible film isolators, in maintenance of immunodeficient strains, 135
Flow cytometry, quantitation of MCH class II antigen expression in murine macrophage by, 145
Fluorescent determination, of intravesicular pH, 302, 303
Fluorescent labeling, of ligand-coated particles, 152–153

G

Gametocyte production, in malaria parasite culture, 18–20

Gene knockouts, in *Toxoplasma gondii* transformation, 57
Gene replacement
in mycobacterium, 120–121
perfect, in *Toxoplasma gondii* transformation, 56
Genetic tools, for future, 58–59
Glycoconjugates
eukaryotic cell, probing, with purified bacterial adhesins, 185–187
soluble, determining relative binding efficiency of, 183–184
Glycoproteins, adsorbed onto hydroxyapatite, bacterial adhesion, 180

H

Histoplasma, phagosomal pH measurement, 304
Homologous recombination
in mycobacterium, 120–121
in *Toxoplasma gondii* transformation, 55
Host cells, compartmentalization of microbes, 314–315

I

ICAM-1, as ligand for *Plasmodium falciparum*-infected erythrocytes, 194–195
Immunoelectron microscopy
of endosomal trafficking in macrophages infected with microbial pathogens, 277–288
processing infected macrophages for, 280–281
Immunofluorescence, indirect
in analysis of phagocytic activity of murine macrophage, 141–142
visual assessment of MCH class II antigen expression in murine macrophage by, 143, 145
Ingestion assays, for phagocytosis, 153–155
Intracellular vesicle/membrane system, of phagocytic cell, 292–296
Intravesicular acid pH generation, 296–298

K

Kupffer cell harvest, 138

L

Laminar flow rack housing, in maintenance of immunodeficient strains, 135

Legionella pneumophila
  intracellular growth mutants, thymineless death enrichment, 249–256
  intracellular survival, 247–258
  laboratory cultivation, 248
  mutants
    enrichment of, from mixed bacterial population, 254–256
    intracellular growth, identification from enriched bacterial pools with "poke plaque" assays, 256–257
  reconstruction studies, 251–253
    genetic manipulation prior to, 250
  tissue culture of UP37 cell-derived macrophages and, 248–249

Leishmania
  phagosomes, 263
    isolation, 269
  transfection experiments with, 65–76
    assay of reporter enzymes in transient transfectants in, 74–76
    electroporation, 71–73
    growth of parasites, 70–71
    plating transfected parasites, 74
    reporter enzymes, 70
    selective plate preparation, 73–74
    with stable transfection, 66–67
    stable transformants selection, 73
    transient assays, 70
    vectors, 67–69

Leukocytes, human peripheral blood, phagocytic function, 147–161

Ligands, for adherence of *P. falciparum*-infected erythrocytes, 194–197

Lymphocytes, T
  in antigen recognition, basic assay, 319–321
  studying phagocytic processing of bacteria for class I/II MHC-restricted antigen recognition by, 313–324

M

Macrophages
  infected with microbial pathogens
    endosomal trafficking
      blocking cryosections and incubation with primary antiserum, 281
      controls, 283–284
      final preparation of grids, 284–286

fluid-phase, routine protocol for analysis, 286–287
      gold-conjugated second antibodies, 282–283
      immunoelectron microscopy, 277–288
      host-pathogen interplay, 278–279
      processing for immunoelectron microscopy, 280–281
  monocyte-derived, assays, 150–151
  murine, *see* Murine macrophage
  for pathogen-containing phagosome study, 264–265
  UP37 cell-derived, tissue culture, 248–249

MADCTW agar recipe, 109

MADCTW liquid recipe, 110

Malaria parasites, *see also Plasmodium falciparum*
  cloning methods, 20
  cryopreservation, 20–21
  cultivation, 7–22
    erythrocytes, 10
    gametocyte production, 18–20
    serum, 8, 10
    serum replacement, 21
    synchronization, 16–18
  culture medium, 8
  culture systems, 10–16
    disk or flask, with manual change of medium, 10–12
    semiautomated, 12–16
  erythrocytic stages, 21–22
  life cycle, preerythrocytic and sporogonic stages of, *in vitro* development of, 22
  Plasmodium falciparum, axenic development of erythrocytic stages, 21

Marker rescue, in *Toxoplasma gondii* transformation, 53

Merozoites, *Plasmodium falciparum, see Plasmodium falciparum*, merozoites of

Microbes, compartmentalization in host cells, 314–315

Microbial pathogens
  biosafety considerations, 1–2
  culture, 3–4
  Department of Health and Human Services guidelines, 2
  federal and institutional requirements, 2–3
  maintenance, 3–4
  obtaining and maintaining, 1–4
  source, 3

Microisolator systems, in maintenance of immunodeficient strains, 135–136

Microscopy, immunoelectron
  of endosomal trafficking in macrophages
      infected with microbial pathogens,
      277–288
  processing infected macrophages for,
      280–281
Monocyte assays, 149–150
Monolayer disruption assays, for *Toxoplasma
    gondii* tachyzoites, 41–42
Murine macrophage
  adherence to solid substrates, 139–140
  alveolar, harvest, 139
  behavior, modulation, 129–145
    animal husbandry, 134–136
    anti-face mounting fluids, 133
    buffer for fluorescent labeling of
        microorganisms, 133
    culture media, 131–133
    maintenance of immunodeficient mice
        strains in, 134–136
    targets for phagocytosis, 133
  bone marrow-derived
    culture media for growth, 132–133
    preparation, 139
  culture media, 131–133
  MHC class II antigen expression in,
      analysis, 143–145
  particles bound versus ingested by,
      differentiation, 142
  peritoneal
    eliciting agents for obtaining, 132
    eliciting and obtaining, 137–138
  phagocytic activity, assays, 140–142
    by indirect immunofluorescence, 141–142
  splenic, harvest, 138
Mutagenesis
  insertional, in *Toxoplasma gondii*
      transformation, 53
    protocol, 54
  mycobacterial, 118–120
  in *Salmonella,* 80–92
    chemical, 80
    deletion or replacement of large DNA
        fragments, 81
    transposon, in *Salmonella, see*
        Transposon mutagenesis, in
        Salmonella
Mutations, transposon-induced, features,
    82–83
Mycobacteriophage infection, 112–114
Mycobacterium, 107–123
  biosafety considerations, 108

complementation of mutants, 122–123
cultural conditions, 111
culture media, 109–111
electroporation of plasmid DNA into,
    116–117
genetic techniques, 112–117
  conjugation, 115
  plasmid transformation of DNA, 115–117
isolation, 111–112
maintenance of stocks, 112
mutagenesis, 118–120
mutant selection and isolation, 121–123
phagosomes, 264
  isolation, 269–271
plasmic preparation from, 117
screening for mutants, 121
strain construction, 120–121

N

Nonhomologous recombination, in
    mycobacterium, 121

O

Opsonized particles, assays, 151–153

P

Parasites, malaria, *see* Malaria parasites
Pathogens, microbial
  biosafety considerations, 1–2
  culture, 3–4
  Department of Health and Human Services
      guidelines, 2
  federal and institutional requirements, 2–3
  macrophages infected with, endosomal
      trafficking in, immunoelectron
      microscopy, 277–288
  maintenance, 3–4
  obtaining and maintaining, 1–4
  source, 3
Periodate oxidation, in biochemical
    characterization of molecular nature of
    bacteria receptors, 173–174
Phage Mu, in *Salmonella* mutagenesis, 87–88
Phage protocols, in mycobacteriophage
    infection, 113–114
Phagocytes, assays, 148–149
Phagocytic processing of bacteria for class
    I/II MHC-restricted antigen recognition
    by T lymphocytes, 313–324

Phagocytosis, 147–161
  assays, 148–155
    ingestion, 153–155
    for opsonized particles, 151–153
    for phagocytes, 148–151
  receptors, 156–159
  stimulation, 159–160
Phagosomal proton pump, 296–298
Phagosome isolation, 160–161
Phagosomes
  constituents, analysis, 271, 273
  Leishmania, isolation, 269
  Mycobacterium, isolation, 269–271
  pathogen-containing, 261–275
    cell lysis conditions, 266–269
    internalization conditions, 265–266
    macrophage, 264–265
    particle adherence conditions, 265–266
    particles for study, 264
    pathogens for study, 263–264
    pH measurement, 289–309
      calibration, 304–306
      difficulties, 304, 307–308
      experimental system, 299–302
      fluorescent dyes, 302, 303
      intravesicular acid pH generation,
        296–298
      local environmental factors affecting,
        302–304
      materials, 299–302
      particle labeling, 299–301
      pathogens, 291–292
      procedures, 302–306
      rationale, 289–291
      system of intracellular vesicles through
        which parasite enters host cell,
        292–296
    shortcomings, 273–275
    storage and handling two-dimensional
      SDS-PAGE data, 273
Phasmids, in mycobacteriophage infection,
  113
Plaque assays, for *Toxoplasma gondii*
  tachyzoites, 39–41
Plasmids
  conjugative, in transposon mutagenesis in
    Salmonella, 85
  nonconjugative, in transposon mutagenesis
    in *Salmonella,* 84–85
Plasmodium falciparum
  asexual life cycle, organization of secretory
    activities at different stages, 234–240

culture
  synchronization, 215–216
  in vitro, 214–215
erythrocyte-free, *in vitro* secretory assays
  with, 221–245
erythrocyte membrane/tubovesicular
    membrane fraction
  distribution of secretory markers in, using
    Western blots, 234–236
  distribution of sphingomyelin synthase
    activity, 236–239
  isolation, 229–230
erythrocytic stages, axenic development, 21
mature asexual stages freed by exposure to
  glycyl-L-serine, 240, 242–244
merozoites
  isolation, 216–217
  purification for analysis of processing of
    merozoitic surface protein-1, 213–
    220
  sphingomyelin synthase in, distribution,
    239
  surface protein-1, assay for secondary
    processing, 217–220
pigmented trophozoites and schizonts,
    release from infected erythrocytes by
    osmotic shock in isoosmolar dipeptide-
    based media, 240, 242–245
released, distribution of sphingomyelin
    synthase activity in, 236–239
rings/trophozoites
  labeling with c_6-NBD-ceramide for
    microscopy, 238
  released
    distribution of secretory markers in,
      using Western blots, 234–236
    preparation, 227–229
    synthesis and secretion of proteins by,
      230–234
secretory processes in plasmodial entry and
    development in erythrocyte, 222–225
sphingomyelin synthase activity in,
    compared by stages, 239–240
stage-specific release, 242
trophozoites
  naked, viability, 242–244
  released, suspension from uninfected and
    surviving intact PRBCs, 242
Plasmodium falciparum-infected erythrocyte
  adhesins, 196–197
  cytoadherence, 193–207
  ligands, 194–197

Poke plaque assays, in identification of intracellular growth *Legionella pneumophila* mutants, 256–257
Proskauer-Beck minimal medium recipe, 110
Proteins
  biosynthetic export, from released *Plasmodium falciparum,* 232–234
  soluble, binding to bacteria, 182–183
  synthesis, by intact ring/trophozoite *Plasmodium falciparum,* 231–232
Protozoan parasite *Toxoplasma gondii, see Toxoplasma gondii*
Pseudo-diploids, in *Toxoplasma gondii* transformation, 56–57
P22 transduction, in transposon mutagenesis in *Salmonella,* 84

R

Recombination
  homologous, in *Toxoplasma gondii* transformation, 55–57
  nonhomologous, in *Toxoplasma gondii* transformation, 52–53
Replication assays, for *Toxoplasma gondii,* 39–43
  direct measurement of doubling time, 43
  incorporation of [³H]-uracil, 42–43
  monolayer disruption, 41–42
  plaque, 39–41
RPMI-1640, in culture of malaria parasites, 8, 9

S

Salmonella, 79–103
  adhesion
    to cells, studying, 93
    to glutaraldehyde-fixed cells, 96–97
  antibiotic-resistant, 94–95
  genetic analysis, 98–101
    mapping strategies, 99–101
      genetic, 99–100
      physical, 100–101
    perspectives, 101–103
  growing for assays, 94
  invasion of cells by, studying, 93
  invasion of epithelial cells by, 97
  macrophage survival assay, 97–98
  mutagenesis, 80–92
    chemical, 80
    deletion or replacement of large DNA fragments, 81

transposon mutagenesis, *see* Transposon mutagenesis, in *Salmonella*
  screening of variants, 92–98
    assay protocols, 96–98
    choice of cell lines, 95–96
    parameters affecting tissue culture assays, 94–95
  sensitivity to detergents, 95
  spinning onto cells, 95
  spontaneous mutants, 80
  survival in macrophages, 93–94
  in vitro models of infection with, 93–94
Sauton's minimal recipe, 110
Serum, in culture of malaria parasites, 8, 10
Shuttle mutagenesis, of mycobacterium, 119–120
Sphingomyelin synthase
  detection in released *Plasmodium falciparum* and EM/TVM fraction, 236–239
  in *Plasmodium falciparum* merozoites compared with rings, trophozoites, and schizont-infected erythrocytes, 239–240
  distribution, 239
Sputum samples, for mycobacteria isolation, 111–112

T

Tachyzoites, *T. gondii,* 30, *31*
  in vitro culture, 32–44
T cell hybridomas
  MHC-I-restricted, generating, 316–319
  MHC-II-restricted, 315–316
Thrombospondin (TSP), as ligand for *Plasmodium falciparum*-infected erythrocytes, 195
Thymineless death enrichment, of *Legionella pneumophila,* 249–256
T lymphocytes
  in antigen recognition, basic assay, 319–321
  studying phagocytic processing of bacteria for class I/II MHC-restricted antigen recognition by, 313–324
Tn5, in *Salmonella* mutagenesis, 88–89
Tn10, in *Salmonella* mutagenesis, 85–87
Toxoplasma gondii, 27–60
  cell biology problems, 59–60
  genetics, 28–29
  life cycle, 29–31
  molecular transformation systems, 44–58
    cloning by complementation, 57–58

gene knockouts, 57
homologous recombination, 55–57
insertional mutagenesis, 53
  protocol, 54
marker rescue, 53
  protocol, 54–55
nonhomologous recombination, 52–53
perfect gene replacement, 56
pseudo-diploids, 56–57
special considerations for working
    with pyrimethamine-resistant
    organisms, 58
stable transgene expression and
    overexpression, 49–52
transient expression, 45, 47–49
vectors, 44, *46*
pathogenesis, problems, 59–60
tachyzoites, 30, *31*
  in vitro culture, 32–44
    cloning by limiting dilution or growth
        under agar, 36–37
    host cells, 32–33
    long-term storage, 37–39
    optimizing production of viable
        parasites, 35–36
    parasite strains, 32
    purification, 35
    replication assays, 39–43
    routine parasite culture, 33–35
    safety issues, 43–44
Transduction
  in mycobacteriophage infection, 112–113
  P22, in transposon mutagenesis in
    Salmonella, 84
Transposon-induced mutations, features,
  82–83

Transposon mutagenesis
  of mycobacterium, 119
  in *Salmonella,* 81–92
    delivery systems, 84–85
    frequently used transposons, 85–89
    generation of P22 HT int lysate, 89–90
    generation of transducing lysate from
        rough strains using F':::P22 HT int,
        90–91
    MudJ, 91–92
    properties of transposons, 81–83
    protocols, 89–92
    P22 transduction, 90
    with Tn10d(Tc), 91
    TnphoA, 92
Transposons
  DNA flanking, cloning and sequencing of,
    in genetic analysis of *Salmonella,*
    98–99
  frequently used, in transposon mutagenesis
    in *Salmonella,* 85–89
  properties, 81–83
Tween-80, 20% recipe, 110

U

UP37 cell-derived macrophages, tissue
  culture, 248–249
[³H]-Uracil incorporation, in *Toxoplasma
  gondii* tachyzoites, 42–43

V

VCAM-1, as ligand for *Plasmodium
  falciparum*-infected erythrocytes, 195

VOLUMES IN SERIES

Founding Series Editor
DAVID M. PRESCOTT

Volume 1 (1964)
Methods in Cell Physiology
Edited by David M. Prescott

Volume 2 (1966)
Methods in Cell Physiology
Edited by David M. Prescott

Volume 3 (1968)
Methods in Cell Physiology
Edited by David M. Prescott

Volume 4 (1970)
Methods in Cell Physiology
Edited by David M. Prescott

Volume 5 (1972)
Methods in Cell Physiology
Edited by David M. Prescott

Volume 6 (1973)
Methods in Cell Physiology
Edited by David M. Prescott

Volume 7 (1973)
Methods in Cell Biology
Edited by David M. Prescott

Volume 8 (1974)
Methods in Cell Biology
Edited by David M. Prescott

Volume 9 (1975)
Methods in Cell Biology
Edited by David M. Prescott

Volume 10 (1975)
Methods in Cell Biology
Edited by David M. Prescott

Volume 11 (1975)
Yeast Cells
Edited by David M. Prescott

Volume 12 (1975)
Yeast Cells
Edited by David M. Prescott

Volume 13 (1976)
Methods in Cell Biology
Edited by David M. Prescott

Volume 14 (1976)
Methods in Cell Biology
Edited by David M. Prescott

Volume 15 (1977)
Methods in Cell Biology
Edited by David M. Prescott

Volume 16 (1977)
Chromatin and Chromosomal Protein Research I
Edited by Gary Stein, Janet Stein, and Lewis J. Kleinsmith

Volume 17 (1978)
Chromatin and Chromosomal Protein Research II
Edited by Gary Stein, Janet Stein, and Lewis J. Kleinsmith

Volume 18 (1978)
Chromatin and Chromosomal Protein Research III
Edited by Gary Stein, Janet Stein, and Lewis J. Kleinsmith

Volume 19 (1978)
Chromatin and Chromosomal Protein Research IV
Edited by Gary Stein, Janet Stein, and Lewis J. Kleinsmith

Volume 20 (1978)
Methods in Cell Biology
Edited by David M. Prescott

Advisory Board Chairman
KEITH R. PORTER

Volume 21A (1980)
Normal Human Tissue and Cell Culture, Part A: Respiratory, Cardiovascular, and Integumentary Systems
Edited by Curtis C. Harris, Benjamin F. Trump, and Gary D. Stoner

Volume 21B (1980)
Normal Human Tissue and Cell Culture, Part B: Endocrine, Urogenital, and Gastrointestinal Systems
Edited by Curtis C. Harris, Benjamin F. Trump, and Gary D. Stoner

Volume 22 (1981)
Three-Dimensional Ultrastructure in Biology
Edited by James N. Turner

Volume 23 (1981)
Basic Mechanisms of Cellular Secretion
Edited by Arthur R. Hand and Constance Oliver

Volume 24 (1982)
The Cytoskeleton, Part A: Cytoskeletal Proteins, Isolation and Characterization
Edited by Leslie Wilson

Volume 25 (1982)
The Cytoskeleton, Part B: Biological Systems and *in Vitro* Models
Edited by Leslie Wilson

Volume 26 (1982)
Prenatal Diagnosis: Cell Biological Approaches
Edited by Samuel A. Latt and Gretchen J. Darlington

Series Editor
LESLIE WILSON

Volume 27 (1986)
Echinoderm Gametes and Embryos
Edited by Thomas E. Schroeder

Volume 28 (1987)
***Dictyostelium discoideum:* Molecular Approaches to Cell Biology**
Edited by James A. Spudich

Volume 29 (1989)
Fluorescence Microscopy of Living Cells in Culture, Part A: Fluorescent Analogs, Labeling Cells, and Basic Microscopy
Edited by Yu-Li Wang and D. Lansing Taylor

Volume 30 (1989)
Fluorescence Microscopy of Living Cells in Culture, Part B: Quantitative Fluorescence Microscopy—Imaging and Spectroscopy
Edited by D. Lansing Taylor and Yu-Li Wang

Volume 31 (1989)
Vesicular Transport, Part A
Edited by Alan M. Tartakoff

Volume 32 (1989)
Vesicular Transport, Part B
Edited by Alan M. Tartakoff

Volume 33 (1990)
Flow Cytometry
Edited by Zbigniew Darzynkiewicz and Harry A. Crissman

Volume 34 (1991)
Vectorial Transport of Proteins into and across Membranes
Edited by Alan M. Tartakoff

Selected from Volumes 31, 32, and 34 (1991)
Laboratory Methods for Vesicular and Vectorial Transport
Edited by Alan M. Tartakoff

Volume 35 (1991)
Functional Organization of the Nucleus: A Laboratory Guide
Edited by Barbara A. Hamkalo and Sarah C. R. Elgin

Volume 36 (1991)
***Xenopus laevis:* Practical Uses in Cell and Molecular Biology**
Edited by Brian K. Kay and H. Benjamin Peng

Series Editors
LESLIE WILSON AND PAUL MATSUDAIRA

Volume 37 (1993)
Antibodies in Cell Biology
Edited by David J. Asai

Volume 38 (1993)
Cell Biological Applications of Confocal Microscopy
Edited by Brian Matsumoto

Volume 39 (1993)
Motility Assays for Motor Proteins
Edited by Jonathan M. Scholey

Volume 40 (1994)
A Practical Guide to the Study of Calcium in Living Cells
Edited by Richard Nuccitelli

Volume 41 (1994)
Flow Cytometry, Second Edition, Part A
Edited by Zbigniew Darzynkiewicz, J. Paul Robinson,
  and Harry A. Crissman

Volume 42 (1994)
Flow Cytometry, Second Edition, Part B
Edited by Zbigniew Darzynkiewicz, J. Paul Robinson,
  and Harry A. Crissman

Volume 43 (1994)
Protein Expression in Animal Cells
Edited by Michael G. Roth

Volume 44 (1994)
***Drosophila melanogaster:* Practical Uses in Cell and Molecular Biology**
Edited by Lawrence S. B. Goldstein and Eric A. Fyrberg

Volume 45 (1994)
Microbes as Tools for Cell Biology
Edited by David G. Russell

Volume 46 (1995) (in preparation)
Cell Death
Edited by Lawrence M. Schwartz and Barbara A. Osborne

ISBN 0-12-564146-X

9 780125 641463

90018